Special Triangles

Name	Characteristic	Examples
Right Triangle	Triangle has a right angle.	
Isosceles Triangle	Triangle has two equal sides.	$AB = BC$ 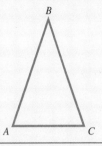
Equilateral Triangle	Triangle has three equal sides.	$AB = BC = CA$ 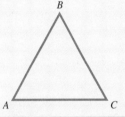
Similar Triangles	Corresponding angles are equal; corresponding sides are proportional.	$A = D,\ B = E,\ C = F$ $$\frac{AB}{DE} = \frac{AC}{DF} = \frac{BC}{EF}$$ 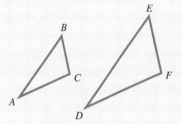

BEGINNING ALGEBRA

MYMATHLAB VERSION

BEGINNING ALGEBRA

MyMathLab version

EIGHTH EDITION

Margaret L. Lial
American River College

John Hornsby
University of New Orleans

Addison
Wesley

Boston San Francisco New York
London Toronto Sydney Tokyo Singapore Madrid
Mexico City Munich Paris Cape Town Hong Kong Montreal

Publisher: Jason Jordan

Acquisitions Editor: Jennifer Crum

Project Manager: Kari Heen

Developmental Editor: Terry McGinnis

Managing Editor: Ron Hampton

Production Supervisor: Sheila Spinney

Production Services: Elm Street Publishing Services, Inc.

Compositor: Typo-Graphics

Art Editor: Jennifer Bagdigian

Art Development: Meredith Nightingale

Artists: Precision Graphics, Jim Bryant, and Darwin Hennings

Marketing Manager: Dona Kenly

Prepress Buyer: Caroline Fell

Manufacturing Coordinator: Evelyn Beaton

Text and Cover Designer: Dennis Schaefer

Cover Illustration: © Peter Siu/SIS

Library of Congress Cataloging-in-Publication Data

Lial, Margaret L.
   Beginning algebra—8th ed. / My mathlab version, Margaret L. Lial, John Hornsby.
     p. cm.
   Includes index.
   ISBN 0-201-74967-X
    1. Algebra.   I. Hornsby, John. II. Title.
   QA152.2.L5 2001
   512.9—dc21

Printed in the U.S.A.

123456789-DOW-05 04 03 02 01

Contents

CHAPTER 6 **Rational Expressions** — **353**

CHAPTER 7 **Equations of Lines, Inequalities, and Functions** — **428**

CHAPTER 8 **Linear Systems** — **467**

CHAPTER 9   Roots and Radicals   518

CHAPTER 10   Quadratic Equations   577

Preface

The eighth edition of *Beginning Algebra* is designed for college students who have never studied algebra or who want to review the basic concepts of algebra before taking additional courses in mathematics, science, business, nursing, or computer science. The primary objective of this text is to familiarize students with mathematical symbols and operations in order to solve first- and second-degree equations and applications that lead to these equations.

This revision of *Beginning Algebra* reflects our ongoing commitment to creating the best possible text and supplements package using the most up-to-date strategies for helping students succeed. One of these strategies, evident in our new Table of Contents and consistent with current teaching practices, involves the early introduction of functions and graphing lines in a rectangular coordinate system. We believe that this pedagogy has a great deal of merit as it provides students with the important "input-output" concept that will be an integral part of later mathematics courses. This organization also allows an early treatment of interesting interpretations of data in the form of line and bar graphs—two pictorial representations that students already see on a daily basis in magazines and newspapers. Chapter 3 introduces three sections on linear equations in two variables, ordered pairs, graphing, and slope, with a gentle introduction to the function concept in the form of input-output relationships. This allows students to read graphs in the chapters that immediately follow and to slowly develop an understanding of the basic idea of a function. Chapter 7 introduces the more involved concepts of forms of equations of lines and inequalities in two variables. The function concept is addressed again here, this time with a discussion of domain, range, and function notation.

If for any reason you choose not to cover these topics as our new edition suggests, it will not be difficult to defer Chapter 3 and combine it with Chapter 7 as in previous editions. You will need to skip the last objective in Section 4.1 (Graphing Simple Polynomials). Also, one or two applied problems in an example or exercise in Chapters 4–6 may refer to the function concept, but even those problems can be used without actually working through Chapter 3.

Other up-to-date pedagogical strategies to foster student success include a strong emphasis on vocabulary and problem solving, an increased number of real world applications in both examples and exercises, and a focus on relevant industry themes throughout the text.

Another strategy for student success, an exciting new CD-ROM called "Pass the Test," debuts with this edition of *Beginning Algebra*. Directly correlated to the text's content, "Pass the Test" helps students master concepts by providing interactive pretests, chapter tests, section reviews, and InterAct Math tutorial exercises. To support an increased emphasis on graphical manipulation, the CD-ROM also includes a graphing tool that can be used for open-ended, student-directed exploration of number lines and coordinate graphs, as well as for exercises relevant to the graphing content throughout the book.

The *Student's Study Guide and Journal,* redesigned and enhanced with an optional journal feature for those who would like to incorporate more writing in their mathematics curriculum, provides an additional strategy for student success.

Although *Beginning Algebra,* Eighth Edition, integrates many new elements, it also retains the time-tested features of previous editions: learning objectives for each section, careful exposition, fully developed examples, Cautions and Notes, and design features that highlight important definitions, rules, and procedures. Since the hallmark of any mathematics text is the quality of its exercise sets, we have carefully developed exercise sets that provide ample opportunity for drill and, at the same time, test conceptual understanding. In preparing this edition we have also addressed the standards of the National Council of Teachers of Mathematics and the American Mathematical Association of Two-Year Colleges, incorporating many new exercises focusing on concepts, writing, graph interpretation, technology use, collaborative work, and analysis of data from a wide variety of sources in the world around us.

CONTENT CHANGES

We have fine-tuned and polished presentations of topics throughout the text based on user and reviewer feedback. Some of the content changes you may notice include the following:

- Operations with real numbers are consolidated from four sections to two sections in Chapter 1.

- We consistently emphasize problem solving using a six-step problem-solving strategy, first introduced in Section 2.3 and continually reinforced in examples and the exercise sets throughout the text. Section 2.6 contains a comprehensive discussion of problem solving.

- New Chapter 3 introduces graphing and slope, along with an intuitive introduction to functions using input-output. Equations with two variables are presented earlier so that the applications used in later chapters (for example, with polynomials and rational expressions) can be more realistic and relevant. Both graphing and functions are continued in later chapters.

- New material is included on graphing parabolas in Section 4.1.

- Division of polynomials is consolidated in Section 4.6.

- Solution set and interval notation are now introduced in *Beginning Algebra,* consistent with the approach in *Intermediate Algebra.*

NEW FEATURES

We believe students and instructors will welcome the following new features:

 Industry Themes To help motivate the material, each chapter features a particular industry that is presented in the chapter opener and revisited in examples and exercises in the chapter. Identified by special icons, these examples and exercises incorporate sourced data, often in the form of graphs and tables. Featured industries include business, health care, entertainment, sports, transportation, and others. (See pages 1, 88, 175, and 428.)

New Examples and Exercises   We have added 25% more real application problems with data sources. These examples and exercises often relate to the industry themes. They are designed to show students how algebra is used to describe and interpret data in everyday life. (See pages 9, 112, 132, 434, and 492.)

The Olympic Committee has come to rely more and more on television rights and major corporate sponsors to finance the games. The pie charts show the funding plans for the first Olympics in Athens and the 1996 Olympics in Atlanta. Use proportions and the figures to answer the questions in Exercises 35 and 36.

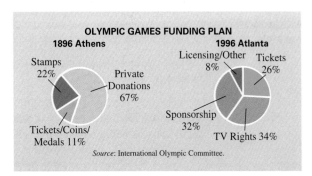

35. In the 1996 Olympics, total revenue of $350 million was raised. There were 10 major sponsors.
   (a) Write a proportion to find the amount of revenue provided by tickets. Solve it.
   (b) What amount was provided by sponsors? Assuming the sponsors contributed equally, how much was provided per sponsor?
   (c) What amount was raised by TV rights?

36. Suppose the amount of revenue raised in the 1896 Olympics was equivalent to the $350 million in 1996.
   (a) Write a proportion for the amount of revenue provided by stamps and solve it.
   (b) What amount (in dollars) would have been provided by private donations?
   (c) In the 1988 Olympics, there were 9 major sponsors, and the total revenue was $95 million. What is the ratio of major sponsors in 1988 to those in 1996? What is the ratio of revenue in 1988 to revenue in 1996?

Technology Insights Exercises   Technology is part of our lives, and we assume that all students of this text have access to scientific calculators. *While graphing calculators are not required for this text,* it is likely that students will go on to courses that use them. For this reason, we have included Technology Insights exercises in selected exercise sets. These exercises illustrate the power of graphing calculators and provide an opportunity for students to interpret typical results seen on graphing calculator screens. (See pages 212, 273, 324, and 506.)

 Mathematical Journal Exercises   While we continue to include conceptual and writing exercises that require short written answers, new journal exercises have been added that ask students to fully explain terminology, procedures, and methods, document their understanding using examples, or make connections between concepts. Instructors who wish to incorporate a journal component in their classes will find these exercises especially useful. For the greatest possible flexibility, both writing exercises and journal exercises are indicated with icons in the Annotated Instructor's Edition, but not in the Student Edition. (See pages 35, 103, 247, and 415.)

Group Activities Appearing at the end of each chapter, these activities allow students to apply the industry theme of the chapter to its mathematical content in a collaborative setting. (See pages 162, 458, and 506.)

Test Your Word Power To help students understand and master mathematical vocabulary, this new feature has been incorporated at the end of each chapter. Key terms from the chapter are presented with four possible definitions in multiple-choice format. Answers and examples illustrating each term are provided at the bottom of the appropriate page. (See pages 163, 277, and 343.)

HALLMARK FEATURES

We have retained the popular features of previous editions of the text. Some of these features are as follows:

Learning Objectives Each section begins with clearly stated numbered objectives, and material in the section is keyed to these objectives. In this way students know exactly what is being covered in each section.

OBJECTIVES

1. Learn the definition of *factor*.
2. Write fractions in lowest terms.
3. Multiply and divide fractions.
4. Add and subtract fractions.
5. Solve applied problems that involve fractions.
6. Interpret data in a circle graph.

Cautions and Notes We often give students warnings of common errors and emphasize important ideas in Cautions and Notes that appear throughout the exposition.

Connections Retained from the previous edition, Connections boxes have been streamlined and now often appear at the beginning or the end of the exposition in selected sections. They continue to provide connections to the real world or to other mathematical concepts, historical background, and thought-provoking questions for writing or class discussion. (See pages 105, 201, 226, and 322.)

Problem Solving Increased emphasis has been given to our six-step problem-solving method to aid students in solving application problems. This method is continually reinforced in examples and exercises throughout the text. (See pages 105, 107, 330, 409, and 497.)

Ample and Varied Exercise Sets Students in beginning algebra require a large number and variety of practice exercises to master the material. This text contains approximately 5800 exercises, including about 1600 review exercises, plus numerous conceptual and writing exercises, journal exercises, and challenging exercises that go beyond the examples. More illustrations, diagrams, graphs, and tables now accompany exercises. Multiple-choice, matching, true/false, and completion exercises help to provide variety. Exercises suitable for calculator use are marked with a calculator icon ▦ in both the Student Edition and the Annotated Instructor's Edition. (See pages 27, 143, 347, and 438.)

Relating Concepts Previously titled Mathematical Connections, these sets of exercises often appear near the end of selected sections. They tie together topics and highlight the relationships among various concepts and skills. For example, they may show how algebra and geometry are related, or how a graph of a linear equation in two variables is related to the solution of the corresponding linear equation in one variable. Instructors have told us that these sets of exercises make great collaborative activities for small groups of students. (See pages 69, 210, 232, 381, and 490.)

Ample Opportunity for Review Each chapter concludes with a Chapter Summary that features Key Terms and Symbols, Test Your Word Power, and a Quick Review of each section's content. Chapter Review Exercises keyed to individual sections are included as well as mixed review exercises and a Chapter Test. Following every chapter after Chapter 1, there is a set of Cumulative Review Exercises that covers material going back to the first chapter. Students always have an opportunity to review material that appears earlier in the text, and this provides an excellent way to prepare for the final examination in the course. (See pages 214–222 and 459–466.)

SUPPLEMENTS

Our extensive supplements package includes the Annotated Instructor's Edition, testing materials, study guides, solutions manuals, CD-ROM software, videotapes, and a Web site. For more information on these and other helpful supplements, contact your Addison Wesley Longman sales representative.

FOR THE INSTRUCTOR

Annotated Instructor's Edition (ISBN 0-321-04128-3)
For immediate access, the Annotated Instructor's Edition provides answers to all text exercises and Group Activities in color in the margin or next to the corresponding exercise, as well as Chalkboard Examples and Teaching Tips. To assist instructors in assigning homework, additional icons not shown in the Student Edition indicate journal exercises 🗐, writing exercises ✒, and challenging exercises ▲.

CHALKBOARD EXAMPLE
Write 90 as the product of prime factors.
Answer: $2 \cdot 3^2 \cdot 5$

TEACHING TIP The term *fraction bar* may be unfamiliar to some students.

Exercises designed for calculator use ▦, ▦ are indicated in both the Student Edition and the Annotated Instructor's Edition.

Instructor's Solutions Manual (ISBN 0-321-06193-4)

The *Instructor's Solutions Manual* provides solutions to all even-numbered exercises, including answer art, and lists of all writing, journal, challenging, Relating Concepts, and calculator exercises.

Answer Book (ISBN 0-321-06194-2)

The *Answer Book* contains answers to all exercises and lists of all writing, journal, challenging, Relating Concepts, and calculator exercises. Instructors may ask the bookstore to order multiple copies of the *Answer Book* for students to purchase.

Printed Test Bank (ISBN 0-321-06192-6)

The *Printed Test Bank* contains short answer and multiple-choice versions of a placement test and final exam; six forms of chapter tests for each chapter, including four open-response (short answer) and two multiple-choice forms; 10 to 20 additional exercises per objective for instructors to use for extra practice, quizzes, or tests; answer keys to all of the above listed tests and exercises; and lists of all writing, journal, challenging, Relating Concepts, and calculator exercises.

 TestGen-EQ with QuizMaster EQ (ISBN 0-321-06132-2)

This fully networkable software presents a friendly graphical interface which enables professors to build, edit, view, print and administer tests. Tests can be printed or easily exported to HTML so they can be posted to the Web for student practice.

FOR THE STUDENT

 Student's Study Guide and Journal (ISBN 0-321-06196-9)

The *Student's Study Guide and Journal* contains a "Chart Your Progress" feature for students to track their scores on homework assignments, quizzes, and tests, additional practice for each learning objective, section summary outlines that give students additional writing opportunities and help with test preparation, and self-tests with answers at the end of each chapter. A manual icon at the beginning of each section in the Student Edition identifies section coverage.

 Student's Solutions Manual (ISBN 0-321-06195-0)

The *Student's Solutions Manual* provides solutions to all odd-numbered exercises (journal and writing exercises excepted). A manual icon at the beginning of each section in the Student Edition identifies section coverage.

 InterAct Math Tutorial Software (ISBN 0-321-06140-3 (Student Version))

This tutorial software correlates with every odd-numbered exercise in the text. The program is highly interactive with sample problems and interactive guided solutions accompanying every exercise. The program recognizes common student errors and provides customized feedback with sophisticated answer recognition capabilities. The management system (InterAct Math Plus) allows instructors to create, administer, and track tests, and to monitor student performance during practice sessions.

 "Real to Reel" Videotapes (0-321-05659-0)

This videotape series provides separate lessons for each section in the book. A videotape icon at the beginning of each section identifies section coverage. All objectives, topics, and problem-solving techniques are covered and content is specific to *Beginning Algebra,* Eighth Edition.

"Pass the Test" Interactive CD-ROM (ISBN 0-321-06204-3)
This CD helps students to master the course content by providing interactive pre-tests, chapter tests, section reviews, and InterAct tutorial exercises. After studying a chapter in class, students take a pre-test to determine what areas in that chapter need additional work. They are then directed to section reviews and tutorial exercises for continued practice. Students continue to take chapter tests and practice their skills until they have mastered the chapter. A unique graphing tool is provided for exploring the relationship between graphs and their algebraic representation.

MathXL (http://www.mathxl.com)
Available on-line with a pre-assigned ID and password by ordering a new copy of *Beginning Algebra,* Eighth Edition, with ISBN 0-201-68155-2, MathXL helps students prepare for tests by allowing them to take practice tests that are similar to the chapter tests in their text. Students also get a personalized study plan that identifies strengths and pinpoints topics where more review is needed. For more information on subscriptions, contact your Addison Wesley Longman sales representative.

 Math Tutor Center
The Addison Wesley Longman Math Tutor Center is staffed by qualified mathematics instructors who provide students with tutoring on text examples, exercises, and problems. Tutoring assistance is provided by telephone, fax, and e-mail and is available five days a week, seven hours a day. A registration number for the tutoring service may be obtained by calling our toll-free customer service number, 1-800-922-0579 and requesting ISBN 0-201-44461-5. Registration for the service is active for one or more of the following time periods depending on the course duration: Fall (8/31–1/31), Spring (1/1–6/30), or Summer (5/1–8/31). The Math Tutor Center service is also available for other Addison Wesley Longman textbooks in developmental math, precalculus math, liberal arts math, applied math, applied calculus, calculus, and introductory statistics. For more information, please contact your Addison Wesley Longman sales representative.

Mathematics Spanish Glossary, Second Edition (ISBN 0-201-72896-6)
This book includes math terms that would be encountered in Basic Math through College Algebra.

ACKNOWLEDGMENTS

For a textbook to last through eight editions, it is necessary for the authors to rely on comments, criticisms, and suggestions of users, nonusers, instructors, and students. We are grateful for the many responses that we have received over the years. We wish to thank the following individuals who reviewed this edition of the text:

Josette Ahlering, *Central Missouri State*

Vickie Aldrich, *Dona Ana Branch Community College*

Lisa Anderson, *Ventura College*

Robert B. Baer, *Miami University–Hamilton*

Julie R. Bonds, *Sonoma State University*

Beverly R. Broomell, *Suffolk Community College*

Cheryl V. Cantwell, *Seminole Community College*

Stanley Carter, *Central Missouri State*

Jeff Clark, *Santa Rosa Junior College*

Ted Corley, *Glendale Community College*

Lisa Delong Cuneo, *Penn State–Dubois Campus*

Marlene Demerjian, *College of the Canyons*

Richard N. Dodge, *Jackson Community College*

Linda Franko, *Cuyahoga Community College*

Linda L. Galloway, *Macon State College*

Theresa A. Geiger, *St. Petersburg Junior College*

Martha Haehl, *Maple Woods Community College*

Melissa Harper, *Embry Riddle Aeronautical University*

W. Hildebrand, *Montgomery College*

Matthew Hudock, *St. Philips College*

Dale W. Hughes, *Johnson County Community College*

Linda Hurst, *Central Texas College*

Nancy Johnson, *Broward Community College–North*

Robert Kaiden, *Lorain County Community College*

Michael Karelius, *American River College*

Margaret Kimbell, *Texas State Technical College*

Linda Kodama, *Kapiolani Community College*

Jeff A. Koleno, *Lorain County Community College*

William R. Livingston, *Missouri Southern State College*

Doug Martin, *Mt. San Antonio College*

Larry Mills, *Johnson County Community College*

Mary Ann Misko, *Gadsden State Community College*

Elsie Newman, *Owens Community College*

Joanne V. Peeples, *El Paso Community College*

Janice Rech, *University of Nebraska–Omaha*

Joyce Saxon, *Morehead University*

Richard Semmler, *Northern Virginia Community College*

LeeAnn Spahr, *Durham Technical Community College*

No author can complete a project of this magnitude without the help of many other individuals. Our sincere thanks go to Jenny Crum of Addison Wesley Longman who coordinated the package of texts of which this book is a part. Other dedicated staff at Addison Wesley Longman who worked long and hard to make this revision a success include Jason Jordan, Kari Heen, Susan Carsten, Meredith Nightingale, and Kathy Manley.

While Terry McGinnis has assisted us for many years "behind the scenes" in producing our texts, she has contributed far more to these revisions than ever. There is no question that these books are improved because of her attention to detail and consistency, and we are most grateful for her work above and beyond the call of duty. Kitty Pellissier continues to do an outstanding job in checking the answers to exercises. Many thanks to Jenny Bagdigian who coordinated the art programs for the books.

Cathy Wacaser of Elm Street Publishing Services provided her usual excellent production work. She is indeed one of the best in the business. As usual, Paul Van Erden created an accurate, useful index. Becky Troutman prepared the Index of Applications. We are also grateful to Tommy Thompson who made suggestions for the feature "For the Student: 10 Ways to Succeed with Algebra," to Vickie Aldrich and Lucy Gurrola who wrote the Group Activity features, and Janis Cimperman of St. Cloud University and Steve C. Ouellette of the Walpole Massachusetts State Public Schools.

To these individuals and all the others who have worked on these books for 30 years, remember that we could not have done it without you. We hope that you share with us our pride in these books.

Margaret L. Lial
John Hornsby

An Introduction to Calculators

There is little doubt that the appearance of handheld calculators nearly three decades ago and the later development of scientific and graphing calculators have changed the methods of learning and studying mathematics forever. Where the study of computations with tables of logarithms and slide rules made up an important part of mathematics courses prior to 1970, today the widespread availability of calculators make their study a topic only of historical significance.

Most consumer models of calculators are inexpensive. At first, however, they were costly. One of the first consumer models available was the Texas Instruments SR-10, which sold for about $150 in 1973. It could perform the four operations of arithmetic and take square roots, but could do very little more.

Today calculators come in a large array of different types, sizes, and prices. *For the course for which this textbook is intended, the most appropriate type is the scientific calculator,* which costs $10–$20.

In this introduction, we explain some of the features of scientific and graphing calculators. However, remember that calculators vary among manufacturers and models, and that while the methods explained here apply to many of them, they may not apply to your specific calculator. For this reason, it is important to remember that *this introduction is only a guide, and is not intended to take the place of your owner's manual.* Always refer to the manual in the event you need an explanation of how to perform a particular operation.

SCIENTIFIC CALCULATORS

Scientific calculators are capable of much more than the typical four-function calculator that you might use for balancing your checkbook. Most scientific calculators use *algebraic logic.* (Models sold by Texas Instruments, Sharp, Casio, and Radio Shack, for example, use algebraic logic.) A notable exception is Hewlett Packard, a company whose calculators use *Reverse Polish Notation* (RPN). In this introduction, we explain the use of calculators with algebraic logic.

ARITHMETIC OPERATIONS

To perform an operation of arithmetic, simply enter the first number, press the operation key ($+$, $-$, $\times$, or $\div$), enter the second number, and then press the $=$ key. For example, to add 4 and 3, use the following keystrokes.

CHANGE SIGN KEY

The key marked $\pm$ allows you to change the sign of a display. This is particularly useful when you wish to enter a negative number. For example, to enter -3, use the following keystrokes.

MEMORY KEY

Scientific calculators can hold a number in memory for later use. The label of the memory key varies among models; two of these are M and STO. $M+$ and $M-$ allow you to add to or subtract from the value currently in memory. The memory recall key, labeled MR, RM, or RCL, allows you to retrieve the value stored in memory.

Suppose that you wish to store the number 5 in memory. Enter 5, then press the key for memory. You can then perform other calculations. When you need to retrieve the 5, press the key for memory recall.

If a calculator has a constant memory feature, the value in memory will be retained even after the power is turned off. Some advanced calculators have more than one memory. It is best to read the owner's manual for your model to see exactly how memory is activated.

CLEARING/CLEAR ENTRY KEYS

These keys allow you to clear the display or clear the last entry entered into the display. They are usually marked C and CE. In some models, pressing the C key once will clear the last entry, while pressing it twice will clear the entire operation in progress.

SECOND FUNCTION KEY

This key is used in conjunction with another key to activate a function that is printed *above* an operation key (and not on the key itself). It is usually marked 2nd. For example, suppose you wish to find the square of a number, and the squaring function (explained in more detail later) is printed above another key. You would need to press 2nd before the desired squaring function can be activated.

SQUARE ROOT KEY

Pressing the square root key, $\sqrt{x}$, will give the square root (or an approximation of the square root) of the number in the display. For example, to find the square root of 36, use the following keystrokes.

3 6 $\sqrt{x}$ 6

The square root of 2 is an example of an irrational number (Chapter 9). The calculator will give an approximation of its value, since the decimal for $\sqrt{2}$ never terminates and never repeats. The number of digits shown will vary among models. To find an approximation of $\sqrt{2}$, use the following keystrokes.

2 $\sqrt{x}$ 1.4142136   An approximation

SQUARING KEY

This key, x^2, allows you to square the entry in the display. For example, to square 35.7, use the following keystrokes.

3 5 . 7 x^2 1274.49

The squaring key and the square root key are often found on the same key, with one of them being a second function (that is, activated by the second function key, described above).

RECIPROCAL KEY

The key marked $1/x$ is the reciprocal key. (When two numbers have a product of 1, they are called *reciprocals*. See Chapter 1.) Suppose that you wish to find the reciprocal of 5. Use the following keystrokes.

5 $1/x$ 0.2

INVERSE KEY

Some calculators have an inverse key, marked $\boxed{\text{INV}}$. Inverse operations are operations that "undo" each other. For example, the operations of squaring and taking the square root are inverse operations. The use of the $\boxed{\text{INV}}$ key varies among different models of calculators, so read your owner's manual carefully.

EXPONENTIAL KEY

The key marked $\boxed{x^y}$ or $\boxed{y^x}$ allows you to raise a number to a power. For example, if you wish to raise 4 to the fifth power (that is, find 4^5, as explained in Chapter 4), use the following keystrokes.

ROOT KEY

Some calculators have this key specifically marked $\boxed{\sqrt[x]{x}}$ or $\boxed{\sqrt[x]{y}}$; with others, the operation of taking roots is accomplished by using the inverse key in conjunction with the exponential key. Suppose, for example, your calculator is of the latter type and you wish to find the fifth root of 1024. Use the following keystrokes.

Notice how this "undoes" the operation explained in the exponential key discussion above.

PI KEY

The number π is an important number in mathematics. It occurs, for example, in the area and circumference formulas for a circle. By pressing the $\boxed{\pi}$ key, you can display the first few digits of π. (Because π is irrational, the display shows only an approximation.) One popular model gives the following display when the $\boxed{\pi}$ key is pressed: $\boxed{3.1415927}$.

METHODS OF DISPLAY

When decimal approximations are shown on scientific calculators, they are either *truncated* or *rounded*. To see how a particular model is programmed, evaluate 1/18 as an example. If the display shows .0555555 (last digit 5), it truncates the display. If it shows .0555556 (last digit 6), it rounds off the display.

When very large or very small numbers are obtained as answers, scientific calculators often express these numbers in scientific notation (Chapter 4). For example, if you multiply 6,265,804 by 8,980,591, the display might look like this:

$$\boxed{5.6270623 \qquad 13}$$.

The "13" at the far right means that the number on the left is multiplied by 10^{13}. This means that the decimal point must be moved 13 places to the right if the answer is to be expressed in its usual form. Even then, the value obtained will only be an approximation: 56,270,623,000,000.

▶ GRAPHING CALCULATORS

Graphing calculators are becoming increasingly popular in mathematics classrooms. While you are not expected to have a graphing calculator to study from this book, we do include a feature in many exercise sets called *Technology Insights* that asks you to interpret typical graphing calculator screens. These exercises can help to prepare you for future courses where graphing calculators may be recommended or even required.

BASIC FEATURES

Graphing calculators provide many features beyond those found on scientific calculators. In addition to the typical keys found on scientific calculators, they have keys that can be used to create graphs, make tables, analyze data, and change settings. One of the major differences between graphing and scientific calculators is that a graphing calculator has a larger viewing screen with graphing capabilities. The screens below illustrate the graphs of $y = x$ and $y = x^2$.

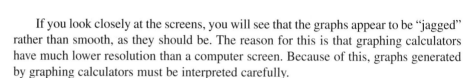

If you look closely at the screens, you will see that the graphs appear to be "jagged" rather than smooth, as they should be. The reason for this is that graphing calculators have much lower resolution than a computer screen. Because of this, graphs generated by graphing calculators must be interpreted carefully.

EDITING INPUT

The screen of a graphing calculator can display several lines of text at a time. This feature allows you to view both previous and current expressions. If an incorrect expression is entered, an error message is displayed. The erroneous expression can be viewed and corrected by using various editing keys, much like a word-processing program. You do not need to enter the entire expression again. Many graphing calculators can also recall past expressions for editing or updating. The screen on the left below shows how two expressions are evaluated. The final line is entered incorrectly, and the resulting error message is shown in the screen on the right.

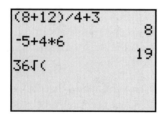

ORDER OF OPERATIONS

Arithmetic operations on graphing calculators are usually entered as they are written in mathematical equations. For example, to evaluate $\sqrt{36}$ on a typical scientific calculator, you would first enter 36 and then press the square root key. As seen above, this is not the correct syntax for a graphing calculator. To find this root, you would first press the square root key, and then enter 36. See the screen on the left at the top of the next page. The order of operations on a graphing calculator is also important, and current models assist the user by inserting parentheses when typical errors might occur. The open parenthesis that follows the square root symbol is automatically entered by the calculator, so that an expression such as $\sqrt{2 \times 8}$ will not be calculated incorrectly as $\sqrt{2} \times 8$. Compare the two entries and their results in the screen on the right.

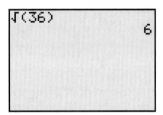

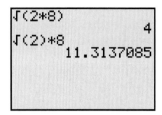

VIEWING WINDOWS

The viewing window for a graphing calculator is similar to the viewfinder in a camera. A camera usually cannot take a photograph of an entire view of a scene. The camera must be centered on some object and can only capture a portion of the available scenery. A camera with a zoom lens can photograph different views of the same scene by zooming in and out. Graphing calculators have similar capabilities. The xy-coordinate plane is infinite. The calculator screen can only show a finite, rectangular region in the plane, and it must be specified before the graph can be drawn. This is done by setting both minimum and maximum values for the x- and y-axes. The scale (distance between tick marks) is usually specified as well. Determining an appropriate viewing window for a graph is often a challenge, and many times it will take a few attempts before a satisfactory window is found.

The screen on the left shows a "standard" viewing window, and the graph of $y = 2x + 1$ is shown on the right. Using a different window would give a different view of the line.

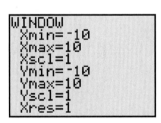

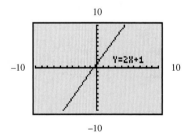

LOCATING POINTS ON A GRAPH: TRACING AND TABLES

Graphing calculators allow you to trace along the graph of an equation, and, while doing this, display the coordinates of points on the graph. See the screen on the left at the top of the next page, which indicates that the point (2, 5) lies on the graph of $y = 2x + 1$. Tables for equations can also be displayed. The screen on the right shows a partial table for this same equation. Note the middle of the screen, which indicates that when $x = 2$, $y = 5$.

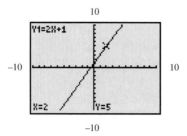

ADDITIONAL FEATURES

There are many features of graphing calculators that go far beyond the scope of this book. These calculators can be programmed, much like computers. Many of them can solve equations at the stroke of a key, analyze statistical data, and perform symbolic algebraic manipulations. Mathematicians from the past would have been amazed by today's calculators. Many important equations in mathematics cannot be solved by hand. However, their solutions can often be approximated using a calculator. Calculators also provide the opportunity to ask "What if . . . ?" more easily. Values in algebraic expressions can be altered and conjectures tested quickly.

FINAL COMMENTS

Despite the power of today's calculators, they cannot replace human thought. **In the entire problem-solving process, your brain is the most important component.** Calculators are only tools, and like any tool, they must be used appropriately in order to enhance our ability to understand mathematics. Mathematical insight may often be the quickest and easiest way to solve a problem; a calculator may neither be needed nor appropriate. By applying mathematical concepts, you can make the decision whether or not to use a calculator.

BEGINNING ALGEBRA

MYMATHLAB VERSION

The Real Number System

Growth in productivity—the amount of goods and services produced for each hour of work—is the most important factor in improving living standards of our country's population. In the latter part of the 1990s, the U.S. construction industry finally began to revive after a long period of decline. Having survived many years of downsizing and layoffs, manufacturing-related industries are again contributing to economic growth. This bar graph shows the increase in construction spending in the latter half of 1996 and all of 1997. During which month and year was construction spending highest? Several exercises in Section 1.2 use this graph. Throughout this chapter, we will see other specific examples and exercises that relate to construction and manufacturing.

Construction/Manufacturing

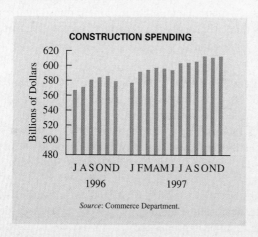

CONSTRUCTION SPENDING

Billions of Dollars

620
600
580
560
540
520
500
480

J A S O N D   J F M A M J J A S O N D
1996   1997

Source: Commerce Department.

1.1 Fractions

As preparation for the study of algebra, this section begins with a brief review of arithmetic. In everyday life the numbers seen most often are the **natural numbers,**

$$1, 2, 3, 4, \ldots ,$$

the **whole numbers,**

$$0, 1, 2, 3, 4, \ldots ,$$

and **fractions,** such as

$$\frac{1}{2}, \quad \frac{2}{3}, \quad \text{and} \quad \frac{15}{7}.$$

The parts of a fraction are named as follows.

$$\text{Fraction bar} \rightarrow \quad \frac{4}{7} \quad \left(\frac{a}{b} = a \div b \right) \quad \begin{array}{l} \leftarrow \text{Numerator} \\ \leftarrow \text{Denominator} \end{array}$$

As we will see later, the fraction bar represents division $\left(\frac{a}{b} = a \div b \right)$ and also serves as a grouping symbol.

CONNECTIONS

A common use of fractions is to measure dimensions of tools, amounts of building materials, and so on. We need to add, subtract, multiply, and divide fractions in order to solve many types of measurement problems.

FOR DISCUSSION OR WRITING
Discuss some situations in your experience where you have needed to perform the operations of addition, subtraction, multiplication, or division on fractions. (*Hint:* To get you started, think of art projects, carpentry projects, adjusting recipes, and working on cars.)

OBJECTIVE 1 **Learn the definition of** *factor.* In the statement $2 \times 9 = 18$, the numbers 2 and 9 are called **factors** of 18. Other factors of 18 include 1, 3, 6, and 18. The result of the multiplication, 18, is called the **product.**

The number 18 is **factored** by writing it as the product of two or more numbers. For example, 18 can be factored in several ways, as $6 \cdot 3$, or $18 \cdot 1$, or $9 \cdot 2$, or $3 \cdot 3 \cdot 2$. In algebra, a raised dot $\cdot$ is often used instead of the $\times$ symbol to indicate multiplication.

A natural number (except 1) is **prime** if it has only itself and 1 as factors. "Factors" are understood here to mean natural number factors. (By agreement, the number 1 is not a prime number.) The first dozen primes are

$$2, 3, 5, 7, 11, 13, 17, 19, 23, 29, 31, 37.$$

A natural number (except 1) that is not prime is a **composite** number.

It is often useful to find all the **prime factors** of a number—those factors that are prime numbers. For example, the only prime factors of 18 are 2 and 3.

EXAMPLE 1   Factoring Numbers

Write the number as the product of prime factors.

(a) 35

Write 35 as the product of the prime factors 5 and 7, or as

$$35 = 5 \cdot 7.$$

(b) 24

One way to begin is to divide by the smallest prime, 2, to get

$$24 = 2 \cdot 12.$$

Now divide 12 by 2 to find factors of 12.

$$24 = 2 \cdot 2 \cdot 6$$

Since 6 can be written as $2 \cdot 3$,

$$24 = 2 \cdot 2 \cdot 2 \cdot 3,$$

where all factors are prime.

 It is not necessary to start with the smallest prime factor, as shown in Example 1(b). In fact, no matter which prime factor we start with, we will *always* obtain the same prime factorization.

OBJECTIVE 2 Write fractions in lowest terms. We use prime numbers to write fractions in *lowest terms*. A fraction is in **lowest terms** when the numerator and denominator have no factors in common (other than 1). By the **basic principle of fractions,** if the numerator and denominator of a fraction are multiplied or divided by the *same* nonzero number, the value of the fraction is unchanged. To write a fraction in lowest terms, use these steps.

Writing a Fraction in Lowest Terms

Step 1 Write the numerator and the denominator as the product of prime factors.

Step 2 Divide the numerator and the denominator by the **greatest common factor,** the product of all factors common to both.

EXAMPLE 2   Writing Fractions in Lowest Terms

Write the fraction in lowest terms.

(a) $\dfrac{10}{15} = \dfrac{2 \cdot 5}{3 \cdot 5} = \dfrac{2 \cdot 1}{3 \cdot 1} = \dfrac{2}{3}$

Since 5 is the greatest common factor of 10 and 15, dividing both numerator and denominator by 5 gives the fraction in lowest terms.

(b) $\dfrac{15}{45} = \dfrac{3 \cdot 5}{3 \cdot 3 \cdot 5} = \dfrac{1 \cdot 1}{3 \cdot 1 \cdot 1} = \dfrac{1}{3}$

The factored form shows that 3 and 5 are the common factors of both 15 and 45. Dividing both 15 and 45 by $3 \cdot 5 = 15$ gives $\frac{15}{45}$ in lowest terms as $\frac{1}{3}$.

We can simplify this process by finding the greatest common factor in the numerator and denominator by inspection. For instance, in Example 2(b), we can use 15 rather than $3 \cdot 5$.

$$\frac{15}{45} = \frac{15}{3 \cdot 15} = \frac{1}{3 \cdot 1} = \frac{1}{3}$$

Errors may occur when writing fractions in lowest terms if the factor 1 is not included. To see this, refer to Example 2(b). In the equation

$$\frac{3 \cdot 5}{3 \cdot 3 \cdot 5} = \frac{?}{3},$$

if 1 is not written in the numerator when dividing common factors, you may make an error. The **?** should be replaced by **1**.

OBJECTIVE ☐3☐ **Multiply and divide fractions.** The basic operations on whole numbers, addition, subtraction, multiplication, and division, also apply to fractions. We multiply two fractions by first multiplying their numerators and then multiplying their denominators. This rule is written in symbols as follows.

Multiplying Fractions

If $\dfrac{a}{b}$ and $\dfrac{c}{d}$ are fractions, then $\quad \dfrac{a}{b} \cdot \dfrac{c}{d} = \dfrac{a \cdot c}{b \cdot d}$.

EXAMPLE 3   **Multiplying Fractions**

Find the product of $\frac{3}{8}$ and $\frac{4}{9}$, and write it in lowest terms.

First, multiply $\frac{3}{8}$ and $\frac{4}{9}$.

$$\frac{3}{8} \cdot \frac{4}{9} = \frac{3 \cdot 4}{8 \cdot 9} \qquad \text{Multiply numerators; multiply denominators.}$$

It is easiest to write a fraction in lowest terms while the product is in factored form. Factor 8 and 9 and then divide out common factors in the numerator and denominator.

$$\frac{3 \cdot 4}{8 \cdot 9} = \frac{3 \cdot 4}{2 \cdot 4 \cdot 3 \cdot 3} = \frac{\mathbf{1} \cdot 3 \cdot 4}{2 \cdot 4 \cdot 3 \cdot 3} \qquad \text{Factor. Introduce a factor of 1.}$$

$$= \frac{1}{2 \cdot 3} \qquad \text{3 and 4 are common factors.}$$

$$= \frac{1}{6} \qquad \text{Lowest terms}$$

Two fractions are **reciprocals** of each other if their product is 1. For example, $\frac{3}{4}$ and $\frac{4}{3}$ are reciprocals since

$$\frac{3}{4} \cdot \frac{4}{3} = \frac{12}{12} = 1.$$

Also, $\frac{7}{11}$ and $\frac{11}{7}$ are reciprocals of each other. We use the reciprocal to divide fractions. To *divide* two fractions, multiply the first fraction by the reciprocal of the second fraction.

Dividing Fractions

For the fractions $\dfrac{a}{b}$ and $\dfrac{c}{d}$, $\dfrac{a}{b} \div \dfrac{c}{d} = \dfrac{a}{b} \cdot \dfrac{d}{c}$.

(To divide by a fraction, multiply by its reciprocal.)

The reason this method works will be explained in Chapter 6. The answer to a division problem is called a **quotient.** For example, the quotient of 20 and 10 is 2, since $20 \div 10 = 2$.

E X A M P L E   4   Dividing Fractions

Find the following quotients, and write them in lowest terms.

(a) $\dfrac{3}{4} \div \dfrac{8}{5} = \dfrac{3}{4} \cdot \dfrac{5}{8} = \dfrac{3 \cdot 5}{4 \cdot 8} = \dfrac{15}{32}$ Multiply by the reciprocal of $\frac{8}{5}$.

(b) $\dfrac{3}{4} \div \dfrac{5}{8} = \dfrac{3}{4} \cdot \dfrac{8}{5} = \dfrac{3 \cdot 8}{4 \cdot 5} = \dfrac{3 \cdot 4 \cdot 2}{4 \cdot 5} = \dfrac{6}{5}$

OBJECTIVE 4 Add and subtract fractions. To find the **sum** of two fractions having the same denominator, add the numerators and keep the same denominator.

Adding Fractions

If $\dfrac{a}{b}$ and $\dfrac{c}{b}$ are fractions, then $\dfrac{a}{b} + \dfrac{c}{b} = \dfrac{a + c}{b}$.

E X A M P L E   5   Adding Fractions with the Same Denominator

Add.

(a) $\dfrac{3}{7} + \dfrac{2}{7} = \dfrac{3 + 2}{7} = \dfrac{5}{7}$ Add numerators and keep the same denominator.

(b) $\dfrac{2}{10} + \dfrac{3}{10} = \dfrac{2 + 3}{10} = \dfrac{5}{10} = \dfrac{1}{2}$

If the fractions to be added do not have the same denominators, the procedure above can still be used, but only *after* the fractions are rewritten with a common denominator. For example, to rewrite $\frac{3}{4}$ as a fraction with a denominator of 32,

$$\frac{3}{4} = \frac{?}{32},$$

find the number that can be multiplied by 4 to give 32. Since $4 \cdot 8 = 32$, use the number 8. By the basic principle, we can multiply the numerator and the denominator by 8.

$$\frac{3}{4} = \frac{3 \cdot 8}{4 \cdot 8} = \frac{24}{32}$$

Finding the Least Common Denominator

To add or subtract fractions with different denominators, find the **least common denominator (LCD)** as follows.

Step 1 Factor both denominators.

Step 2 For the LCD, use every factor that appears in any factored form. If a factor is repeated, use the largest number of repeats in the LCD.

The next example shows this procedure.

E X A M P L E   6   Adding Fractions with Different Denominators

Add the following fractions.

(a) $\dfrac{4}{15} + \dfrac{5}{9}$

To find the least common denominator, first factor both denominators.

$$15 = 5 \cdot 3 \qquad \text{and} \qquad 9 = 3 \cdot 3$$

Since 5 and 3 appear as factors, and 3 is a factor of 9 twice, the LCD is

$$5 \cdot 3 \cdot 3 \qquad \text{or} \qquad 45.$$

Write each fraction with 45 as denominator.

$$\frac{4}{15} = \frac{4 \cdot 3}{15 \cdot 3} = \frac{12}{45} \qquad \text{and} \qquad \frac{5}{9} = \frac{5 \cdot 5}{9 \cdot 5} = \frac{25}{45}$$

Now add the two equivalent fractions.

$$\frac{4}{15} + \frac{5}{9} = \frac{12}{45} + \frac{25}{45} = \frac{37}{45}$$

(b) $3\dfrac{1}{2} + 2\dfrac{3}{4}$

These numbers are called mixed numbers. A **mixed number** is understood to be the sum of a whole number and a fraction. We can add mixed numbers using either of two methods.

Method 1

Rewrite both numbers as follows.

$$3\frac{1}{2} = 3 + \frac{1}{2} = \frac{3}{1} + \frac{1}{2} = \frac{6}{2} + \frac{1}{2} = \frac{6+1}{2} = \frac{7}{2}$$

$$2\frac{3}{4} = 2 + \frac{3}{4} = \frac{8}{4} + \frac{3}{4} = \frac{8+3}{4} = \frac{11}{4}$$

Now add. The common denominator is 4.

$$3\frac{1}{2} + 2\frac{3}{4} = \frac{7}{2} + \frac{11}{4} = \frac{14}{4} + \frac{11}{4} = \frac{25}{4} \qquad \text{or} \qquad 6\frac{1}{4}$$

Method 2

Write $3\frac{1}{2}$ as $3\frac{2}{4}$. Then add vertically.

$$3\frac{1}{2} \qquad\qquad 3\frac{2}{4}$$

$$\underline{+\;2\frac{3}{4}} \quad\rightarrow\quad \underline{+\;2\frac{3}{4}}$$

$$5\frac{5}{4}$$

Since $\frac{5}{4} = 1\frac{1}{4}$,

$$5\frac{5}{4} = 5 + 1\frac{1}{4} = 6\frac{1}{4}, \quad\text{or}\quad \frac{25}{4}.$$

To multiply and divide mixed numbers, follow the same general procedure shown in Example 6(b), Method 1. First change to fractions, then perform the operation, and then convert back to a mixed number if desired. For example,

$$3\frac{1}{2} \cdot 2\frac{3}{4} = \frac{7}{2} \cdot \frac{11}{4} = \frac{77}{8} \quad\text{or}\quad 9\frac{5}{8}$$

$$3\frac{1}{2} \div 2\frac{3}{4} = \frac{7}{2} \div \frac{11}{4} = \frac{7}{2} \cdot \frac{4}{11} = \frac{14}{11} \quad\text{or}\quad 1\frac{3}{11}.$$

The **difference** between two numbers is found by subtraction. For example, $9 - 5 = 4$ so the difference between 9 and 5 is 4. Subtraction of fractions is similar to addition. Just subtract the numerators instead of adding them; again, keep the same denominator.

Subtracting Fractions

$$\frac{a}{b} - \frac{c}{b} = \frac{a - c}{b}$$

E X A M P L E 7   **Subtracting Fractions**

Subtract. Write the differences in lowest terms.

(a) $\dfrac{15}{8} - \dfrac{3}{8} = \dfrac{15 - 3}{8}$ *Subtract numerators; keep the same denominator.*

$$= \frac{12}{8} = \frac{3}{2} \qquad \text{*Lowest terms*}$$

(b) $\dfrac{7}{18} - \dfrac{4}{15}$

Here, $18 = 2 \cdot 3 \cdot 3$ and $15 = 3 \cdot 5$, so the LCD is $2 \cdot 3 \cdot 3 \cdot 5 = 90$.

$$\frac{7}{18} - \frac{4}{15} = \frac{7 \cdot 5}{2 \cdot 3 \cdot 3 \cdot 5} - \frac{4 \cdot 2 \cdot 3}{2 \cdot 3 \cdot 3 \cdot 5} = \frac{35}{90} - \frac{24}{90} = \frac{11}{90}$$

(c) $\dfrac{15}{32} - \dfrac{11}{45}$

Since $32 = 2 \cdot 2 \cdot 2 \cdot 2 \cdot 2$ and $45 = 3 \cdot 3 \cdot 5$, there are no common factors, and the LCD is $32 \cdot 45 = 1440$.

$$\frac{15}{32} - \frac{11}{45} = \frac{15 \cdot 45}{32 \cdot 45} - \frac{11 \cdot 32}{45 \cdot 32} \qquad \text{\textit{Get a common denominator.}}$$

$$= \frac{675}{1440} - \frac{352}{1440}$$

$$= \frac{323}{1440} \qquad \text{\textit{Subtract.}}$$

OBJECTIVE 5 **Solve applied problems that involve fractions.** Applied problems often require work with fractions. For example, when a carpenter reads diagrams and plans, he or she often must work with fractions whose denominators are 2, 4, 8, 16, or 32, as shown in the next example.

EXAMPLE 8   **Adding Fractions to Solve a Manufacturing (Woodworking) Problem**

The diagram in Figure 1 appears in the book *Woodworker's 39 Sure-Fire Projects*. It is the front view of a corner bookcase/desk. Add the fractions shown in the diagram to find the height of the bookcase/desk.

We must add the following measures (in inches):

$$\frac{3}{4}, \quad 4\frac{1}{2}, \quad 9\frac{1}{2}, \quad \frac{3}{4}, \quad 9\frac{1}{2}, \quad \frac{3}{4}, \quad 4\frac{1}{2}.$$

Begin by changing $4\frac{1}{2}$ to $4\frac{2}{4}$ and $9\frac{1}{2}$ to $9\frac{2}{4}$, since the common denominator is 4. Then, use Method 2 from Example 6(b).

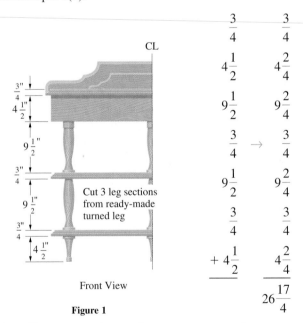

Front View

Figure 1

Since $\frac{17}{4} = 4\frac{1}{4}$, $26\frac{17}{4} = 26 + 4\frac{1}{4} = 30\frac{1}{4}$. The height is $30\frac{1}{4}$ inches. It is best to give answers as mixed numbers in applications like this.

OBJECTIVE ▢6 Interpret data in a circle graph. A **circle graph** or **pie chart** is often used to give a pictorial representation of data. A circle is used to indicate the total of all the categories represented. The circle is divided into sectors, or wedges (like pieces of pie) whose sizes show the relative magnitudes of the categories. The sum of all the fractional parts must be 1 (for 1 whole circle).

 E X A M P L E 9 Using a Pie Chart to Interpret Information

The pie chart in Figure 2 shows the job categories of African Americans employed in 1993 (age 16 or older).

Figure 2

(a) In a group of 150,000 such employees, about how many would we expect to be employed in precision production/repair?

To find the answer, we multiply the fraction indicated in the chart for the category $\left(\frac{2}{25}\right)$ by the number of people in the group (150,000):

$$\frac{2}{25} \cdot 150,000 = \frac{300,000}{25} = 12,000.$$

About 12,000 people in the group are employed in precision production/repair.

(b) *Estimate* the number employed in service.

The fraction $\frac{23}{100}$ is approximately $\frac{1}{4}$. Therefore, a good estimate for this number is

$$\frac{1}{4} \cdot 150,000 = 37,500.$$

1.1 EXERCISES

Decide whether each statement is true or false. If it is false, say why.

1. In the fraction $\frac{3}{7}$, 3 is the numerator and 7 is the denominator.

2. The mixed number equivalent of $\frac{41}{5}$ is $8\frac{1}{5}$.

3. The fraction $\frac{17}{51}$ is in lowest terms.

4. The reciprocal of $\dfrac{8}{2}$ is $\dfrac{4}{1}$.

5. The product of 8 and 2 is 10.

6. The difference between 12 and 2 is 6.

Identify each number as prime, composite, or neither. If the number is composite, write it as the product of prime factors. See Example 1.

7. 19	**8.** 31	**9.** 64	**10.** 99
11. 3458	**12.** 1025	**13.** 1	**14.** 0
15. 30	**16.** 40	**17.** 500	**18.** 700
19. 124	**20.** 120	**21.** 29	**22.** 83

Write each fraction in lowest terms. See Example 2.

23. $\dfrac{8}{16}$ **24.** $\dfrac{4}{12}$ **25.** $\dfrac{15}{18}$ **26.** $\dfrac{16}{20}$

27. $\dfrac{15}{45}$ **28.** $\dfrac{16}{64}$ **29.** $\dfrac{144}{120}$ **30.** $\dfrac{132}{77}$

31. One of the following is the correct way to write $\dfrac{16}{24}$ in lowest terms. Which one is it?

    (a) $\dfrac{16}{24} = \dfrac{8+8}{8+16} = \dfrac{8}{16} = \dfrac{1}{2}$ **(b)** $\dfrac{16}{24} = \dfrac{4 \cdot 4}{4 \cdot 6} = \dfrac{4}{6}$

    (c) $\dfrac{16}{24} = \dfrac{8 \cdot 2}{8 \cdot 3} = \dfrac{2}{3}$ **(d)** $\dfrac{16}{24} = \dfrac{14+2}{21+3} = \dfrac{2}{3}$

32. For the fractions $\dfrac{p}{q}$ and $\dfrac{r}{s}$, which one of the following can serve as a common denominator?

    (a) $q \cdot s$ **(b)** $q + s$ **(c)** $p \cdot r$ **(d)** $p + r$

Find each product or quotient, and write it in lowest terms. See Examples 3 and 4.

33. $\dfrac{4}{5} \cdot \dfrac{6}{7}$ **34.** $\dfrac{5}{9} \cdot \dfrac{10}{7}$ **35.** $\dfrac{1}{10} \cdot \dfrac{12}{5}$ **36.** $\dfrac{6}{11} \cdot \dfrac{2}{3}$

37. $\dfrac{15}{4} \cdot \dfrac{8}{25}$ **38.** $\dfrac{4}{7} \cdot \dfrac{21}{8}$ **39.** $2\dfrac{2}{3} \cdot 5\dfrac{4}{5}$ **40.** $3\dfrac{3}{5} \cdot 7\dfrac{1}{6}$

41. $\dfrac{5}{4} \div \dfrac{3}{8}$ **42.** $\dfrac{7}{6} \div \dfrac{9}{10}$ **43.** $\dfrac{32}{5} \div \dfrac{8}{15}$ **44.** $\dfrac{24}{7} \div \dfrac{6}{21}$

45. $\dfrac{3}{4} \div 12$ **46.** $\dfrac{2}{5} \div 30$ **47.** $2\dfrac{5}{8} \div 1\dfrac{15}{32}$ **48.** $2\dfrac{3}{10} \div 7\dfrac{4}{5}$

49. Write a summary explaining how to multiply and divide two fractions. Give examples.

50. Write a summary explaining how to add and subtract two fractions. Give examples.

Find each sum or difference, and write it in lowest terms. See Examples 5–7.

51. $\dfrac{7}{12} + \dfrac{1}{12}$ **52.** $\dfrac{3}{16} + \dfrac{5}{16}$ **53.** $\dfrac{5}{9} + \dfrac{1}{3}$ **54.** $\dfrac{4}{15} + \dfrac{1}{5}$

55. $3\dfrac{1}{8} + \dfrac{1}{4}$ **56.** $5\dfrac{3}{4} + \dfrac{2}{3}$ **57.** $\dfrac{7}{12} - \dfrac{1}{9}$ **58.** $\dfrac{11}{16} - \dfrac{1}{12}$

59. $6\dfrac{1}{4} - 5\dfrac{1}{3}$ **60.** $8\dfrac{4}{5} - 7\dfrac{4}{9}$ **61.** $\dfrac{5}{3} + \dfrac{1}{6} - \dfrac{1}{2}$ **62.** $\dfrac{7}{15} + \dfrac{1}{6} - \dfrac{1}{10}$

The following chart appears on a package of Quaker Quick Grits.

	Microwave		Stove Top	
Servings	1	1	4	6
Water	$\frac{3}{4}$ cup	1 cup	3 cups	4 cups
Grits	3 Tbsp	3 Tbsp	$\frac{3}{4}$ cup	1 cup
Salt (optional)	dash	dash	$\frac{1}{4}$ tsp	$\frac{1}{2}$ tsp

Use the chart to answer the questions in Exercises 63 and 64.

63. How many cups of water would be needed for 8 microwave servings?

64. How many tsp of salt would be needed for 5 stove top servings? (*Hint:* 5 is halfway between 4 and 6.)

Work each problem. See Example 8.

65. On Tuesday, February 10, 1998, Earthlink stock on the NASDAQ exchange closed the day at $4\frac{5}{8}$ (dollars) ahead of where it had opened. It closed at $38\frac{5}{8}$ (dollars). What was its opening price?

66. A report in *USA Today* on February 10, 1998, stated that Teva Pharmaceutical skidded $9\frac{9}{16}$ (dollars) to $37\frac{1}{2}$ (dollars) after the Israeli drugmaker said fourth-quarter net income was likely to be below expectations. What was its price before the skid?

67. A hardware store sells a 40-piece socket wrench set. The measure of the largest socket is $\frac{3}{4}$ inch, while the measure of the smallest socket is $\frac{3}{16}$ inch. What is the difference between these measures?

68. Two sockets in a socket wrench set have measures of $\frac{9}{16}$ inch and $\frac{3}{8}$ inch. What is the difference between these two measures?

69. A motel owner has decided to expand his business by buying a piece of property next to the motel. The property has an irregular shape, with five sides as shown in the figure. Find the total distance around the piece of property. (This is called the *perimeter* of the figure.)

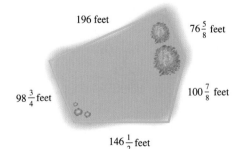

196 feet, $76\frac{5}{8}$ feet, $100\frac{7}{8}$ feet, $146\frac{1}{2}$ feet, $98\frac{3}{4}$ feet

70. Find the perimeter of the triangle in the figure.

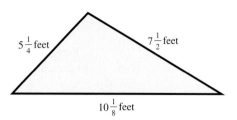

$5\frac{1}{4}$ feet $7\frac{1}{2}$ feet

$10\frac{1}{8}$ feet

71. A piece of board is $15\frac{5}{8}$ inches long. If it must be divided into 3 pieces of equal length, how long must each piece be?

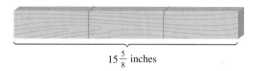

$15\frac{5}{8}$ inches

72. If one serving of a macaroni and cheese meal requires $\frac{1}{8}$ cup of chopped onions, how many cups of onions will $7\frac{1}{2}$ servings require?

73. Tex's favorite recipe for barbecue sauce calls for $2\frac{1}{3}$ cups of tomato sauce. The recipe makes enough barbecue sauce to serve 7 people. How much tomato sauce is needed for 1 serving?

74. A cake recipe calls for $1\frac{3}{4}$ cups of sugar. A caterer has $15\frac{1}{2}$ cups of sugar on hand. How many cakes can he make?

The pie chart gives fractional job categories of white employees in the workforce in 1993 (age 16 or older). Use the chart to answer the questions in Exercises 75 and 76. See Example 9.

75. In a random sample of 5000 such employees, about how many would we expect to be employed in service occupations?

76. In a random sample of 7500 such employees, about how many would we expect to be employed in farming/forestry/fishing occupations?

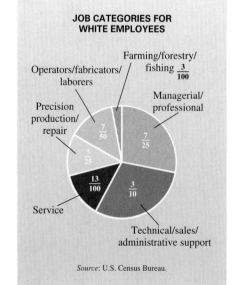

JOB CATEGORIES FOR WHITE EMPLOYEES

Operators/fabricators/ laborers

Farming/forestry/ fishing $\frac{3}{100}$

Managerial/ professional

Precision production/ repair

$\frac{7}{50}$ $\frac{7}{25}$

$\frac{3}{25}$

$\frac{13}{100}$ $\frac{3}{10}$

Service

Technical/sales/ administrative support

Source: U.S. Census Bureau.

77. At the conclusion of the AWL softball league season, batting statistics for five players were as follows.

Player	At-bats	Hits	Home Runs
Stephanie Baldock	40	9	2
Jennifer Crum	36	12	3
Jason Jordan	11	5	1
Greg Tobin	16	8	0
David Perry	20	10	2

Answer each of the following, using estimation skills as necessary.

(a) Which player got a hit in exactly $\frac{1}{3}$ of his or her at-bats?

(b) Which player got a hit in just less than $\frac{1}{2}$ of his or her at-bats?

(c) Which player got a home run in just less than $\frac{1}{10}$ of his or her at-bats?

(d) Which player got a hit in just less than $\frac{1}{4}$ of his or her at-bats?

(e) Which two players got hits in exactly the same fractional parts of their at-bats? What was the fractional part, expressed in lowest terms?

78. For each of the following, write a fraction in lowest terms that represents the region described.

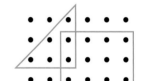

(a) the dots in the rectangle as a part of the dots in the entire figure

(b) the dots in the triangle as a part of the dots in the entire figure

(c) the dots in the overlapping region of the triangle and the rectangle as a part of the dots in the triangle alone

(d) the dots in the overlapping region of the triangle and the rectangle as a part of the dots in the rectangle alone

79. Estimate the best approximation for the following sum:

$$\frac{14}{26} + \frac{98}{99} + \frac{100}{51} + \frac{90}{31} + \frac{13}{27}.$$

(a) 6   **(b)** 7   **(c)** 5   **(d)** 8

80. In the local softball league, the first five games produced the following results: David Horwitz got 8 hits in 20 at-bats, and Chalon Bridges got 12 hits in 30 at-bats. David claims that he did just as well as Chalon. Is he correct? Why or why not?

1.2 Exponents, Order of Operations, and Inequality

OBJECTIVE 1 Use exponents. In a multiplication problem, the same factor may appear several times. For example, in the product

$$3 \cdot 3 \cdot 3 \cdot 3 = 81,$$

the factor 3 appears four times. In algebra, repeated factors are written with an *exponent.* For example, in $3 \cdot 3 \cdot 3 \cdot 3$, the number 3 appears as a factor four times, so the product is written as 3^4, and is read "3 to the fourth power."

$$3 \cdot 3 \cdot 3 \cdot 3 = 3^4$$

The number 4 is the **exponent** or **power** and 3 is the **base** in the **exponential expression** 3^4. A natural number exponent, then, tells how many times the base is used as a factor. A number raised to the first power is simply that number. For example, $5^1 = 5$ and $\left(\frac{1}{2}\right)^1 = \frac{1}{2}$.

EXAMPLE 1 Evaluating an Exponential Expression

Find the values of the following.

(a) 5^2 $5 \cdot 5 = 25$

    5 is used as a factor 2 times.

Read 5^2 as "5 squared."

(b) 6^3 $6 \cdot 6 \cdot 6 = 216$

    6 is used as a factor 3 times.

Read 6^3 as "6 cubed."

(c) $2^5 = 2 \cdot 2 \cdot 2 \cdot 2 \cdot 2 = 32$ 2 is used as a factor 5 times.

Read 2^5 as "2 to the fifth power."

(d) $\left(\frac{2}{3}\right)^3 = \frac{2}{3} \cdot \frac{2}{3} \cdot \frac{2}{3} = \frac{8}{27}$ $\frac{2}{3}$ is used as a factor 3 times.

OBJECTIVE 2 Use the order of operations rules. Many problems involve more than one operation. To indicate the order in which the operations should be performed, we often use *grouping symbols.* If no grouping symbols are used, we apply the order of operations rules discussed below.

Consider the expression $5 + 2 \cdot 3$. To show that the multiplication should be performed before the addition, parentheses can be used to write

$$5 + (2 \cdot 3) = 5 + 6 = 11.$$

If addition is to be performed first, the parentheses should group $5 + 2$ as follows.

$$(5 + 2) \cdot 3 = 7 \cdot 3 = 21$$

Other grouping symbols used in more complicated expressions are brackets [ ], braces { }, and fraction bars. (For example, in $\frac{8-2}{3}$, the expression $8 - 2$ is considered to be grouped in the numerator.)

To work problems with more than one operation, use the following **order of operations.** This order is used by most calculators and computers.

Order of Operations

If grouping symbols are present, simplify within them, innermost first (and above and below fraction bars separately), in the following order.

Step 1 Apply all exponents.

Step 2 Do any multiplications or divisions in the order in which they occur, working from left to right.

Step 3 Do any additions or subtractions in the order in which they occur, working from left to right.

If no grouping symbols are present, start with Step 1.

A dot has been used to show multiplication; another way to show multiplication is with parentheses. For example, 3(7), (3)7, and (3)(7) each mean $3 \cdot 7$ or 21. The next example shows the use of parentheses for multiplication.

E X A M P L E   2   **Using the Order of Operations**

Find the values of the following.

(a) $9(6 + 11)$

Using the order of operations given above, work first inside the parentheses.

$$9(6 + 11) = 9(17) \qquad \text{Work inside parentheses.}$$
$$= 153 \qquad \text{Multiply.}$$

(b) $6 \cdot 8 + 5 \cdot 2$

Do any multiplications, working from left to right, and then add.

$$6 \cdot 8 + 5 \cdot 2 = 48 + 10 \qquad \text{Multiply.}$$
$$= 58 \qquad \text{Add.}$$

(c) $2(5 + 6) + 7 \cdot 3 = 2(11) + 7 \cdot 3 \qquad \text{Work inside parentheses.}$
$$= 22 + 21 \qquad \text{Multiply.}$$
$$= 43 \qquad \text{Add.}$$

(d) $9 - 2^3 + 5$

Find 2^3 first.

$$9 - 2^3 + 5 = 9 - 2 \cdot 2 \cdot 2 + 5 \qquad \text{Use the exponent.}$$
$$= 9 - 8 + 5 \qquad \text{Multiply.}$$
$$= 1 + 5 \qquad \text{Subtract.}$$
$$= 6 \qquad \text{Add.}$$

(e) $72 \div 2 \cdot 3 + 4 \cdot 2^3$

$$72 \div 2 \cdot 3 + 4 \cdot 2^3 = 72 \div 2 \cdot 3 + 4 \cdot 8 \qquad \text{Use the exponent.}$$
$$= 36 \cdot 3 + 4 \cdot 8 \qquad \text{Perform the division.}$$
$$= 108 + 32 \qquad \text{Perform the multiplications.}$$
$$= 140 \qquad \text{Add.}$$

Notice that the multiplications and divisions are performed from left to right *as they appear;* then the additions and subtractions should be done from left to right, *as they appear.*

OBJECTIVE **3**   **Use more than one grouping symbol.** An expression with double (or *nested*) parentheses, such as $2(8 + 3(6 + 5))$, can be confusing. For clarity, square brackets, [ ], often are used in place of one pair of parentheses. Fraction bars also act as grouping symbols. The next example explains these situations.

E X A M P L E   3   Using Brackets and Fraction Bars as Grouping Symbols
Simplify each expression.

(a) $2[8 + 3(6 + 5)]$
Work first within the parentheses, and then simplify inside the brackets until a single number remains.

$$2[8 + 3(6 + 5)] = 2[8 + 3(11)]$$
$$= 2[8 + 33]$$
$$= 2[41]$$
$$= 82$$

(b) $\dfrac{4(5 + 3) + 3}{2(3) - 1}$

The expression can be written as the quotient

$$[4(5 + 3) + 3] \div [2(3) - 1],$$

which shows that the fraction bar groups the numerator and denominator separately. Simplify both numerator and denominator, then divide, if possible.

$$\frac{4(5 + 3) + 3}{2(3) - 1} = \frac{4(8) + 3}{2(3) - 1} \qquad \text{Work inside parentheses.}$$

$$= \frac{32 + 3}{6 - 1} \qquad \text{Multiply.}$$

$$= \frac{35}{5} \qquad \text{Add and subtract.}$$

$$= 7 \qquad \text{Divide.}$$

   Parentheses and fraction bars are used as grouping symbols to indicate an expression that represents a single number. That is why we must first simplify within parentheses and above and below fraction bars.

OBJECTIVE **4**   **Know the meanings of $\neq$, $<$, $>$, $\leq$, and $\geq$.** So far, we have used the symbols for the operations of arithmetic and the symbol for equality ($=$). The equality symbol with a slash through it, $\neq$, means "is not equal to." For example,

$$7 \neq 8$$

indicates that 7 is not equal to 8.

Note that for *any* value of x, −2x and 2x are additive inverses; this is why we can use the inverse property in this simplification.

NOTE The detailed procedure shown in Example 8 is seldom, if ever, used in practice. We include the example to show how the properties of this section apply, even though steps may be skipped when actually doing the simplification.

OBJECTIVE 5 Use the distributive property. The everyday meaning of the word *distribute* is "to give out from one to several." An important property of real number operations involves this idea.

Look at the following statements.

$$2(5 + 8) = 2(13) = 26$$
$$2(5) + 2(8) = 10 + 16 = 26$$

Since both expressions equal 26,

$$2(5 + 8) = 2(5) + 2(8).$$

This result is an example of the **distributive property,** the only property involving *both* addition and multiplication. With this property, a product can be changed to a sum or difference.

The distributive property says that multiplying a number *a* by a sum of numbers *b* + *c* gives the same result as multiplying *a* by *b* and *a* by *c* and then adding the two products.

Distributive Property

$$a(b + c) = ab + ac \quad \text{and} \quad (b + c)a = ba + ca$$

As the arrows show, the *a* outside the parentheses is "distributed" over the *b* and *c* inside. Another form of the distributive property is valid for subtraction.

$$a(b - c) = ab - ac \quad \text{and} \quad (b - c)a = ba - ca$$

The distributive property also can be extended to more than two numbers.

$$a(b + c + d) = ab + ac + ad$$

NOTE The distributive property can be used "in reverse." For example, we can write

$$ac + bc = (a + b)c.$$

EXAMPLE 9 Using the Distributive Property

Use the distributive property to rewrite each expression.

(a) $5(9 + 6) = 5 \cdot 9 + 5 \cdot 6$ Distributive property
$$= 45 + 30 \quad \text{Multiply.}$$
$$= 75 \quad \text{Add.}$$

(b) $4(x + 5 + y) = 4x + 4 \cdot 5 + 4y$ Distributive property
$$= 4x + 20 + 4y \quad \text{Multiply.}$$

(c) $-2(x + 3) = -2x + (-2)(3)$ Distributive property
          $= -2x - 6$ Multiply.

(d) $3(k - 9) = 3k - 3 \cdot 9$ Distributive property
        $= 3k - 27$ Multiply.

(e) $8(3r + 11t + 5z) = 8(3r) + 8(11t) + 8(5z)$ Distributive property
     $= (8 \cdot 3)r + (8 \cdot 11)t + (8 \cdot 5)z$ Associative property
     $= 24r + 88t + 40z$ Multiply.

(f) $6 \cdot 8 + 6 \cdot 2 = 6(8 + 2)$ Distributive property
     $= 6(10) = 60$ Add, then multiply.

(g) $4x - 4m = 4(x - m)$ Distributive property

(h) $6x - 12 = 6 \cdot x - 6 \cdot 2 = 6(x - 2)$ Distributive property

The symbol $-a$ may be interpreted as $-1 \cdot a$. Similarly, when a negative sign precedes an expression within parentheses, it may also be interpreted as a factor of -1. The distributive property is used to remove parentheses from expressions such as $-(2y + 3)$. We do this by first writing $-(2y + 3)$ as $-1 \cdot (2y + 3)$.

$$-(2y + 3) = -1 \cdot (2y + 3)$$
$$= -1 \cdot (2y) + (-1) \cdot (3) \quad \text{Distributive property}$$
$$= -2y - 3 \quad \text{Multiply.}$$

EXAMPLE 10   Using the Distributive Property to Remove Parentheses
Write without parentheses.

(a) $-(7r - 8) = -1(7r) + (-1)(-8)$ Distributive property
     $= -7r + 8$ Multiply.

(b) $-(-9w + 2) = 9w - 2$

The properties discussed here are the basic properties that justify how we do algebra. You should know them by name because we will be referring to them frequently. Here is a summary of these properties.

Properties of Addition and Multiplication

For any real numbers a, b, and c, the following properties hold.

Commutative Properties $a + b = b + a$ $ab = ba$

Associative Properties $(a + b) + c = a + (b + c)$
       $(ab)c = a(bc)$

Identity Properties   There is a real number 0 such that
$$a + 0 = a \quad \text{and} \quad 0 + a = a.$$
There is a real number 1 such that
$$a \cdot 1 = a \quad \text{and} \quad 1 \cdot a = a.$$

Properties of Addition and Multiplication (continued)

Inverse Properties   For each real number a, there is a single real number $-a$ such that

$$a + (-a) = 0 \quad \text{and} \quad (-a) + a = 0.$$

For each nonzero real number a, there is a single real number $\frac{1}{a}$ such that

$$a \cdot \frac{1}{a} = 1 \quad \text{and} \quad \frac{1}{a} \cdot a = 1.$$

Distributive Property $a(b + c) = ab + ac \qquad (b + c)a = ba + ca$

1.7 EXERCISES

Match each item in Column I with the correct choice(s) from Column II. Choices may be used once, more than once, or not at all.

I

1. Identity element for addition
2. Identity element for multiplication
3. Additive inverse of a
4. Multiplicative inverse, or reciprocal, of the nonzero number a
5. The number that is its own additive inverse
6. The two numbers that are their own multiplicative inverses
7. The only number that has no multiplicative inverse
8. An example of the associative property
9. An example of the commutative property
10. An example of the distributive property

II

A. $(5 \cdot 4) \cdot 3 = 5 \cdot (4 \cdot 3)$
B. 0
C. $-a$
D. -1
E. $5 \cdot 4 \cdot 3 = 60$
F. 1
G. $(5 \cdot 4) \cdot 3 = 3 \cdot (5 \cdot 4)$
H. $5(4 + 3) = 5 \cdot 4 + 5 \cdot 3$
I. $\dfrac{1}{a}$

Decide whether each statement is an example of the commutative, associative, identity, inverse, or distributive property. See Examples 1, 2, 3, 5, 6, 7, and 9.

EXERCISES

11. $7 + 18 = 18 + 7$
12. $13 + 12 = 12 + 13$
13. $5(13 \cdot 7) = (5 \cdot 13) \cdot 7$
14. $-4(2 \cdot 6) = (-4 \cdot 2) \cdot 6$
15. $-6 + (12 + 7) = (-6 + 12) + 7$
16. $(-8 + 13) + 2 = -8 + (13 + 2)$
17. $-6 + 6 = 0$
18. $12 + (-12) = 0$
19. $\left(\dfrac{2}{3}\right)\left(\dfrac{3}{2}\right) = 1$
20. $\left(\dfrac{5}{8}\right)\left(\dfrac{8}{5}\right) = 1$
21. $2.34 + 0 = 2.34$
22. $-8.456 + 0 = -8.456$
23. $(4 + 17) + 3 = 3 + (4 + 17)$
24. $(-8 + 4) + (-12) = -12 + (-8 + 4)$
25. $6(x + y) = 6x + 6y$
26. $14(t + s) = 14t + 14s$
27. $-\dfrac{5}{9} = -\dfrac{5}{9} \cdot \dfrac{3}{3} = -\dfrac{15}{27}$
28. $\dfrac{13}{12} = \dfrac{13}{12} \cdot \dfrac{7}{7} = \dfrac{91}{84}$
29. $5(2x) + 5(3y) = 5(2x + 3y)$
30. $3(5t) - 3(7r) = 3(5t - 7r)$

31. The following conversation actually took place between one of the authors of this book and his son, Jack, when Jack was four years old:

DADDY: "Jack, what is 3 + 0?"

JACK: "3."

DADDY: "Jack, what is 4 + 0?"

JACK: "4. And Daddy, *string* plus zero equals *string*!"

What property of addition did Jack recognize?

32. The distributive property holds for multiplication with respect to addition. Is there a distributive property for addition with respect to multiplication? If not, give an example to show why.

33. Write a paragraph explaining in your own words the following properties of addition and multiplication: commutative, associative, identity, inverse.

34. Write a paragraph explaining in your own words the distributive property of multiplication with respect to addition. Give examples.

Use the indicated property to write a new expression that is equal to the given expression. Then simplify the new expression if possible. See Examples 1, 2, 5, 7, and 9.

35. $r + 7$;   commutative

36. $t + 9$;   commutative

37. $s + 0$;   identity

38. $w + 0$;   identity

39. $-6(x + 7)$;   distributive

40. $-5(y + 2)$;   distributive

41. $(w + 5) + (-3)$;   associative

42. $(b + 8) + (-10)$;   associative

Use the properties of this section to simplify each expression. See Examples 7 and 8.

43. $6t + 8 - 6t + 3$

44. $9r + 12 - 9r + 1$

45. $\frac{2}{3}x - 11 + 11 - \frac{2}{3}x$

46. $\frac{1}{5}y + 4 - 4 - \frac{1}{5}y$

47. $\left(\frac{9}{7}\right)(-.38)\left(\frac{7}{9}\right)$

48. $\left(\frac{4}{5}\right)(-.73)\left(\frac{5}{4}\right)$

49. $t + (-t) + \frac{1}{2}(2)$

50. $w + (-w) + \frac{1}{4}(4)$

51. Evaluate $25 - (6 - 2)$ and evaluate $(25 - 6) - 2$. Do you think subtraction is associative?

52. Evaluate $180 \div (15 \div 3)$ and evaluate $(180 \div 15) \div 3$. Do you think division is associative?

53. Suppose that a student shows you the following work.

$$-3(4 - 6) = -3(4) - 3(6) = -12 - 18 = -30$$

The student has made a very common error. Explain the student's mistake, and work the problem correctly.

54. Explain how the procedure of changing $\frac{3}{4}$ to $\frac{9}{12}$ requires the use of the multiplicative identity element, 1.

Use the distributive property to rewrite each expression. Simplify if possible. See Example 9.

55. $5x + x$

56. $6q + q$

57. $4(t + 3)$

58. $5(w + 4)$

59. $-8(r + 3)$

60. $-11(x + 4)$

61. $-5(y - 4)$

62. $-9(g - 4)$

63. $-\frac{4}{3}(12y + 15z)$

64. $-\frac{2}{5}(10b + 20a)$

65. $8 \cdot z + 8 \cdot w$

66. $4 \cdot s + 4 \cdot r$

67. $7(2v) + 7(5r)$

68. $13(5w) + 13(4p)$

69. $8(3r + 4s - 5y)$

EXERCISES

70. $2(5u - 3v + 7w)$ **71.** $q + q + q$ **72.** $m + m + m + m$

73. $-5x + x$ **74.** $-9p + p$

Use the distributive property to write each expression without parentheses. See Example 10.

75. $-(4t + 3m)$ **76.** $-(9x + 12y)$ **77.** $-(-5c - 4d)$

78. $-(-13x - 15y)$ **79.** $-(-3q + 5r - 8s)$ **80.** $-(-4z + 5w - 9y)$

81. The operations of "getting out of bed" and "taking a shower" are not commutative. Give an example of another pair of everyday operations that are not commutative.

82. The phrase "dog biting man" has two different meanings, depending on how the words are associated:

(dog biting) man   dog (biting man)

Give another example of a three-word phrase that has different meanings depending on how the words are associated.

■ RELATING CONCEPTS (EXERCISES 83-86)

In Section 1.6 we used a pattern to see that the product of two negative numbers is a positive number. In the group of exercises that follows, we show another justification for determining the sign of the product of two negative numbers.

Work Exercises 83–86 in order.

83. Evaluate the expression $-3[5 + (-5)]$ by using the rules for order of operations.

84. Write the expression in Exercise 83 using the distributive property. Do not simplify the products.

85. The product -3×5 should be one of the terms you wrote when answering Exercise 84. Based on the results in Section 1.6, what is this product?

86. In Exercise 83, you should have obtained 0 as an answer. Now, consider the following, using the results of Exercises 83 and 85.

$$-3[5 + (-5)] = -3(5) + (-3)(-5)$$
$$0 = -15 + ?$$

The question mark represents the product $(-3)(-5)$. When added to -15, it must give a sum of 0. Therefore, how must we interpret $(-3)(-5)$?

Did you make the connection that a rule can be obtained in more than one way, with consistent results from each method?

1.8 Simplifying Expressions

OBJECTIVES

1 Simplify expressions.

2 Identify terms and numerical coefficients.

3 Identify like terms.

OBJECTIVE 1 Simplify expressions. In this section we show how to simplify expressions using the properties of addition and multiplication introduced in the previous section.

EXAMPLE 1 Simplifying Expressions

Simplify the following expressions.

(a) $4x + 8 + 9$

Since $8 + 9 = 17$,

$$4x + 8 + 9 = 4x + 17.$$

4 Combine like terms.

5 Simplify expressions from word phrases.

(b) $4(3m - 2n)$

Use the distributive property first.

$$4(3m - 2n) = 4(3m) - 4(2n) \qquad \text{Arrows denote distributive property.}$$
$$= (4 \cdot 3)m - (4 \cdot 2)n \qquad \text{Associative property}$$
$$= 12m - 8n$$

(c)
$$6 + 3(4k + 5) = 6 + 3(4k) + 3(5) \qquad \text{Distributive property}$$
$$= 6 + (3 \cdot 4)k + 3(5) \qquad \text{Associative property}$$
$$= 6 + 12k + 15$$
$$= 6 + 15 + 12k \qquad \text{Commutative property}$$
$$= 21 + 12k$$

(d)
$$5 - (2y - 8) = 5 - 1 \cdot (2y - 8) \qquad \text{Replace } - \text{ with } -1.$$
$$= 5 - 2y + 8 \qquad \text{Distributive property}$$
$$= 5 + 8 - 2y \qquad \text{Commutative property}$$
$$= 13 - 2y$$

 NOTE In Example 1, parts (c) and (d), a different use of the commutative property would have resulted in answers of $12k + 21$ and $-2y + 13$. These answers also would be acceptable.

The steps using the commutative and associative properties will not be shown in the rest of the examples, but you should be aware that they are usually involved.

OBJECTIVE 2 Identify terms and numerical coefficients. A **term** is a number, a variable, or a product or quotient of numbers and variables raised to powers.* Examples of terms include

$$-9x^2, \quad 15y, \quad -3, \quad 8m^2n, \quad \frac{2}{p}, \quad \text{and} \quad k.$$

The **numerical coefficient** of the term $9m$ is 9, the numerical coefficient of $-15x^3y^2$ is -15, the numerical coefficient of x is 1, and the numerical coefficient of 8 is 8. In the expression $\frac{x}{3}$, the numerical coefficient of x is $\frac{1}{3}$. Do you see why?

CAUTION It is important to be able to distinguish between *terms* and *factors*. For example, in the expression $8x^3 + 12x^2$, there are two *terms*, $8x^3$ and $12x^2$. On the other hand, in the one-term expression $(8x^3)(12x^2)$, $8x^3$ and $12x^2$ are *factors*.

Several examples of terms and their numerical coefficients follow.

*Another name for certain terms, **monomial,** is introduced in Chapter 4.

Term	Numerical Coefficient
$-7y$	-7
$8p$	8
$34r^3$	34
$-26x^5yz^4$	-26
$-k$	-1
$\dfrac{x}{7}$	$\dfrac{1}{7}$

OBJECTIVE 3 Identify like terms. Terms with exactly the same variables that have the same exponents are **like terms.** For example, $9m$ and $4m$ have the same variable and are like terms. Also, $6x^3$ and $-5x^3$ are like terms. The terms $-4y^3$ and $4y^2$ have different exponents and are **unlike terms.**

Here are some additional examples.

$$5x \text{ and } -12x \qquad 3x^2y \text{ and } 5x^2y \qquad \text{Like terms}$$
$$4xy^2 \text{ and } 5xy \qquad -7w^3z^3 \text{ and } 2xz^3 \qquad \text{Unlike terms}$$

OBJECTIVE 4 Combine like terms. Recall the distributive property:

$$x(y + z) = xy + xz.$$

As seen in the previous section, this statement can also be written "backward" as

$$xy + xz = x(y + z).$$

This form of the distributive property may be used to find the sum or difference of like terms. For example,

$$3x + 5x = (3 + 5)x = 8x.$$

This process is called **combining like terms.**

 Remember that *only like terms may be combined.* For example, $5x^2 + 2x \neq 7x^3$.

EXAMPLE 2 Combining Like Terms

Combine like terms in the following expressions.

(a) $9m + 5m$

Use the distributive property as given above.

$$9m + 5m = (9 + 5)m = 14m$$

(b) $6r + 3r + 2r = (6 + 3 + 2)r = 11r$ Distributive property

(c) $4x + x = 4x + 1x = (4 + 1)x = 5x$ (Note: $x = 1x$.)

(d) $16y^2 - 9y^2 = (16 - 9)y^2 = 7y^2$

(e) $32y + 10y^2$ cannot be combined because $32y$ and $10y^2$ are unlike terms. The distributive property cannot be used here to combine coefficients.

When an expression involves parentheses, the distributive property is used both "forward" and "backward" to combine like terms, as shown in the following example.

E X A M P L E 3   Simplifying Expressions Involving Like Terms

Combine like terms in the following expressions.

(a) $14y + 2(6 + 3y) = 14y + 2(6) + 2(3y)$ Distributive property

$\qquad\qquad\qquad\quad = 14y + 12 + 6y$ Multiply.

$\qquad\qquad\qquad\quad = 20y + 12$ Combine like terms.

(b) $9k - 6 - 3(2 - 5k) = 9k - 6 - 3(2) - 3(-5k)$ Distributive property

$\qquad\qquad\qquad\qquad\quad = 9k - 6 - 6 + 15k$ Multiply.

$\qquad\qquad\qquad\qquad\quad = 24k - 12$ Combine like terms.

(c) $-(2 - r) + 10r = -1(2 - r) + 10r$ Replace $-$ with -1.

$\qquad\qquad\qquad\quad = -1(2) - 1(-r) + 10r$ Distributive property

$\qquad\qquad\qquad\quad = -2 + 1r + 10r$ Multiply.

$\qquad\qquad\qquad\quad = -2 + 11r$ Combine like terms.

(d) $5(2a - 6) - 3(4a - 9) = 10a - 30 - 12a + 27$ Distributive property

$\qquad\qquad\qquad\qquad\quad = -2a - 3$ Combine like terms.

Example 3(d) shows that the commutative property can be used with subtraction by treating the subtracted terms as the addition of their additive inverses.

 Examples 2 and 3 suggest that like terms may be combined by adding or subtracting the coefficients of the terms and keeping the same variable factors.

O B J E C T I V E 5   Simplify expressions from word phrases. Earlier we saw how to translate words, phrases, and statements into expressions and equations. Now we can simplify translated expressions by combining like terms.

E X A M P L E 4   Converting Words to a Mathematical Expression

Convert to a mathematical expression, and simplify: The sum of 9, five times a number, four times the number, and six times the number.

The word "sum" indicates that the terms should be added. Use x to represent the number. Then the phrase translates as follows.

$\qquad\qquad 9 + 5x + 4x + 6x$ Write as a mathematical expression.

$\qquad\qquad = 9 + 15x$ Combine like terms.

 In Example 4, we are dealing with an expression to be simplified, *not* an equation to be solved.

1.8 EXERCISES

Decide whether each statement is true or false.

1. $6t + 5t^2 = 11t^3$

2. $9xy^2 - 3x^2y = 6xy$

3. $8r^2 + 3r - 12r^2 + 4r = -4r^2 + 7r$

4. $4 + 3t^3 = 7t^3$

In Exercises 5–8, choose the letter of the correct response.

5. Which one of the following is true for all real numbers x?
  (a) $6 + 2x = 8x$ **(b)** $6 - 2x = 4x$
  (c) $6x - 2x = 4x$ **(d)** $3 + 8(4x - 6) = 11(4x - 6)$

6. Which one of the following is an example of a pair of like terms?
  (a) $6t, 6w$ **(b)** $-8x^2y, 9xy^2$ **(c)** $5ry, 6yr$ **(d)** $-5x^2, 2x^3$

7. Which one of the following is an example of a term with numerical coefficient 5?
  (a) $5x^3y^7$ **(b)** x^5 **(c)** $\dfrac{x}{5}$ **(d)** 5^2xy^3

8. Which one of the following is a correct translation for "six times a number, subtracted from the product of eleven and the number" (if x represents the number)?
  (a) $6x - 11x$ **(b)** $11x - 6x$ **(c)** $(11 + x) - 6x$ **(d)** $6x - (11 + x)$

Simplify each expression. See Example 1.

9. $4r + 19 - 8$

10. $7t + 18 - 4$

11. $5 + 2(x - 3y)$

12. $8 + 3(s - 6t)$

13. $-2 - (5 - 3p)$

14. $-10 - (7 - 14r)$

Give the numerical coefficient of each term.

15. $-12k$ **16.** $-23y$ **17.** $5m^2$ **18.** $-3n^6$ **19.** xw

20. pq **21.** $-x$ **22.** $-t$ **23.** 74 **24.** 98

25. Give an example of a pair of like terms in the variable x, such that one of them has a negative numerical coefficient, one has a positive numerical coefficient, and their sum has a positive numerical coefficient.

26. Give an example of a pair of unlike terms such that each term has x as the only variable factor.

Identify each group of terms as like *or* unlike.

27. $8r, -13r$ **28.** $-7a, 12a$ **29.** $5z^4, 9z^3$ **30.** $8x^5, -10x^3$

31. $4, 9, -24$ **32.** $7, 17, -83$ **33.** x, y **34.** t, s

35. There is an old saying, "You can't add apples and oranges." Explain how this saying can be applied to the goal of Objective 4 in this section.

36. Explain how the distributive property is used in combining $6t + 5t$ to get $11t$.

Simplify each expression by combining like terms. See Examples 1–3.

37. $4k + 3 - 2k + 8 + 7k - 16$

38. $9x + 7 - 13x + 12 + 8x - 15$

39. $-\dfrac{4}{3} + 2t + \dfrac{1}{3}t - 8 - \dfrac{8}{3}t$

40. $-\dfrac{5}{6} + 8x + \dfrac{1}{6}x - 7 - \dfrac{7}{6}$

41. $-5.3r + 4.9 - 2r + .7 + 3.2r$

42. $2.7b + 5.8 - 3b + .5 - 4.4b$

43. $2y^2 - 7y^3 - 4y^2 + 10y^3$

44. $9x^4 - 7x^6 + 12x^4 + 14x^6$

45. $13p + 4(4 - 8p)$

46. $5x + 3(7 - 2x)$

47. $-4(y - 7) - 6$

48. $-5(t - 13) - 4$

49. $-5(5y - 9) + 3(3y + 6)$

50. $-3(2t + 4) + 8(2t - 4)$

51. $-4(-3k + 3) - (6k - 4) - 2k + 1$ **52.** $-5(8j + 2) - (5j - 3) - 3j + 17$

53. $-7.5(2y + 4) - 2.9(3y - 6)$ **54.** $8.4(6t - 6) + 2.4(9 - 3t)$

Convert each phrase into a mathematical expression. Use x as the variable. Combine like terms when possible. See Example 4.

55. Five times a number, added to the sum of the number and three

56. Six times a number, added to the sum of the number and six

57. A number multiplied by -7, subtracted from the sum of 13 and six times the number

58. A number multiplied by 5, subtracted from the sum of 14 and eight times the number

59. Six times a number added to -4, subtracted from twice the sum of three times the number and 4 (*Hint: Twice* means two times.)

60. Nine times a number added to 6, subtracted from triple the sum of 12 and 8 times the number (*Hint: Triple* means three times.)

61. Write the expression $9x - (x + 2)$ using words, as in Exercises 55–60.

62. Write the expression $2(3x + 5) - 2(x + 4)$ using words, as in Exercises 55–60.

RELATING CONCEPTS (EXERCISES 63–70)

Work Exercises 63–70 in order. They will help prepare you for graphing later in the text.

63. Evaluate the expression $x + 2$ for the values of x shown in the chart.

x	$x + 2$
0	
1	
2	
3	

64. Based on your results from Exercise 63, complete the following statement: For every increase of 1 unit for x, the value of $x + 2$ increases by _____ unit(s).

65. Repeat Exercise 63 for these expressions:
   (a) $x + 1$ **(b)** $x + 3$ **(c)** $x + 4$

66. Based on your results from Exercises 63 and 65, make a conjecture (an educated guess) about what happens to the value of an expression of the form $x + b$ for any value of b, as x increases by 1 unit.

67. Repeat Exercise 63 for these expressions:
   (a) $2x + 2$ **(b)** $3x + 2$ **(c)** $4x + 2$

68. Based on your results from Exercise 67, complete the following statement: For every increase of 1 unit for x, the value of $mx + 2$ increases by _____ units.

69. Repeat Exercise 63 and compare your results to those in Exercise 67 for these expressions:
   (a) $2x + 7$ **(b)** $3x + 5$ **(c)** $4x + 1$

70. Based on your results from Exercises 63–69, complete the following statement: For every increase of 1 unit for x, the value of $mx + b$ increases by _____ units.

Did you make the connection that an increase of 1 for x yields an increase of m for $mx + b$?

CHAPTER 1 GROUP ACTIVITY

⊞ Comparing Shapes of Houses

Objective: Use arithmetic skills to make comparisons.

People throughout the world live in different shaped homes. As the population of the earth continues to grow, issues of housing and heating become increasingly important. This activity will explore perimeters and areas of some of these different shaped homes. Floor plans for three different homes are given below.

A. As a group, look at the dimensions of the given floor plans. Considering only the dimensions, which plan do you think has the greatest area?

B. Now have each student in your group pick one floor plan. For each plan, find the following. (Round all answers to the nearest whole number. In Plan one, let $\pi = 3.14$.)
   1. The area of the plan
   2. The perimeter or circumference (the distance around the outside) of the plan

C. Share your findings with the group and answer the following questions.
   1. What did you determine about the areas of the three floor plans?
   2. Which plan has the smallest perimeter? Which has the largest perimeter?
   3. Why do you think houses with round floor plans might be more energy efficient?
   4. What advantages do you think houses with square or rectangular floor plans have? Why do you think floor plans with these shapes are most common for homes today?

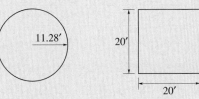 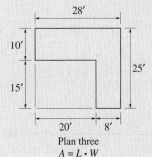

Plan one
$A = \pi r^2$
$C = 2\pi r$

Plan two
$A = s^2$

Plan three
$A = L \cdot W$

CHAPTER 1 SUMMARY

KEY TERMS

1.1 natural numbers	mixed number	signed numbers	identity property
whole numbers	difference	integers	identity element for
numerator	**1.2** exponent (power)	graph	addition
denominator	base	coordinate	identity element for
factor	exponential	rational numbers	multiplication
product	expression	set-builder notation	inverse property
factored	grouping symbols	irrational numbers	distributive property
prime	**1.3** variable	real numbers	**1.8** term
composite	algebraic expression	additive inverse	numerical coefficient
greatest common	equation	(opposite)	like and unlike terms
factor	solution	absolute value	combining like terms
lowest terms	set	**1.6** multiplicative	
reciprocal	element	inverse	
quotient	**1.4** number line	(reciprocal)	
sum	negative numbers	**1.7** commutative property	
least common	positive numbers	associative property	
denominator (LCD)			

NEW SYMBOLS

a^n	n factors of a	$\leq$	is less than or equal to (read from left to right)	$|x|$	absolute value of x
[ ]	square brackets (used as grouping symbols)	$\geq$	is greater than or equal to (read from left to right)	$\dfrac{1}{x}$ or $1/x$	the multiplicative inverse, or reciprocal, of the nonzero number x
$=$	is equal to	{ }	set braces	$a(b), (a)(b), a \cdot b,$ or ab	a times b
$\neq$	is not equal to	$\{x \mid x \text{ has a certain property}\}$			
$<$	is less than (read from left to right)		set-builder notation	$\dfrac{a}{b}$ or a/b	a divided by b
$>$	is greater than (read from left to right)	$-x$	the additive inverse, or opposite, of x		

TEST YOUR WORD POWER

See how well you have learned the vocabulary in this chapter. Answers, with examples, are given at the bottom of the next page.

1. A **factor** is
(a) the answer in an addition problem
(b) the answer in a multiplication problem
(c) one of two or more numbers that are added to get another number
(d) one of two or more numbers that are multiplied to get another number.

2. A number is **prime** if
(a) it cannot be factored
(b) it has just one factor
(c) it has only itself and 1 as factors
(d) it has at least two different factors.

3. An **exponent** is
(a) a symbol that tells how many numbers are being multiplied

(b) a number raised to a power
(c) a number that tells how many times a factor is repeated
(d) one of two or more numbers that are multiplied.

4. A **variable** is
(a) a symbol used to represent an unknown number
(b) a value that makes an equation true
(c) a solution of an equation
(d) the answer in a division problem.

5. An **integer** is
(a) a positive or negative number
(b) a natural number, its opposite, or zero
(c) any number that can be graphed on a number line
(d) the quotient of two numbers.

6. A **coordinate** is
(a) the number that corresponds to a point on a number line
(b) the graph of a number
(c) any point on a number line
(d) the distance from 0 on a number line.

7. The **absolute value** of a number is
(a) the graph of the number
(b) the reciprocal of the number
(c) the opposite of the number
(d) the distance between 0 and the number on a number line.

8. A **term** is
(a) a numerical factor
(b) a number or a product or quotient of numbers and variables raised to powers
(c) one of several variables with the same exponents
(d) a sum of numbers and variables raised to powers.

9. A **numerical coefficient** is
(a) the numerical factor in a term
(b) the number of terms in an expression
(c) a variable raised to a power
(d) the variable factor in a term.

QUICK REVIEW

CONCEPTS	EXAMPLES

1.1 FRACTIONS

Operations with Fractions

Addition/Subtraction:

1. To add/subtract fractions with the same denominator, add/subtract the numerators and keep the same denominator.

2. To add/subtract fractions with different denominators, find the LCD and write each fraction with this LCD. Then follow the procedure above.

Multiplication: Multiply numerators and multiply denominators.

Division: Multiply the first fraction by the reciprocal of the second fraction.

Perform the operations.

$$\frac{2}{5} + \frac{7}{5} = \frac{2+7}{5} = \frac{9}{5}$$

$$\frac{2}{3} - \frac{1}{2} = \frac{4}{6} - \frac{3}{6} \quad \text{6 is the LCD.}$$
$$= \frac{4-3}{6} = \frac{1}{6}$$

$$\frac{4}{3} \cdot \frac{5}{6} = \frac{20}{18} = \frac{10}{9}$$

$$\frac{6}{5} \div \frac{1}{4} = \frac{6}{5} \cdot \frac{4}{1} = \frac{24}{5}$$

1.2 EXPONENTS, ORDER OF OPERATIONS, AND INEQUALITY

Order of Operations
Simplify within parentheses or above and below fraction bars first, in the following order.

1. Apply all exponents.

2. Do any multiplications or divisions from left to right.

3. Do any additions or subtractions from left to right. If no grouping symbols are present, start with Step 1.

Simplify $36 - 4(2^2 + 3)$.

$$36 - 4(2^2 + 3) = 36 - 4(4 + 3)$$
$$= 36 - 4(7)$$
$$= 36 - 28$$
$$= 8$$

Answers to Test Your Word Power
1. (d) *Example:* Since $2 \times 5 = 10$, the numbers 2 and 5 are factors of 10; other factors of 10 are $-10, -5, -2, -1, 1,$ and 10. **2.** (c) *Examples:* 2, 3, 11, 41, 53 **3.** (c) *Example:* In 2^3, the number 3 is the exponent (or power), so 2 is a factor three times; $2^3 = 2 \cdot 2 \cdot 2 = 8$. **4.** (a) *Examples: a, b, c* **5.** (b) *Examples:* $-9, 0, 6$ **6.** (a) *Example:* The point graphed three units to the right of 0 on a number line has coordinate 3. **7.** (d) *Examples:* $|2| = 2$ and $|-2| = 2$ **8.** (b) *Examples:* $6, \frac{x}{2}, -4ab^2$ **9.** (a) *Example:* The term 3 has numerical coefficient 3, 8z has numerical coefficient 8, and $-10x^4y$ has numerical coefficient -10.

CONCEPTS	EXAMPLES

1.3 VARIABLES, EXPRESSIONS, AND EQUATIONS

Evaluate an expression with a variable by substituting a given number for the variable.

Evaluate $2x + y^2$ if $x = 3$ and $y = -4$.

$$2x + y^2 = 2(3) + (-4)^2$$
$$= 6 + 16$$
$$= 22$$

Values of a variable that make an equation true are solutions of the equation.

Is 2 a solution of $5x + 3 = 18$?

$$5(2) + 3 = 18 \quad ?$$
$$13 = 18 \quad \text{False}$$

2 is not a solution.

1.4 REAL NUMBERS AND THE NUMBER LINE

The Ordering of Real Numbers
a is less than b if a is to the left of b on the number line.

Graph -2, 0, and 3.

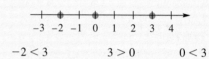

$$-2 < 3 \qquad 3 > 0 \qquad 0 < 3$$

The additive inverse of x is $-x$.

$$-(5) = -5 \qquad -(-7) = 7 \qquad -0 = 0$$

The absolute value of x, $|x|$, is the distance between x and 0 on the number line.

$$|13| = 13 \qquad |0| = 0 \qquad |-5| = 5$$

1.5 ADDITION AND SUBTRACTION OF REAL NUMBERS

Rules for Addition
To add two numbers with the same sign, add their absolute values. The sum has that same sign.

Add.

$$9 + 4 = 13$$
$$-8 + (-5) = -13$$

To add two numbers with different signs, subtract their absolute values. The sum has the sign of the number with larger absolute value.

$$7 + (-12) = -5$$
$$-5 + 13 = 8$$

Definition of Subtraction

$$x - y = x + (-y)$$

Subtract.

$$5 - (-2) = 5 + 2 = 7$$

Rules for Subtraction
1. Change the subtraction symbol to the addition symbol.

$$-3 - 4 = -3 + (-4) = -7$$

2. Change the sign of the number being subtracted.

$$-2 - (-6) = -2 + 6 = 4$$

3. Add, using the rules for addition.

$$13 - (-8) = 13 + 8 = 21$$

1.6 MULTIPLICATION AND DIVISION OF REAL NUMBERS

Rules for Multiplying and Dividing Signed Numbers
The product (or quotient) of two numbers having the *same sign* is *positive;* the product (or quotient) of two numbers having *different signs* is *negative.*

Multiply or divide.

$$6 \cdot 5 = 30 \qquad (-7)(-8) = 56 \qquad \frac{20}{4} = 5$$

$$\frac{-24}{-6} = 4 \qquad (-6)(5) = -30 \qquad (6)(-5) = -30$$

$$\frac{-18}{9} = -2 \qquad \frac{49}{-7} = -7$$

CONCEPTS	EXAMPLES

Definition of Division

$$\frac{x}{y} = x \cdot \frac{1}{y}, \quad y \neq 0$$

$$\frac{10}{2} = 10 \cdot \frac{1}{2} = 5$$

Division by 0 is undefined.

$$\frac{5}{0} \text{ is undefined.}$$

0 divided by a nonzero number equals 0.

$$\frac{0}{5} = 0$$

1.7 PROPERTIES OF REAL NUMBERS

Commutative

$$a + b = b + a$$
$$ab = ba$$

$$7 + (-1) = -1 + 7$$
$$5(-3) = (-3)5$$

Associative

$$(a + b) + c = a + (b + c)$$
$$(ab)c = a(bc)$$

$$(3 + 4) + 8 = 3 + (4 + 8)$$
$$[(-2)(6)](4) = (-2)[(6)(4)]$$

Identity

$$a + 0 = a \qquad 0 + a = a$$
$$a \cdot 1 = a \qquad 1 \cdot a = a$$

$$-7 + 0 = -7 \qquad 0 + (-7) = -7$$
$$9 \cdot 1 = 9 \qquad 1 \cdot 9 = 9$$

Inverse

$$a + (-a) = 0 \qquad -a + a = 0$$
$$a \cdot \frac{1}{a} = 1 \qquad \frac{1}{a} \cdot a = 1 \quad (a \neq 0)$$

$$7 + (-7) = 0 \qquad -7 + 7 = 0$$
$$-2\left(-\frac{1}{2}\right) = 1 \qquad -\frac{1}{2}(-2) = 1$$

Distributive

$$a(b + c) = ab + ac$$
$$(b + c)a = ba + ca$$
$$a(b - c) = ab - ac$$

$$5(4 + 2) = 5(4) + 5(2)$$
$$(4 + 2)5 = 4(5) + 2(5)$$
$$9(5 - 4) = 9(5) - 9(4)$$

1.8 SIMPLIFYING EXPRESSIONS

Only like terms may be combined.

Simplify $-3y^2 + 6y^2 + 14y^2 = 17y^2$.

$$4(3 + 2x) - 6(5 - x)$$
$$= 12 + 8x - 30 + 6x \qquad \text{Distributive property}$$
$$= 14x - 18$$

CHAPTER 1 REVIEW EXERCISES

[1.1] *Perform each operation.**

1. $\dfrac{8}{5} \div \dfrac{32}{15}$

2. $\dfrac{3}{8} + 3\dfrac{1}{2} - \dfrac{3}{16}$

---

*For help with any of these exercises, refer to the section given in brackets.

3. The pie chart illustrates how 800 people responded to a survey that asked "Do you believe that there was a conspiracy to assassinate John F. Kennedy?" What fractional part of the group did not have an opinion?

4. Based on the chart in Exercise 3, how many people responded "yes"?

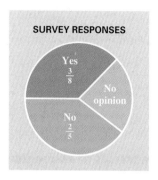

SURVEY RESPONSES

Yes
$\frac{3}{8}$

No opinion

No
$\frac{2}{5}$

[1.2]   *Find the value of each exponential expression.*

5. 5^4 **6.** $\left(\dfrac{3}{5}\right)^3$ **7.** $(.02)^5$ **8.** $(.001)^3$

Find the value of each expression.

9. $8 \cdot 5 - 13$

10. $7[3 + 6(3^2)]$

11. $\dfrac{9(4^2 - 3)}{4 \cdot 5 - 17}$

12. $\dfrac{6(5 - 4) + 2(4 - 2)}{3^2 - (4 + 3)}$

Tell whether each statement is true or false.

13. $12 \cdot 3 - 6 \cdot 6 \le 0$ **14.** $3[5(2) - 3] > 20$ **15.** $9 \le 4^2 - 8$

Write each word statement in symbols.

16. Thirteen is less than seventeen.

17. Five plus two is not equal to ten.

18. Americans are literally bombarded by mail-order catalogs on a daily basis. The bar graph shows the estimated number of catalogs mailed to consumers and businesses during the years 1983–1996.

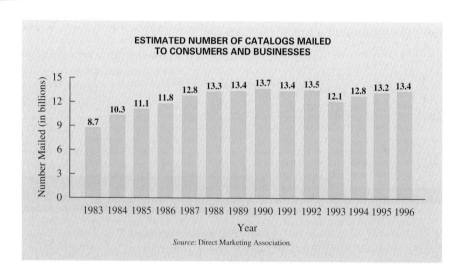

ESTIMATED NUMBER OF CATALOGS MAILED TO CONSUMERS AND BUSINESSES

Source: Direct Marketing Association.

(a) In which years were *fewer than* 13.4 billion catalogs mailed?

(b) In which years were *at least* 12.8 billion mailed?

(c) How many *total* catalogs were mailed in the five years having the largest numbers of mailings?

[1.3] *Find the numerical value of each expression if x = 6 and y = 3.*

19. $2x + 6y$ **20.** $4(3x - y)$ **21.** $\dfrac{x}{3} + 4y$ **22.** $\dfrac{x^2 + 3}{3y - x}$

Change each word phrase to an algebraic expression. Use x as the variable to represent the number.

23. Six added to a number
24. A number subtracted from eight
25. Nine subtracted from six times a number
26. Three-fifths of a number added to 12

Decide whether the given number is a solution of the given equation.

27. $5x + 3(x + 2) = 22;\quad 2$ **28.** $\dfrac{t + 5}{3t} = 1;\quad 6$

Change each word statement to an equation. Use x as the variable. Then find the solution from the set {0, 2, 4, 6, 8, 10}.

29. Six less than twice a number is 10. **30.** The product of a number and 4 is 8.

[1.4] *Graph each group of numbers on a number line.*

31. $-4, -\dfrac{1}{2}, 0, 2.5, 5$ **32.** $-2, |-3|, -3, |-1|$

Classify each number, using the sets natural numbers, whole numbers, integers, rational numbers, irrational numbers, real numbers.

33. $\dfrac{4}{3}$ **34.** $\sqrt{6}$

Select the smaller number in each pair.

35. $-10, 5$ **36.** $-8, -9$ **37.** $-\dfrac{2}{3}, -\dfrac{3}{4}$ **38.** $0, -|23|$

Decide whether each statement is true or false.

39. $12 > -13$ **40.** $0 > -5$ **41.** $-9 < -7$ **42.** $-13 \geq -13$

For the following, (a) find the opposite of the number and (b) find the absolute value of the number.

43. -9 **44.** 0 **45.** 6 **46.** $-\dfrac{5}{7}$

Simplify each number by removing absolute value symbols.

47. $|-12|$ **48.** $-|3|$ **49.** $-|-19|$ **50.** $-|9 - 2|$

[1.5] *Perform the indicated operations.*

51. $-10 + 4$ **52.** $14 + (-18)$ **53.** $-8 + (-9)$
54. $\dfrac{4}{9} + \left(-\dfrac{5}{4}\right)$ **55.** $-13.5 + (-8.3)$ **56.** $(-10 + 7) + (-11)$
57. $[-6 + (-8) + 8] + [9 + (-13)]$ **58.** $(-4 + 7) + (-11 + 3) + (-15 + 1)$
59. $-7 - 4$ **60.** $-12 - (-11)$
61. $5 - (-2)$ **62.** $-\dfrac{3}{7} - \dfrac{4}{5}$
63. $2.56 - (-7.75)$ **64.** $(-10 - 4) - (-2)$
65. $(-3 + 4) - (-1)$ **66.** $-(-5 + 6) - 2$

Write a numerical expression for each phrase, and simplify the expression.

67. 19 added to the sum of -31 and 12

68. 13 more than the sum of -4 and -8

69. The difference between -4 and -6

70. Five less than the sum of 4 and -8

Find the solution of the equation from the set $\{-3, -2, -1, 0, 1, 2, 3\}$ by guessing or by trial and error.

71. $x + (-2) = -4$

72. $12 + x = 11$

Solve each problem.

73. Like many people, Kareem Dunlap neglects to keep up his checkbook balance. When he finally balanced his account, he found the balance was $-\$23.75$, so he deposited $\$50.00$. What is his new balance?

74. The low temperature in Yellowknife, in the Canadian Northwest Territories, one January day was $-26°$F. It rose $16°$ that day. What was the high temperature?

75. Eric owed his brother $\$28$. He repaid $\$13$ but then borrowed another $\$14$. What positive or negative amount represents his present financial status?

76. If the temperature drops $7°$ below its previous level of $-3°$, what is the new temperature?

77. A football team gained 3 yards on the first play from scrimmage, lost 12 yards on the second play, and then gained 13 yards on the third play. How many yards did the team gain or lose altogether?

78. In 1985, the construction industry had 4,480,000 employees. By 1990, this number had increased by 759,000 employees, but it experienced a decrease of 530,000 employees by 1994. How many employees were there in 1994? (*Source:* U.S. Bureau of the Census.)

[1.6] *Perform the indicated operations.*

79. $(-12)(-3)$

80. $15(-7)$

81. $\left(-\dfrac{4}{3}\right)\left(-\dfrac{3}{8}\right)$

82. $(-4.8)(-2.1)$

83. $5(8 - 12)$

84. $(5 - 7)(8 - 3)$

85. $2(-6) - (-4)(-3)$ **86.** $3(-10) - 5$

87. $\dfrac{-36}{-9}$

88. $\dfrac{220}{-11}$

89. $-\dfrac{1}{2} \div \dfrac{2}{3}$

90. $-33.9 \div (-3)$

91. $\dfrac{-5(3) - 1}{8 - 4(-2)}$

92. $\dfrac{5(-2) - 3(4)}{-2[3 - (-2)] - 1}$

93. $\dfrac{10^2 - 5^2}{8^2 + 3^2 - (-2)}$

94. $\dfrac{(.6)^2 + (.8)^2}{(-1.2)^2 - (-.56)}$

Evaluate each expression if $x = -5$, $y = 4$, and $z = -3$.

95. $6x - 4z$

96. $5x + y - z$

97. $5x^2$

98. $z^2(3x - 8y)$

Write a numerical expression for each phrase, and simplify the expression.

99. Nine less than the product of -4 and 5

100. Five-sixths of the sum of 12 and -6

101. The quotient of 12 and the sum of 8 and -4

102. The product of -20 and 12, divided by the difference between 15 and -15

Write each sentence in symbols, using x as the variable, and find the solution by guessing or by trial and error. All solutions come from the list of integers between -12 and 12.

103. 8 times a number is -24.

104. The quotient of a number and 3 is -2.

105. The payrolls and average salaries of 5 of the 28 major league baseball teams on opening day of 1998 are listed here.

Team	Payroll	Average Salary
Baltimore Orioles	$68,988,134	$2,555,116
New York Yankees	63,460,567	2,440,791
Cleveland Indians	59,583,500	2,127,982
Atlanta Braves	59,536,000	2,126,286
Texas Rangers	55,304,595	1,975,164

Source: The Associated Press.

(a) What is the average of the five payrolls? Round to the nearest dollar.

(b) What is the average of the five average salaries? Round to the nearest dollar.

106. The bar graph shows the 1993 sales in millions of dollars of four of the largest brands in the United States. What was the average of these sales?

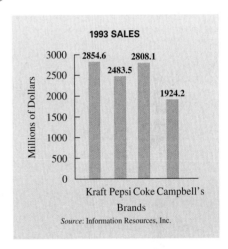

[1.7] *Decide whether each statement is an example of the commutative, associative, identity, inverse, or distributive property.*

107. $6 + 0 = 6$

108. $5 \cdot 1 = 5$

109. $-\dfrac{2}{3}\left(-\dfrac{3}{2}\right) = 1$

110. $17 + (-17) = 0$

111. $5 + (-9 + 2) = [5 + (-9)] + 2$

112. $w(xy) = (wx)y$

113. $3x + 3y = 3(x + y)$

114. $(1 + 2) + 3 = 3 + (1 + 2)$

Use the distributive property to rewrite each expression. Simplify if possible.

115. $7y + y$ **116.** $-12(4 - t)$ **117.** $3(2s) + 3(5y)$ **118.** $-(-4r + 5s)$

119. Evaluate $25 - (5 - 2)$ and $(25 - 5) - 2$. Use this example to explain why subtraction is not associative.

120. Evaluate $180 \div (15 \div 5)$ and $(180 \div 15) \div 5$. Use this example to explain why division is not associative.

[1.8] *Combine terms whenever possible.*

121. $2m + 9m$

122. $15p^2 - 7p^2 + 8p^2$

123. $5p^2 - 4p + 6p + 11p^2$

124. $-2(3k - 5) + 2(k + 1)$

125. $7(2m + 3) - 2(8m - 4)$

126. $-(2k + 8) - (3k - 7)$

Perform the indicated operations.

127. $[(-2) + 7 - (-5)] + [-4 - (-10)]$

128. $\left(-\dfrac{5}{6}\right)^2$

129. $\dfrac{6(-4) + 2(-12)}{5(-3) + (-3)}$

130. $\dfrac{3}{8} - \dfrac{5}{12}$

131. $\dfrac{8^2 + 6^2}{7^2 + 1^2}$

132. $-16(-3.5) - 7.2(-3)$

133. $2\dfrac{5}{6} - 4\dfrac{1}{3}$

134. $-8 + [(-4 + 17) - (-3 - 3)]$

135. $-\dfrac{12}{5} \div \dfrac{9}{7}$

136. $(-8 - 3) - 5(2 - 9)$

137. $5x^2 - 12y^2 + 3x^2 - 9y^2$

138. $-4(2t + 1) - 8(-3t + 4)$

139. Write a sentence or two explaining the special considerations involving zero when dividing.

140. "Two negatives give a positive" is often heard from students. Is this correct? Use more precise language in explaining what this means.

141. Use x as the variable and write an expression for "the product of 5 and the sum of a number and 7." Then use the distributive property to rewrite the expression.

142. The highest temperature ever recorded in Albany, New York, was 99°F, while the lowest was 112° less than the highest. What was the lowest temperature ever recorded in Albany? (*Source: The World Almanac and Book of Facts, 1998.*)

The year-to-year percentage change in net corporate income for companies in the DJ-Global U.S. Index is given in the accompanying graph for the years 1994 through the second quarter of 1997.

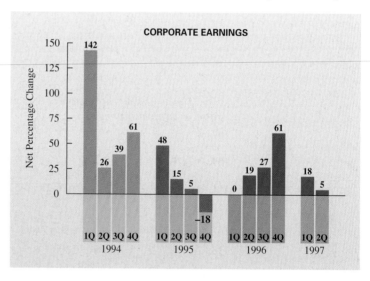

143. What signed number represents the change from the fourth quarter in 1994 to the fourth quarter in 1995?

144. What signed number represents the change from the third quarter in 1995 to the third quarter in 1996?

*The order of exercises in this final group does not correspond to the order in which topics occur in the chapter. This random ordering should help you prepare for the chapter test in yet another way.

78. When a number is multiplied by 4, the result is 6. Find the number.

79. When a number is divided by -5, the result is 2. Find the number.

80. If twice a number is divided by 5, the result is 4. Find the number.

2.2 More on Solving Linear Equations

OBJECTIVES

1 Learn and use the four steps for solving a linear equation.

2 Solve equations with fractions or decimals as coefficients.

3 Recognize equations with no solutions or infinitely many solutions.

4 Write expressions for two related unknown quantities.

OBJECTIVE **1** Learn and use the four steps for solving a linear equation. To solve linear equations in general, follow these steps.

Solving a Linear Equation

Step 1 **Simplify each side separately.** Use the distributive property to clear parentheses and combine terms, as needed.

Step 2 **Isolate the variable terms on one side.** If necessary, use the addition property to get all variable terms on one side of the equation and all numbers on the other.

Step 3 **Isolate the variable.** Use the multiplication property, if necessary, to get the equation in the form x = a number.

Step 4 **Check.** Check the solution by substituting into the *original* equation.

EXAMPLE 1 Using the Four Steps to Solve an Equation

Solve $3r + 4 - 2r - 7 = 4r + 3$.

Step 1 $3r + 4 - 2r - 7 = 4r + 3$

$\qquad\qquad r - 3 = 4r + 3$ Combine like terms.

Step 2 $\quad r - 3 - r = 4r + 3 - r$ Use the addition property of equality.
Subtract r.

$\qquad\qquad\quad -3 = 3r + 3$

$\qquad\quad -3 - 3 = -3 + 3r + 3$ Add -3.

$\qquad\qquad\quad -6 = 3r$

Step 3 $\qquad\quad \dfrac{-6}{3} = \dfrac{3r}{3}$ Use the multiplication property of equality. Divide by 3.

$\qquad\quad -2 = r \quad \text{or} \quad r = -2$

Step 4 Substitute -2 for r in the original equation.

$\qquad\qquad 3r + 4 - 2r - 7 = 4r + 3$

$\qquad 3(-2) + 4 - 2(-2) - 7 = 4(-2) + 3 \qquad ? \qquad$ Let $r = -2$.

$\qquad\qquad -6 + 4 + 4 - 7 = -8 + 3 \qquad ? \qquad$ Multiply.

$\qquad\qquad\qquad\qquad -5 = -5 \qquad\qquad$ True

The solution set of the equation is $\{-2\}$.

In Step 2 of Example 1, the terms were added and subtracted so that the variable term ended up on the right. Choosing differently would lead to the variable term being on the left side of the equation. Usually there is no advantage either way.

E X A M P L E 2   Using the Four Steps to Solve an Equation

Solve $4(k - 3) - k = k - 6$.

Step 1   Before combining like terms, use the distributive property to simplify $4(k - 3)$.

$$4(k - 3) - k = k - 6$$
$$4 \cdot k - 4 \cdot 3 - k = k - 6 \qquad \text{Distributive property}$$
$$4k - 12 - k = k - 6$$
$$3k - 12 = k - 6 \qquad \text{Combine like terms.}$$

Step 2
$$-k + 3k - 12 = -k + k - 6 \qquad \text{Add } -k.$$
$$2k - 12 = -6$$
$$2k - 12 + 12 = -6 + 12 \qquad \text{Add 12.}$$
$$2k = 6$$

Step 3
$$\frac{2k}{2} = \frac{6}{2} \qquad \text{Divide by 2.}$$
$$k = 3$$

Step 4   Check your answer by substituting 3 for k in the original equation. Remember to do the work inside the parentheses first.

$$4(k - 3) - k = k - 6$$
$$4(3 - 3) - 3 = 3 - 6 \quad ? \qquad \text{Let } k = 3.$$
$$4(0) - 3 = 3 - 6 \quad ?$$
$$0 - 3 = 3 - 6 \quad ?$$
$$-3 = -3 \qquad \text{True}$$

The solution set of the equation is {3}.

E X A M P L E 3   Using the Four Steps to Solve an Equation

Solve $8a - (3 + 2a) = 3a + 1$.

Step 1   Simplify.

$$8a - (3 + 2a) = 3a + 1$$
$$8a - 3 - 2a = 3a + 1 \qquad \text{Distributive property}$$
$$6a - 3 = 3a + 1 \qquad \text{Combine terms.}$$

Step 2
$$-3a + 6a - 3 = -3a + 3a + 1 \qquad \text{Add } -3a.$$
$$3a - 3 = 1$$
$$3a - 3 + 3 = 1 + 3 \qquad \text{Add 3.}$$
$$3a = 4$$

Step 3
$$\frac{3a}{3} = \frac{4}{3} \qquad \text{Divide by 3.}$$

$$a = \frac{4}{3}$$

Step 4   Check the solution in the original equation.

$$8a - (3 + 2a) = 3a + 1$$

$$8\left(\frac{4}{3}\right) - \left[3 + 2\left(\frac{4}{3}\right)\right] = 3\left(\frac{4}{3}\right) + 1 \qquad ? \qquad \text{Let } a = \tfrac{4}{3}.$$

$$\frac{32}{3} - \left[3 + \frac{8}{3}\right] = 4 + 1 \qquad ?$$

$$\frac{32}{3} - \left[\frac{9}{3} + \frac{8}{3}\right] = 5 \qquad ?$$

$$\frac{32}{3} - \frac{17}{3} = 5 \qquad ?$$

$$5 = 5 \qquad \text{True}$$

The check shows that $\left\{\frac{4}{3}\right\}$ is the solution set.

Be very careful with signs when solving equations like the one in Example 3. When a subtraction sign appears immediately in front of a quantity in parentheses, such as in the expression

$$8 - (3 + 2a),$$

 remember that the $-$ sign acts like a factor of -1 and affects the sign of *every* term within the parentheses. Thus,

$$8 - (3 + 2a) = 8 + (-1)(3 + 2a) = 8 - 3 - 2a.$$

↑   ↑

Change to $-$ in *both* terms.

┌ **E X A M P L E   4**   **Using the Four Steps to Solve an Equation**

Solve $4(8 - 3t) = 32 - 8(t + 2)$.

Step 1   Use the distributive property.

$$4(8 - 3t) = 32 - 8(t + 2) \qquad \text{Given equation}$$
$$32 - 12t = 32 - 8t - 16 \qquad \text{Distributive property}$$
$$32 - 12t = 16 - 8t$$

Step 2 $$32 - 12t + 12t = 16 - 8t + 12t \qquad \text{Add } 12t.$$
$$32 = 16 + 4t$$
$$32 - 16 = 16 + 4t - 16 \qquad \text{Subtract 16.}$$
$$16 = 4t$$

Step 3 $$\frac{16}{4} = \frac{4t}{4} \qquad \text{Divide by 4.}$$
$$4 = t \quad \text{or} \quad t = 4$$

Step 4   Check the solution.

$$4(8 - 3t) = 32 - 8(t + 2)$$
$$4(8 - 3 \cdot 4) = 32 - 8(4 + 2) \qquad ? \qquad \text{Let } t = 4.$$
$$4(8 - 12) = 32 - 8(6) \qquad ?$$
$$4(-4) = 32 - 48 \qquad ?$$
$$-16 = -16 \qquad\qquad \text{True}$$

Since 4 satisfies the equation, the solution set is {4}.

OBJECTIVE 2   **Solve equations with fractions or decimals as coefficients.** We can clear an equation of fractions by multiplying both sides by the least common denominator (LCD) of all denominators in the equation. It is a good idea to do this immediately after using the distributive property to clear parentheses. Most students make fewer errors working with integer coefficients.

EXAMPLE 5   **Solving an Equation with Fractions as Coefficients**

Solve $\dfrac{2}{3}x - \dfrac{1}{2}x = -\dfrac{1}{6}x - 2$.

The LCD of all the fractions in the equation is 6. Start by multiplying both sides of the equation by 6.

$$\frac{2}{3}x - \frac{1}{2}x = -\frac{1}{6}x - 2$$

$$6\left(\frac{2}{3}x - \frac{1}{2}x\right) = 6\left(-\frac{1}{6}x - 2\right) \qquad \text{Multiply by 6.}$$

$$6\left(\frac{2}{3}x\right) + 6\left(-\frac{1}{2}x\right) = 6\left(-\frac{1}{6}x\right) + 6(-2) \qquad \text{Distributive property}$$

$$4x - 3x = -x - 12$$

Now use the four steps to solve this simpler equivalent equation.

Step 1 $x = -x - 12$ Combine like terms.

Step 2 $x + x = x - x - 12$ Add x.

$$2x = -12$$

Step 3 $\dfrac{2x}{2} = \dfrac{-12}{2}$ Divide by 2.

$$x = -6$$

Step 4   Check the answer.

$$\frac{2}{3}(-6) - \frac{1}{2}(-6) = -\frac{1}{6}(-6) - 2 \qquad ? \qquad \text{Let } x = -6.$$

$$-4 + 3 = 1 - 2 \qquad\qquad ?$$

$$-1 = -1 \qquad\qquad\qquad \text{True}$$

The solution set of the equation is {−6}.

Set up a proportion. One ratio in the proportion can involve the number of packs, and the other can involve the costs. Make sure that the corresponding numbers appear in the numerator and the denominator.

$$\frac{\text{Cost of 3}}{\text{Cost of 10}} = \frac{3}{10}$$

$$\frac{.87}{x} = \frac{3}{10}$$

$3x = .87(10)$ Cross products

$3x = 8.7$

$x = 2.90$ Divide by 3.

The 10 packs should cost $2.90. As shown earlier, the proportion could also be written as $\frac{3}{.87} = \frac{10}{x}$, which would give the same cross products.

Many people would solve the problem in Example 4 mentally as follows: Three packs cost $.87, so one pack costs $.87/3 = $.29. Then ten packs will cost 10($.29) = $2.90. If you do the problem this way, you are using proportions and probably not even realizing it!

An important application that uses proportions is *unit pricing*—deciding which size of an item offered in different sizes produces the best price per unit. For example, suppose you can buy 36 ounces of pancake syrup for $3.89. To find the price per unit, set up the proportion

$$\frac{36 \text{ ounces}}{1 \text{ ounce}} = \frac{3.89}{x}$$

and solve for *x*.

$36x = 3.89$ Cross products

$x = \dfrac{3.89}{36}$ Divide by 36.

$x \approx .108$ Use a calculator.

Thus, the price for 1 ounce is $.108, or about 11 cents. Notice that the unit price is the ratio of the cost for 36 ounces, $3.89, to the number of ounces, 36, which means that the unit price for an item is found by dividing the cost by the number of units.

EXAMPLE 5   Determining Unit Price to Obtain the Best Buy

Besides the 36-ounce size discussed above, the local supermarket carries two other sizes of a popular brand of pancake syrup, priced as follows.

Size	Price
36-ounce	$3.89
24-ounce	$2.79
12-ounce	$1.89

Which size is the best buy? That is, which size has the lowest unit price?

To find the best buy, divide the price by the number of units to get the price per ounce. Each result in the following table was found by using a calculator and rounding the answer to three decimal places.

Size	Unit Cost (dollars per ounce)	
36-ounce	$\dfrac{\$3.89}{36} = \$.108$	← The best buy
24-ounce	$\dfrac{\$2.79}{24} = \$.116$	
12-ounce	$\dfrac{\$1.89}{12} = \$.158$	

Since the 36-ounce size produces the lowest price per unit, it would be the best buy. (*Be careful:* Sometimes the largest container *does not* produce the lowest price per unit.)

OBJECTIVE 5   Solve direct variation problems. Suppose that gasoline costs $1.50 per gallon. Then 1 gallon costs $1.50, 2 gallons cost 2($1.50) = $3.00, 3 gallons cost 3($1.50) = $4.50, and so on. Using a proportion,

$$\frac{\text{Cost of 1 gallon}}{\text{Total cost}} = \frac{1 \text{ gallon}}{x \text{ gallons}}$$

$$x \cdot \text{Cost of 1 gallon} = 1 \cdot \text{Total cost.}$$

Thus, the total cost is obtained by multiplying the number of gallons by the price per gallon. In general, if k equals the price per gallon and x equals the number of gallons, then the total cost y is equal to kx. Notice that as number of gallons increases, total cost increases.

The preceding discussion is an example of variation. As in the gasoline example, two variables *vary directly* if one is a multiple of the other.

Direct Variation

y **varies directly** as x if there exists a number k such that

$$y = kx.$$

Another way to say this is that y **is directly proportional to** x.

EXAMPLE 6   Using Direct Variation to Find the Cost of Gasoline

If 6 gallons of gasoline cost $8.52, find the cost of 20 gallons.

Here, the total cost of the gasoline, y, varies directly as (or is directly proportional to) the number of gallons purchased, x. This means there is a number k such that $y = kx$. To find y when $x = 20$, use the fact that 6 gallons cost $8.52.

$$y = kx$$
$$8.52 = k(6) \qquad \text{Let } x = 6,\ y = 8.52.$$
$$1.42 = k \qquad \text{Divide by 6.}$$

Since $k = 1.42$,

$$y = 1.42x.$$

When $x = 20$,

$$y = 1.42(20) = 28.4,$$

so the cost to purchase 20 gallons of gasoline is $28.40.

2.5 EXERCISES

Determine the ratio and write it in lowest terms. See Example 1.

1. 25 feet to 40 feet
2. 16 miles to 48 miles
3. 18 dollars to 72 dollars
4. 300 people to 250 people
5. 144 inches to 6 feet
6. 60 inches to 2 yards
7. 5 days to 40 hours
8. 75 minutes to 2 hours

9. Which one of the following ratios is not the same as the ratio 2 to 5?
   (a) .4   (b) 4 to 10   (c) 20 to 50   (d) 5 to 2
10. Give three ratios that are equivalent to the ratio 4 to 3.
11. Explain the distinction between *ratio* and *proportion*. Give examples.
12. Suppose that someone told you to use cross products in order to multiply fractions. How would you explain to the person what is wrong with his or her thinking?

Decide whether each proportion is true or false. See Example 2.

13. $\dfrac{5}{35} = \dfrac{8}{56}$
14. $\dfrac{4}{12} = \dfrac{7}{21}$
15. $\dfrac{120}{82} = \dfrac{7}{10}$

16. $\dfrac{27}{160} = \dfrac{18}{110}$
17. $\dfrac{\frac{1}{2}}{5} = \dfrac{1}{10}$
18. $\dfrac{\frac{1}{3}}{6} = \dfrac{1}{18}$

Solve each equation. See Example 3.

19. $\dfrac{k}{4} = \dfrac{175}{20}$
20. $\dfrac{49}{56} = \dfrac{z}{8}$
21. $\dfrac{x}{6} = \dfrac{18}{4}$

22. $\dfrac{z}{80} = \dfrac{20}{100}$
23. $\dfrac{3y - 2}{5} = \dfrac{6y - 5}{11}$
24. $\dfrac{2p + 7}{3} = \dfrac{p - 1}{4}$

Solve each problem by setting up and solving a proportion. See Example 4.

25. A chain saw requires a mixture of 2-cycle engine oil and gasoline. According to the directions on a bottle of Oregon 2-cycle Engine Oil, for a 50 to 1 ratio requirement, approximately 2.5 fluid ounces of oil are required for 1 gallon of gasoline. For 2.75 gallons, how many fluid ounces of oil are required?

26. The directions on the bottle mentioned in Exercise 25 indicate that if the ratio requirement is 24 to 1, approximately 5.5 ounces of oil are required for 1 gallon of gasoline. If gasoline is to be mixed with 22 ounces of oil, how much gasoline is to be used?

27. In 1998, the average exchange rate between U.S. dollars and United Kingdom pounds was 1 pound to $1.6762. Margaret went to London and exchanged her U.S. currency for U.K. pounds, and received 400 pounds. How much in U.S. money did Margaret exchange?

28. If 3 U.S. dollars can be exchanged for 4.5204 Swiss francs, how many Swiss francs can be obtained for $49.20? (Round to the nearest hundredth.)

29. If 6 gallons of premium unleaded gasoline cost $3.72, how much would it cost to completely fill a 15-gallon tank?

30. If sales tax on a $16.00 compact disc is $1.32, how much would the sales tax be on a $120.00 compact disc player?

31. The distance between Kansas City, Missouri, and Denver is 600 miles. On a certain wall map, this is represented by a length of 2.4 feet. On the map, how many feet would there be between Memphis and Philadelphia, two cities that are actually 1000 miles apart?

32. The distance between Singapore and Tokyo is 3300 miles. On a certain wall map, this distance is represented by 11 inches. The actual distance between Mexico City and Cairo is 7700 miles. How far apart are they on the same map?

33. Biologists tagged 250 fish in Willow Lake on October 5. On a later date they found 7 tagged fish in a sample of 350. Estimate the total number of fish in Willow Lake to the nearest hundred.

34. On May 13 researchers at Argyle Lake tagged 420 fish. When they returned a few weeks later, their sample of 500 fish contained 9 that were tagged. Give an approximation of the fish population in Argyle Lake to the nearest hundred.

 The Olympic Committee has come to rely more and more on television rights and major corporate sponsors to finance the games. The pie charts show the funding plans for the first Olympics in Athens and the 1996 Olympics in Atlanta. Use proportions and the figures to answer the questions in Exercises 35 and 36.

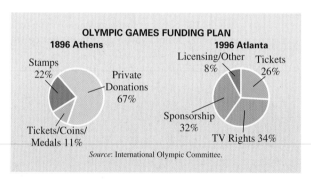

OLYMPIC GAMES FUNDING PLAN

1896 Athens
Stamps 22%
Private Donations 67%
Tickets/Coins/ Medals 11%

1996 Atlanta
Licensing/Other 8%
Tickets 26%
Sponsorship 32%
TV Rights 34%

Source: International Olympic Committee.

35. In the 1996 Olympics, total revenue of $350 million was raised. There were 10 major sponsors.
 (a) Write a proportion to find the amount of revenue provided by tickets. Solve it.
 (b) What amount was provided by sponsors? Assuming the sponsors contributed equally, how much was provided per sponsor?
 (c) What amount was raised by TV rights?

36. Suppose the amount of revenue raised in the 1896 Olympics was equivalent to the $350 million in 1996.
 (a) Write a proportion for the amount of revenue provided by stamps and solve it.
 (b) What amount (in dollars) would have been provided by private donations?
 (c) In the 1988 Olympics, there were 9 major sponsors, and the total revenue was $95 million. What is the ratio of major sponsors in 1988 to those in 1996? What is the ratio of revenue in 1988 to revenue in 1996?

A supermarket was surveyed to find the prices charged for items in various sizes. Find the best buy (based on price per unit) for each particular item. See Example 5.

37. Trash bags
 20 count: $3.09
 30 count: $4.59

38. Black pepper
 1-ounce size: $ .99
 2-ounce size: $1.65
 4-ounce size: $4.39

39. Breakfast cereal
    15-ounce size: $2.99
    25-ounce size: $4.49
    31-ounce size: $5.49

40. Cocoa mix
    8-ounce size: $1.39
    16-ounce size: $2.19
    32-ounce size: $2.99

41. Tomato ketchup
    14-ounce size: $ .89
    32-ounce size: $1.19
    64-ounce size: $2.95

42. Cut green beans
    8-ounce size: $ .45
    16-ounce size: $ .49
    50-ounce size: $1.59

Two triangles are **similar** if they have the same shape (but not necessarily the same size). Similar triangles have sides that are proportional. The figure shows two similar triangles. Notice that the ratios of the corresponding sides are all equal to $\frac{3}{2}$:

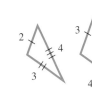

$$\frac{3}{2} = \frac{3}{2} \qquad \frac{4.5}{3} = \frac{3}{2} \qquad \frac{6}{4} = \frac{3}{2}.$$

If we know that two triangles are similar, we can set up a proportion to solve for the length of an unknown side.

Use a proportion to find the length x, given that the pair of triangles are similar.

43.

44.

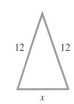

45.

46.

*For the problems in Exercises 47 and 48, (**a**) draw a sketch consisting of two right triangles, depicting the situation described, and (**b**) solve the problem. (*Source: The Guinness Book of World Records.)*

47. An enlarged version of the chair used by George Washington at the Constitutional Convention casts a shadow 18 feet long at the same time a vertical pole 12 feet high casts a shadow 4 feet long. How tall is the chair?

48. One of the tallest candles ever constructed was exhibited at the 1897 Stockholm Exhibition. If it cast a shadow 5 feet long at the same time a vertical pole 32 feet high cast a shadow 2 feet long, how tall was the candle?

The Consumer Price Index, issued by the U.S. Bureau of Labor Statistics, provides a means of determining the purchasing power of the U.S. dollar from one year to the next. Using the period from 1982 to 1984 as a measure of 100.0, the Consumer Price Index from 1990 to 1995 is shown here.

Year	Consumer Price Index
1990	130.7
1991	136.2
1992	140.3
1993	144.5
1994	148.2
1995	152.4

Source: Bureau of Labor Statistics.

EXERCISES

To use the Consumer Price Index to predict a price in a particular year, we can set up a proportion and compare it with a known price in another year, as follows:

$$\frac{\text{Price in year } A}{\text{Index in year } A} = \frac{\text{Price in year } B}{\text{Index in year } B}.$$

Use the Consumer Price Index figures above to find the amount that would be charged for the use of the same amount of electricity that cost $225 in 1990. Give your answer to the nearest dollar.

49. in 1992   **50.** in 1993   **51.** in 1994   **52.** in 1995

53. The Consumer Price Index figures for shelter for the years 1981 and 1991 are 90.5 and 146.3. If shelter for a particular family cost $3000 in 1981, what would be the comparable cost in 1991? Give your answer to the nearest dollar.

54. Due to a volatile fuel oil market in the early 1980s, the price of fuel decreased during the first three quarters of the decade. The Consumer Price Index figures for 1982 and 1986 were 105.0 and 74.1. If it cost you $21.50 to fill your tank with fuel oil in 1982, how much would it have cost to fill the same tank in 1986? Give your answer to the nearest cent.

RELATING CONCEPTS (EXERCISES 55–58)

In Section 2.2 we learned that to make the solution process easier, if an equation involves fractions, we can multiply both sides of the equation by the least common denominator of all the fractions in the equation. A proportion consists of two fractions equal to each other, so a proportion is a special case of this kind of equation.

Work Exercises 55–58 in order to see how the process of solving by cross products is justified.

55. In the equation $\frac{x}{6} = \frac{2}{5}$, what is the least common denominator of the two fractions?

56. Solve the equation in Exercise 55 as follows:
   (a) Multiply both sides by the LCD. What is the equation that you obtain?
   (b) Solve for x by dividing both sides by the coefficient of x. What is the solution?

57. Solve the equation in Exercise 55 as follows:
   (a) Set the cross products equal. What is the equation you obtain?
   (b) Repeat part (b) of Exercise 56.

58. Compare your results from Exercises 56(b) and 57(b). What do you notice?

Did you make the connection that solving a proportion by cross products is justified by the general method of solving an equation with fractions?

Solve each variation problem. See Example 6.

59. The interest on an investment varies directly as the rate of simple interest. If the interest is $48 when the interest rate is 5%, find the interest when the rate is 4.2%.

60. For a given base, the area of a triangle varies directly as its height. Find the area of a triangle with a height of 6 inches, if the area is 10 square inches when the height is 4 inches.

61. The distance a spring stretches is directly proportional to the force applied. If a force of 30 pounds stretches a spring 16 inches, how far will a force of 50 pounds stretch the spring?

62. The perimeter of a square is directly proportional to the length of its side. What is the perimeter of a square with a 5-centimeter side, if a square with a 12-centimeter side has a perimeter of 48 centimeters?

EXERCISES

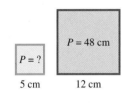

63. According to *The Guinness Book of World Records,* the longest recorded voyage in a paddleboat is 2226 miles in 103 days; the boat was propelled down the Mississippi River by the foot power of two boaters. Assuming a constant rate, how far would they have gone in 120 days? (*Hint:* Distance varies directly as time.)

2.6 More about Problem Solving

OBJECTIVES

1. Use percent in problems involving rates.
2. Solve problems involving mixtures.
3. Solve problems involving simple interest.
4. Solve problems involving denominations of money.
5. Solve problems involving distance, rate, and time.

OBJECTIVE 1 Use percent in problems involving rates. Recall that percent means "per hundred." Thus, percents are ratios where the second number is always 100. For example, 50% represents the ratio of 50 to 100 and 27% represents the ratio of 27 to 100.

PROBLEM SOLVING

Percents are often used in problems involving mixing different concentrations of a substance or different interest rates. In each case, to get the amount of pure substance or the interest, we multiply.

Mixture Problems	Interest Problems (annual)
base × rate (%) = percentage	principal × rate (%) = interest
$b \times r = p$	$p \times r = I$

In an equation, the percent always appears as a decimal. For example, 35% is written as .35, not 35.

EXAMPLE 1   Using Percent to Find a Percentage

(a) If a chemist has 40 liters of a 35% acid solution, then the amount of pure acid in the solution is

$$40 \quad \times \quad .35 \quad = \quad 14 \text{ liters.}$$

Amount of solution   Rate of concentration   Amount of pure acid

(b) If $1300 is invested for one year at 7% simple interest, the amount of interest earned in the year is

$$\$1300 \quad \times \quad .07 \quad = \quad \$91.$$

Principal   Interest rate   Interest earned

PROBLEM SOLVING

In the examples that follow, we use charts to organize the information in the problems. A chart enables us to more easily set up the equation for the problem, which is usually the most difficult step. The six steps described in Section 2.3 are used but are not always specifically numbered.

OBJECTIVE 2   Solve problems involving mixtures. In the next example, we use percent to solve a mixture problem.

EXAMPLE 2   Solving a Mixture Problem

A chemist needs to mix 20 liters of 40% acid solution with some 70% solution to get a mixture that is 50% acid. How many liters of the 70% solution should be used?

Step 1   Let x = the number of liters of 70% solution that are needed.

Step 2   Recall from part (a) of Example 1 that the amount of pure acid in this solution will be given by the product of the percent of strength and the number of liters of solution, or

liters of pure acid in x liters of 70% solution = $.70x$.

The amount of pure acid in the 20 liters of 40% solution is

liters of pure acid in the 40% solution = $.40(20) = 8$.

The new solution will contain $20 + x$ liters of 50% solution. The amount of pure acid in this solution is

liters of pure acid in the 50% solution = $.50(20 + x)$.

The given information is summarized in the chart below.

Liters of Mixture	Rate (as a decimal)	Liters of Pure Acid
x	.70	.70(x)
20	.40	.40(20)
20 + x	.50	.50(20 + x)

Step 3   The number of liters of pure acid in the 70% solution added to the number of liters of pure acid in the 40% solution will equal the number of liters of pure acid in the final mixture, so the equation is

Pure acid in 70%	plus	pure acid in 40%	is	pure acid in 50%.
↓	↓	↓	↓	↓
$.70x$	$+$	$.40(20)$	$=$	$.50(20 + x)$.

Step 4   Clear parentheses, then multiply by 100 to clear decimals.

$$.70x + .40(20) = .50(20) + .50x \qquad \text{Distributive property}$$
$$70x + 40(20) = 50(20) + 50x \qquad \text{Multiply by 100.}$$

Solve for *x*.

$$70x + 800 = 1000 + 50x$$
$$20x + 800 = 1000 \qquad \text{Subtract } 50x.$$
$$20x = 200 \qquad \text{Subtract } 800.$$
$$x = 10 \qquad \text{Divide by 20.}$$

Steps 5 and 6   Check this solution to see that the chemist needs to use 10 liters of 70% solution.

OBJECTIVE 3 Solve problems involving simple interest. The next example uses the formula for simple interest, $I = prt$. Remember that when $t = 1$, the formula becomes $I = pr$, as shown in the Problem-Solving box at the beginning of this section. Once again the idea of multiplying the total amount (principal) by the rate (rate of interest) gives the percentage (amount of interest).

EXAMPLE 3   Solving a Simple Interest Problem

Elizabeth Thornton receives an inheritance. She plans to invest part of it at 9% and $2000 more than this amount in a less secure investment at 10%. To earn $1150 per year in interest, how much should she invest at each rate?

Let x = the amount invested at 9% (in dollars);

$x + 2000$ = the amount invested at 10% (in dollars).

Use a chart to arrange the information given in the problem.

Amount Invested in Dollars	Rate of Interest	Interest for One Year
x	.09	$.09x$
$x + 2000$	.10	$.10(x + 2000)$

We multiply amount by rate to get the interest earned. Since the total interest is to be $1150, the equation is

Interest at 9%	plus	interest at 10%	is	total interest.
↓	↓	↓	↓	↓
$.09x$	$+$	$.10(x + 2000)$	$=$	1150.

Clear parentheses; then clear decimals.

$$.09x + .10x + .10(2000) = 1150 \qquad \text{Distributive property}$$
$$9x + 10x + 10(2000) = 115{,}000 \qquad \text{Multiply by 100.}$$

Now solve for x.

$$9x + 10x + 20{,}000 = 115{,}000$$
$$19x + 20{,}000 = 115{,}000 \qquad \text{Combine terms.}$$
$$19x = 95{,}000 \qquad \text{Subtract 20,000.}$$
$$x = 5000 \qquad \text{Divide by 19.}$$

She should invest $5000 at 9% and $5000 + $2000 = $7000 at 10%.

 Although decimals were cleared in Examples 2 and 3, the equations also can be solved without doing so.

OBJECTIVE **4**   Solve problems involving denominations of money.   If a cash drawer contains 37 quarters, the total value of the coins is

$$37 \qquad \times \qquad \$.25 \qquad = \qquad \$9.25.$$

Number of coins   Denomination   Total value

PROBLEM SOLVING

Problems that involve different denominations of money or items with different monetary values are very similar to mixture and interest problems. The basic relationship is the same.

Money Problems

Number × Value of one = Total value

A chart is helpful for these problems, too.

EXAMPLE 4   Solving a Problem about Money

A bank teller has 25 more five-dollar bills than ten-dollar bills. The total value of the money is $200. How many of each denomination of bill does he have?

We must find the number of each denomination of bill that the teller has.

$$\text{Let} \qquad x = \text{the number of ten-dollar bills;}$$
$$x + 25 = \text{the number of five-dollar bills.}$$

Organize the given information in a chart.

Number of Bills	Denomination in Dollars	Dollar Value
x	10	10x
x + 25	5	5(x + 25)

Multiplying the number of bills by the denomination gives the monetary value. The value of the tens added to the value of the fives must be $200:

Value of fives	plus	value of tens	is	$200.
↓	↓	↓	↓	↓
$5(x + 25)$	$+$	$10x$	$=$	$200.$

Solve this equation.

$$5x + 125 + 10x = 200 \qquad \text{Distributive property}$$
$$15x + 125 = 200 \qquad \text{Combine terms.}$$
$$15x = 75 \qquad \text{Subtract 125.}$$
$$x = 5 \qquad \text{Divide by 15.}$$

Since x represents the number of tens, the teller has 5 tens and $5 + 25 = 30$ fives. Check that the value of this money is $5(\$10) + 30(\$5) = \$200$.

EXAMPLE 5   Solving a Problem about Ticket Prices

Lindsay Davenport, the winner of the 1998 U.S. Open women's tennis championship, plays World Team Tennis for the Sacramento Capitals. Tickets are $35 for box seats and $25 for all other seats. There are 5 times as many other seats as box seats. How many of each kind of ticket must the promoter sell for ticket sales revenue to reach $16,000?

Choose a variable and use it to write the revenue from each kind of ticket.

$$\text{Let} \quad x = \text{the number of box seats;}$$
$$5x = \text{the number of other seats.}$$

Again, organize the information in a chart.

Number of Seats	Ticket Price	Revenue
x	35	$35x$
$5x$	25	$125x$

Total sales revenue is

Revenue from box seats	plus	revenue from other seats	is	total revenue.
↓	↓	↓	↓	↓
$35x$	$+$	$125x$	$=$	$16,000.$

To solve the equation, combine the terms on the left side of the equation, then divide.

$$160x = 16,000$$
$$x = 100 \qquad \text{Divide by 160.}$$

Thus, there are 100 box seats and $5(100) = 500$ other seats. Check that this produces revenue of $100(\$35) + 500(\$25) = \$16,000$, as required.

OBJECTIVE 5   Solve problems involving distance, rate, and time. If an automobile travels at an average rate of 50 miles per hour for two hours, then it travels $50 \times 2 = 100$ miles. This is an example of the basic relationship between distance, rate, and time:

$$\text{distance} = \text{rate} \times \text{time,}$$

given by the formula $d = rt$. By solving, in turn, for r and t in the formula, we obtain two other equivalent forms of the formula. The three forms are given below.

Distance, Rate, Time Relationship

$$d = rt \qquad r = \frac{d}{t} \qquad t = \frac{d}{r}$$

The following examples illustrate the uses of these formulas.

EXAMPLE 6 Finding Distance, Rate, or Time

(a) The speed of sound is 1088 feet per second at sea level at 32°F. In 5 seconds under these conditions, sound travels

$$\underset{\text{Rate}}{1088} \quad \times \quad \underset{\text{Time}}{5} \quad = \quad \underset{\text{Distance}}{5440 \text{ feet.}}$$

Here, we found distance given rate and time, using $d = rt$.

(b) The winner of the first Indianapolis 500 race (in 1911) was Ray Harroun, driving a Marmon Wasp at an average speed of 74.59 miles per hour.* To complete the 500 miles, it took him

$$\underset{\text{Rate}}{\overset{\text{Distance} \rightarrow}{\frac{500}{74.59}}} = 6.70 \text{ hours} \quad \text{(rounded).} \quad \leftarrow \text{Time}$$

Here, we found time given rate and distance, using $t = \dfrac{d}{r}$. To convert .7 hour to minutes, multiply by 60: .7(60) = 42 minutes, so the race took Harroun 6 hours, 42 minutes to complete.

(c) In the 1996 Olympic Games in Atlanta, Claudia Poll of Costa Rica won the women's 200-meter freestyle swimming event in 1 minute, 58.16 seconds or 60 + 58.16 = 118.16 seconds.* Her rate was

$$\underset{\text{Time} \rightarrow}{\overset{\text{Distance} \rightarrow}{\frac{200}{118.16}}} = 1.69 \text{ meters per second} \quad \text{(rounded).} \quad \leftarrow \text{Rate}$$

Here, we found rate given distance and time, using $r = \dfrac{d}{t}$.

*Source: The Universal Almanac, 1997, John W. Wright, General Editor.

EXAMPLE 3   Using the Multiplication Property of Inequality

(a) Solve the inequality $3r < -18$.

Simplify this inequality by using the multiplication property of inequality and dividing both sides by 3. Since 3 is a positive number, the direction of the inequality symbol does not change.

$$3r < -18$$
$$\frac{3r}{3} < \frac{-18}{3} \qquad \text{Divide by 3.}$$
$$r < -6$$

The solution set is $(-\infty, -6)$. The graph is shown in Figure 16.

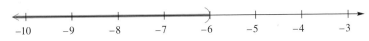

Figure 16

(b) Solve the inequality $-4t \geq 8$.

Here both sides of the inequality must be divided by -4, a negative number, which *does* change the direction of the inequality symbol.

$$-4t \geq 8$$
$$\frac{-4t}{-4} \leq \frac{8}{-4} \qquad \text{Divide by } -4; \text{ symbol is reversed.}$$
$$t \leq -2$$

The solution set is $(-\infty, -2]$. The solutions are graphed in Figure 17.

Figure 17

CAUTION   Even though the number on the right side of the inequality in Example 3(a) is negative (-18), *do not reverse the direction of the inequality symbol.* Reverse the symbol only when multiplying or dividing by a negative number, as shown in Example 3(b).

OBJECTIVE 4   Solve linear inequalities. A **linear inequality** is an inequality that can be written in the form $ax + b < 0$, for real numbers a and b, with $a \neq 0$. ($<$ may be replaced with $>$, $\leq$, or $\geq$ in this definition.) To solve a linear inequality, follow these steps.

Solving a Linear Inequality

Step 1   **Simplify each side separately.**   Use the properties to clear parentheses and combine terms.

Step 2   **Isolate the variable terms on one side.**   Use the addition property of inequality to simplify the inequality to the form $ax < b$ or $ax > b$, where a and b are real numbers.

Step 3   **Isolate the variable.**   Use the multiplication property of inequality to simplify the inequality to the form $x < d$ or $x > d$, where d is a real number.

Notice how these steps are used in the next example.

E X A M P L E   4   Solving a Linear Inequality

Solve the inequality $3z + 2 - 5 > -z + 7 + 2z$.

Step 1   Combine like terms and simplify.

$$3z + 2 - 5 > -z + 7 + 2z$$
$$3z - 3 > z + 7$$

Step 2   Use the addition property of inequality.

$$3z - 3 + 3 > z + 7 + 3 \qquad \text{Add 3.}$$
$$3z > z + 10$$
$$3z - z > z + 10 - z \qquad \text{Subtract } z.$$
$$2z > 10$$

Step 3   Use the multiplication property of inequality.

$$\frac{2z}{2} > \frac{10}{2} \qquad \text{Divide by 2.}$$
$$z > 5$$

Since 2 is positive, the direction of the inequality symbol was not changed in the third step. The solution set is $(5, \infty)$. Its graph is shown in Figure 18.

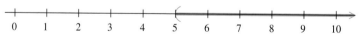

Figure 18

E X A M P L E   5   Solving a Linear Inequality

Solve $5(k - 3) - 7k \geq 4(k - 3) + 9$.

Step 1   Simplify and combine like terms.

$$5(k - 3) - 7k \geq 4(k - 3) + 9$$
$$5k - 15 - 7k \geq 4k - 12 + 9 \qquad \text{Distributive property}$$
$$-2k - 15 \geq 4k - 3 \qquad \text{Combine like terms.}$$

Step 2   Use the addition property.

$$-2k - 15 - 4k \geq 4k - 3 - 4k \qquad \text{Subtract } 4k.$$
$$-6k - 15 \geq -3$$
$$-6k - 15 + 15 \geq -3 + 15 \qquad \text{Add } 15.$$
$$-6k \geq 12$$

Step 3   Divide both sides by -6, a negative number. Change the direction of the inequality symbol.

$$\frac{-6k}{-6} \leq \frac{12}{-6} \qquad \text{Divide by } -6; \text{ symbol is reversed.}$$
$$k \leq -2$$

The solution set is $(-\infty, -2]$. Its graph is shown in Figure 19.

Figure 19

OBJECTIVE 5   Solve applied problems by using inequalities.   Until now, the applied problems that we have studied have all led to equations.

PROBLEM SOLVING

Inequalities can be used to solve applied problems involving phrases that suggest inequality. The following chart gives some of the more common such phrases along with examples and translations.

Phrase	Example	Inequality
Is more than	A number *is more than* 4	$x > 4$
Is less than	A number *is less than* -12	$x < -12$
Is at least	A number *is at least* 6	$x \geq 6$
Is at most	A number *is at most* 8	$x \leq 8$

Do not confuse statements like "5 is more than a number" with the phrase "5 more than a number." The first of these is expressed as "$5 > x$" while the second is expressed with addition, as "$x + 5$."

The next example shows an application of algebra that is important to anyone who has ever asked himself or herself "What score can I make on my next test and have a (particular grade) in this course?" It uses the idea of finding the average of a number of grades. In general, to find the average of *n* numbers, add the numbers, and divide by *n*.

EXAMPLE 6   Finding an Average Test Score

Brent has test grades of 86, 88, and 78 on his first three tests in geometry. If he wants an average of at least 80 after his fourth test, what are the possible scores he can make on his fourth test?

Let x = Brent's score on his fourth test. To find his average after 4 tests, add the test scores and divide by 4.

$$\underset{\text{Average}}{\underset{\downarrow}{}} \; \underset{\substack{\text{is at} \\ \text{least}}}{\underset{\downarrow}{}} \; 80.$$

$$\frac{86 + 88 + 78 + x}{4} \geq 80$$

$$\frac{252 + x}{4} \geq 80 \qquad \text{Add the known scores.}$$

$$4\left(\frac{252 + x}{4}\right) \geq 4(80) \qquad \text{Multiply by 4.}$$

$$252 + x \geq 320$$

$$252 - 252 + x \geq 320 - 252 \qquad \text{Subtract 252.}$$

$$x \geq 68 \qquad \text{Combine terms.}$$

He must score 68 or more on the fourth test to have an average of *at least* 80.

 Errors often occur when the phrases "at least" and "at most" appear in applied problems. Remember that

at least   translates as   **greater than or equal to**

and

at most   translates as   **less than or equal to.**

OBJECTIVE 6   Solve three-part inequalities. Inequalities that say that one number is *between* two other numbers are *three-part inequalities*. For example,

$$-3 < 5 < 7$$

says that 5 is between -3 and 7. Three-part inequalities can also be solved by using the addition and multiplication properties of inequality. The idea is to get the inequality in the form

a number $< x <$ **another number,**

using "is less than." The solution set can then easily be graphed.

EXAMPLE 7   Solving Three-Part Inequalities

(a) Solve $4 \leq 3x - 5 < 6$ and graph the solutions.

This inequality is equivalent to the statement $4 \leq 3x - 5$ and $3x - 5 < 6$. Using the inequality properties, we could solve each of these simple inequalities separately, and then combine the solutions with "and." We get the same result if we simply work with all three parts at once. Working separately, the first step would be to add 5 on each side. Do this to all three parts.

$$4 \leq 3x - 5 < 6$$
$$4 + 5 \leq 3x - 5 + 5 < 6 + 5 \qquad \text{Add 5.}$$
$$9 \leq 3 < 11$$

Now divide each part by the positive number 3.

$$\frac{9}{3} \leq \frac{3x}{3} < \frac{11}{3} \qquad \text{Divide by 3.}$$
$$3 \leq x < \frac{11}{3}$$

The solution set is $\left[3, \frac{11}{3}\right)$. Its graph is shown in Figure 20.

Figure 20

(b) Solve $-4 \leq \frac{2}{3}m - 1 < 8$ and graph the solutions.

Recall from Section 2.2 that fractions as coefficients in equations can be eliminated by multiplying both sides by the least common denominator of the fractions. The same is true for inequalities. One way to begin is to multiply all three parts by 3.

$$-4 \leq \frac{2}{3}m - 1 < 8$$
$$3(-4) \leq 3\left(\frac{2}{3}m - 1\right) < 3(8) \qquad \text{Multiply by 3.}$$
$$-12 \leq 2m - 3 < 24 \qquad \text{Distributive property}$$

Now add 3 to each part.

$$-12 + 3 \leq 2m - 3 + 3 < 24 + 3 \qquad \text{Add 3.}$$
$$-9 \leq 2m < 27$$

Finally, divide by 2 to get

$$-\frac{9}{2} \leq m < \frac{27}{2}.$$

The solution set is $\left[-\frac{9}{2}, \frac{27}{2}\right)$. Its graph is shown in Figure 21.

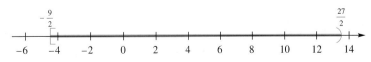

Figure 21

This inequality could also have been solved by first adding 1 to each part, and then multiplying each part by $\frac{3}{2}$.

CONNECTIONS

Many mathematical models involve inequalities rather than equations. This is often the case in economics. For example, a company that produces videocassettes has found that revenue from the sales of the cassettes is $5 per cassette less sales costs of $100. Production costs are $125 plus $4 per cassette. Profit ($P$) is given by revenue ($R$) less cost ($C$), so the company must find the production level x that makes

$$P = R - C > 0.$$

FOR DISCUSSION OR WRITING

Write an expression for revenue letting x represent the production level (number of cassettes to be produced). Write an expression for production costs using x. Write an expression for profit and solve the inequality shown above. Describe the solution in terms of the problem.

2.7 EXERCISES

1. Explain how to determine whether to use a parenthesis or a square bracket at the endpoint when graphing an inequality on a number line.

2. How does the graph of $t \geq -7$ differ from the graph of $t > -7$?

Write an inequality involving the variable x that describes each set of numbers graphed. See Example 1.

3.

        $-4\ -3\ -2\ -1\ \ 0\ \ 1\ \ 2\ \ 3$

4.

        $-4\ -3\ -2\ -1\ \ 0\ \ 1\ \ 2\ \ 3\ \ 4$

5.

        $-2\ -1\ \ 0\ \ 1\ \ 2\ \ 3\ \ 4\ \ 5$

6.

        $-2\ -1\ \ 0\ \ 1\ \ 2\ \ 3\ \ 4\ \ 5$

Write each inequality in interval notation and graph the interval on a number line. See Example 1.

    7. $k \leq 4$ **8.** $r \leq -11$ **9.** $x < -3$ **10.** $y < 3$

    11. $t > 4$ **12.** $m > 5$ **13.** $8 \leq x \leq 10$ **14.** $3 \leq x \leq 5$

    15. $0 < y \leq 10$ **16.** $-3 \leq x < 5$

17. Why is it *wrong* to write $3 < x < -2$ to indicate that x is between -2 and 3?

18. If $p < q$ and $r < 0$, which one of the following statements is *false*?
    (a) $pr < qr$ **(b)** $pr > qr$ **(c)** $p + r < q + r$ **(d)** $p - r < q - r$

Solve each inequality and graph the solution set. See Example 2.

    19. $z - 8 \geq -7$ **20.** $p - 3 \geq -11$ **21.** $2k + 3 \geq k + 8$

    22. $3x + 7 \geq 2x + 11$ **23.** $3n + 5 < 2n - 6$ **24.** $5x - 2 < 4x - 5$

25. Under what conditions must the inequality symbol be reversed when solving an inequality?

26. Explain the steps you would use to solve the inequality $-5x > 20$.

27. Your friend tells you that when solving the inequality $6x < -42$ he reversed the direction of the inequality because of the presence of -42. How would you respond?

28. By what number must you *multiply* both sides of $.2x > 6$ to get just x on the left side?

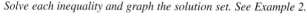

Solve each inequality. Write the solution set in interval notation and graph it. See Example 3.

29. $3x < 18$　　　**30.** $5x < 35$　　　**31.** $2y \geq -20$　　　**32.** $6m \geq -24$

33. $-8t > 24$　　　**34.** $-7x > 49$　　　**35.** $-x \geq 0$　　　**36.** $-k < 0$

37. $-\dfrac{3}{4}r < -15$　　　　　　　　　　**38.** $-\dfrac{7}{8}t < -14$

39. $-.02x \leq .06$　　　　　　　　　　**40.** $-.03v \geq -.12$

Solve each inequality. Write the solution set in interval notation and graph it. See Examples 4 and 5.

41. $5r + 1 \geq 3r - 9$　　　　　　**42.** $6t + 3 < 3t + 12$

43. $6x + 3 + x < 2 + 4x + 4$　　　　　　**44.** $-4w + 12 + 9w \geq w + 9 + w$

45. $-x + 4 + 7x \leq -2 + 3x + 6$　　　　　　**46.** $14y - 6 + 7y > 4 + 10y - 10$

47. $5(x + 3) - 6x \leq 3(2x + 1) - 4x$　　　　　　**48.** $2(x - 5) + 3x < 4(x - 6) + 1$

49. $\dfrac{2}{3}(p + 3) > \dfrac{5}{6}(p - 4)$　　　　　　**50.** $\dfrac{7}{9}(y - 4) \leq \dfrac{4}{3}(y + 5)$

51. $4x - (6x + 1) \leq 8x + 2(x - 3)$　　　　　　**52.** $2y - (4y + 3) > 6y + 3(y + 4)$

53. $5(2k + 3) - 2(k - 8) > 3(2k + 4) + k - 2$

54. $2(3z - 5) + 4(z + 6) \geq 2(3z + 2) + 3z - 15$

Write a three-part inequality involving the variable x that describes each set of numbers graphed. See Example 1(b).

55.

56.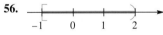

57.

58.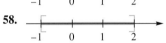

Solve each inequality. Write the solution set in interval notation and graph it. See Example 7.

59. $-5 \leq 2x - 3 \leq 9$　　　　　　**60.** $-7 \leq 3x - 4 \leq 8$

61. $5 < 1 - 6m < 12$　　　　　　**62.** $-1 \leq 1 - 5q \leq 16$

63. $10 < 7p + 3 < 24$　　　　　　**64.** $-8 \leq 3r - 1 \leq -1$

65. $-12 \leq \dfrac{1}{2}z + 1 \leq 4$　　　　　　**66.** $-6 \leq 3 + \dfrac{1}{3}a \leq 5$

67. $1 \leq 3 + \dfrac{2}{3}p \leq 7$　　　　　　**68.** $2 < 6 + \dfrac{3}{4}y < 12$

69. $-7 \leq \dfrac{5}{4}r - 1 \leq -1$　　　　　　**70.** $-12 \leq \dfrac{3}{7}a + 2 \leq -4$

RELATING CONCEPTS (EXERCISES 71-76)

The methods for solving linear equations and linear inequalities are quite similar. In Exercises 71–76, we show how the solutions of an inequality are closely connected to the solution of the corresponding equation.

Work these exercises in order.

71. Solve the equation $3x + 2 = 14$ and graph the solution set as a single point on the number line.

72. Solve the inequality $3x + 2 > 14$ and graph the solution set as an interval on the number line. How does this result compare to the one in Exercise 71?

(continued)

RELATING CONCEPTS (EXERCISES 71–76) (CONTINUED)

73. Solve the inequality $3x + 2 < 14$ and graph the solution set as an interval on the number line. How does this result compare to the one in Exercise 72?

74. If you were to graph all the solution sets from Exercises 71–73 on the same number line, what would the graph be? (This is called the *union* of all the solution sets.)

75. Based on your results from Exercises 71–74, if you were to graph the union of the solution sets of

$$-4x + 3 = -1, \qquad -4x + 3 > -1, \qquad \text{and} \qquad -4x + 3 < -1,$$

what do you think the graph would be?

76. Comment on the following statement: *Equality* is the boundary between *less than* and *greater than.*

Did you make the connection that the value that satisfies an equation separates the values that satisfy the corresponding less than and greater than inequalities?

Solve each problem by writing and solving an inequality. See Example 6.

77. Inkie Landry has grades of 76 and 81 on her first two algebra tests. If she wants an average of at least 80 after her third test, what possible scores can she make on her third test?

78. Mabimi Pampo has grades of 96 and 86 on his first two geometry tests. What possible scores can he make on his third test so that his average is at least 90?

79. The formula for converting Fahrenheit temperature to Celsius is

$$C = \frac{5}{9}(F - 32).$$

If the Celsius temperature on a certain summer day in Toledo is never more than 30°, how would you describe the corresponding Fahrenheit temperatures?

80. The formula for converting Celsius temperature to Fahrenheit is

$$F = \frac{9}{5}C + 32.$$

The Fahrenheit temperature of Key West, Florida, has never exceeded 95°. How would you describe this using Celsius temperature?

81. A product will break even or produce a profit if the revenue R from selling the product is at least equal to the cost C of producing it. Suppose that the cost C (in dollars) to produce x units of bicycle helmets is $C = 50x + 5000$, while the revenue R (in dollars) collected from the sale of x units is $R = 60x$. For what values of x does the product break even or produce a profit?

82. (See Exercise 81.) If the cost to produce x units of basketball cards is $C = 100x + 6000$ (in dollars), and the revenue collected from selling x units is $R = 500x$ (in dollars), for what values of x does the product break even or produce a profit?

83. For what values of x would the rectangle have perimeter of at least 400?

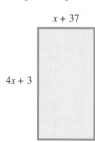

$x + 37$

$4x + 3$

84. For what values of x would the triangle have perimeter of at least 72?

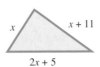

85. A long-distance phone call costs \$2.00 for the first three minutes plus \$.30 per minute for each minute or fractional part of a minute after the first three minutes. If x represents the number of minutes of the length of the call after the first three minutes, then $2 + .30x$ represents the cost of the call. If Jorge has \$5.60 to spend on a call, what is the maximum total time he can use the phone?

86. If the call described in Exercise 85 costs between \$5.60 and \$6.50, what are the possible total time lengths for the call?

RELATING CONCEPTS (EXERCISES 87–98)

The words *and* and *or* are very important in the context of inequalities. We use them to combine simple inequalities. For instance, the statement $-1 \le x < 3$ means $-1 \le x$ *and* $x < 3$, so that only values of x that satisfy *both* conditions are in the solution set, the interval $[-1, 3)$. On the other hand, the statement $x < -1$ *or* $x > 3$, that uses the connective *or*, will be true if $x < -1$ or if $x > 3$. Only one of the conditions must be true for the statement to be true. The two intervals that make this *or* statement true are combined, using the union symbol $\cup$, as $(-\infty, -1) \cup (3, \infty)$.

Work Exercises 87–98 in order.

87. Is the statement "Today is Tuesday *or* today is Wednesday" true if today is actually Wednesday? Is it true if it is actually Tuesday?

88. Is the statement "Today is Tuesday *and* today is Wednesday" true if today is actually Wednesday? Is it true if it is actually Tuesday?

89. Is the statement $x > 3$ and $x < 8$ true for $x = 5$? For $x = 10$? For $x = 3$?

90. Graph $x > 3$.

91. Graph $x < 8$.

92. Compare the graphs from Exercises 90 and 91. Give the values of x in interval notation that make *both* inequalities true.

93. Graph the solution set of $x > 3$ and $x < 8$. How does the graph compare to your answer to Exercise 92?

94. Give the solution set of $x < -4$ and $x < 2$ in interval notation and graph it.

95. What is the solution set of $3 - x > 6$ and $2x > -4$?

96. Graph the solution set of $x > 3$ or $x < 8$. Compare the graph with your answers to Exercises 90 and 91.

97. Give the solution set of $x < -4$ or $x < 2$ in interval notation and graph it. Compare with the answer to Exercise 94.

98. Give the solution set in interval notation and graph the solution set of $3 - x > 6$ or $2x > -4$. Compare with the answer to Exercise 95.

Did you make the connection that when *and* is used to connect two inequalities, both statements must be true for the combined statement to be true, but when *or* is used as the connective, the combined statement is true if either one or both statements are true?

CHAPTER 2 GROUP ACTIVITY

▦ Are You a Race-Walker?

Objective: Use proportions to calculate walking speeds.

Materials: Students will need a large area for walking. Each group will need a stopwatch.

Race-walking at speeds exceeding 8 miles per hour is a high fitness, long-distance competitive sport. The table below contains the gold medal winners of the 1996 Olympic race-walking competition. Complete the table by applying the proportion given below to find the race-walker's steps per minute.

Use 10 km ≈ 6.21 miles. (Round all answers except those for steps per minute to the nearest thousandths. Round steps per minute to the nearest whole number.)

$$\frac{70 \text{ steps per minute}}{2 \text{ miles per hour}} = \frac{x \text{ steps per minute}}{y \text{ miles per hour}}$$

Event	Gold Medal Winner	Country	Time in Hours: Minutes: Seconds	Time in Minutes	Time in Hours	y Miles per Hour	x Steps per Minute
10 km Walk, Women	Yelena Nikolayeva	Russia	0:41:49				
20 km Walk, Men	Jefferson Perez	Ecuador	1:20:07				
50 km Walk, Men	Robert Korzeniowski	Poland	3:43:30				

Source: 1999 World Almanac.

A. Using a stopwatch, take turns counting how many steps each member of the group takes in one minute while walking at a normal pace. Record the results in the chart below. Then do it again at a fast pace. Record these results.

Name	Normal Pace		Fast Pace	
	x Steps per Minute	y Miles per Hour	x Steps per Minute	y Miles per Hour

B. Use the proportion above to convert the numbers from part A to miles per hour and complete the chart.

1. Find the average speed for the group at a normal pace and at a fast pace.

2. What is the minimum number of steps per minute you would have to take to be a race-walker?

3. At a fast pace did anyone in the group walk fast enough to be a race-walker? Explain how you decided.

CHAPTER 2 SUMMARY

KEY TERMS

2.1 linear equation in one variable solution set equivalent equations **2.2** empty (null) set **2.3** degree complementary angles	supplementary angles consecutive integers **2.4** perimeter circumference area vertical angles straight angle	**2.5** ratio proportion cross products vary directly as (is proportional to) **2.7** interval on the number line
		interval notation linear inequality three-part inequality

NEW SYMBOLS

$\emptyset$ empty set

$1°$ one degree

a to b, $a{:}b$, or $\dfrac{a}{b}$ the ratio of a to b

(a, b) interval notation for $a < x < b$

$[a, b]$ interval notation for $a \le x \le b$

∞ infinity

$-\infty$ negative infinity

$(-\infty, \infty)$ set of all real numbers

TEST YOUR WORD POWER

See how well you have learned the vocabulary in this chapter. Answers, with examples, are given at the bottom of the page.

1. A **solution set** is the set of numbers that
(a) make an expression undefined
(b) make an equation false
(c) make an equation true
(d) make an expression equal to 0.

2. The **empty set** is a set
(a) with 0 as its only element
(b) with an infinite number of elements
(c) with no elements
(d) of ideas.

3. Complementary angles are angles
(a) formed by two parallel lines
(b) whose sum is 90°
(c) whose sum is 180°
(d) formed by perpendicular lines.

4. Supplementary angles are angles
(a) formed by two parallel lines
(b) whose sum is 90°
(c) whose sum is 180°
(d) formed by perpendicular lines.

5. A **ratio**
(a) compares two quantities using a quotient
(b) says that two quotients are equal
(c) is a product of two quantities
(d) is a difference between two quantities.

6. A **proportion**
(a) compares two quantities using a quotient
(b) says that two quotients are equal
(c) is a product of two quantities
(d) is a difference between two quantities.

7. An **inequality** is
(a) a statement that two algebraic expressions are equal
(b) a point on a number line
(c) an equation with no solutions
(d) a statement with algebraic expressions related by $<$, $\le$, $>$, or $\ge$.

8. Interval notation is
(a) a portion of a number line
(b) a special notation for describing a point on a number line
(c) a way to use symbols to describe an interval on a number line
(d) a notation to describe unequal quantities.

Answers to Test Your Word Power

1. (c) *Example:* {8} is the solution set of $2x + 5 = 21$. **2.** (c) *Example:* The empty set $\emptyset$ is the solution set of $5x + 3 = 5x + 4$.
3. (b) *Example:* Angles with measures 35° and 55° are complementary angles. **4.** (c) *Example:* Angles with measures 112° and 68° are supplementary angles. **5.** (a) *Example:* $\dfrac{7 \text{ inches}}{12 \text{ inches}}$ **6.** (b) *Example:* $\dfrac{7}{12} = \dfrac{2}{3}$ **7.** (d) *Example:* $x < 5$.
8. (c) *Examples:* $(-\infty, 5]$, $(1, \infty)$, $[-3, 3)$ $7 + 2y \ge 11$, $-5 < 2z - 1 \le 3$

QUICK REVIEW

CONCEPTS	EXAMPLES

2.1 THE ADDITION AND MULTIPLICATION PROPERTIES OF EQUALITY

The same expression may be added to (or subtracted from) each side of an equation without changing the solution.

Solve $x - 6 = 12$.

$$x - 6 + 6 = 12 + 6 \qquad \text{Add 6.}$$
$$x = 18 \qquad \text{Combine terms.}$$

Solution set: $\{18\}$

Each side of an equation may be multiplied (or divided) by the same nonzero expression without changing the solution.

Solve $\dfrac{3}{4}x = -9$.

$$\frac{4}{3} \cdot \frac{3}{4}x = \frac{4}{3}(-9) \qquad \text{Multiply by } \tfrac{4}{3}.$$
$$x = -12$$

Solution set: $\{-12\}$

2.2 MORE ON SOLVING LINEAR EQUATIONS

Solving a Linear Equation

Solve the equation $2x + 3(x + 1) = 38$.

1. Clear parentheses and combine like terms to simplify each side.

$$2x + 3x + 3 = 38 \qquad \text{Clear parentheses.}$$
$$5x + 3 = 38 \qquad \text{Combine like terms.}$$

2. Get the variable term on one side, a number on the other.

$$5x + 3 - 3 = 38 - 3 \qquad \text{Subtract 3.}$$
$$5x = 35 \qquad \text{Combine terms.}$$

3. Get the equation into the form $x =$ a number.

$$\frac{5x}{5} = \frac{35}{5} \qquad \text{Divide by 5.}$$
$$x = 7$$

4. Check by substituting the result into the original equation.

$$2x + 3(x + 1) = 38 \qquad \text{Check.}$$
$$2(7) + 3(7 + 1) = 38 \quad ? \qquad \text{Let } x = 7.$$
$$14 + 24 = 38 \quad ? \qquad \text{Multiply.}$$
$$38 = 38 \qquad \text{True}$$

Solution set: $\{7\}$

2.3 AN INTRODUCTION TO APPLICATIONS OF LINEAR EQUATIONS

Solving an Applied Problem Using the Six-Step Method

One number is 5 more than another. Their sum is 21. Find both numbers.

Step 1 Choose a variable to represent the unknown.

Let x be the smaller number.

Step 2 Determine expressions for any other unknown quantities, using the variable. Draw figures or diagrams and use charts if they apply.

Let $x + 5$ be the larger number.

Step 3 Write an equation.

Step 4 Solve the equation.

$$x + (x + 5) = 21$$
$$2x + 5 = 21 \qquad \text{Combine terms.}$$
$$2x + 5 - 5 = 21 - 5 \qquad \text{Subtract 5.}$$
$$2x = 16 \qquad \text{Combine terms.}$$
$$x = 8 \qquad \text{Divide by 2.}$$

Linear Equations in Two Variables

3

Calvin Coolidge, 30th president of the United States, once said "The chief business of America is business." Small businesses account for 99 percent of the 19 million nonfarm businesses in the United States today. Small businesses employ 55 percent of the private workforce, make 44 percent of all sales in America, and produce 38 percent of the nation's gross national product.*

Business

3.1 Linear Equations in Two Variables

3.2 Graphing Linear Equations in Two Variables

3.3 The Slope of a Line

The number of new consumer packaged-goods products introduced in the United States from 1991–1996 is illustrated in the line graph. The graph shows that, with the exception of 1994, this number has increased from year to year at a steady pace. What might account for the one-year decrease? We will return to this graph in the exercises for Section 3.1.

NEW CONSUMER PRODUCTS

Source: Marketing Intelligence Service, Ltd.

Mathematics plays an important role in the business world. Preparing a business plan, pricing, borrowing money, determining market share, and tracking costs, revenue, and profit all require a good understanding of mathematics. Throughout this chapter many of the examples and exercises will further illustrate the use of mathematics in business management.

*Data from *The Universal Almanac*, 1997, John W. Wright, General Editor.

3.1 Linear Equations in Two Variables

Graphs are prevalent in our society. Pie charts (circle graphs) and bar graphs were introduced in Chapter 1, and we have seen many examples of them in the first two chapters of this book. It is important to be able to interpret graphs correctly.

OBJECTIVE **1** Interpret graphs. We begin with a bar graph where we must estimate the heights of the bars.

EXAMPLE 1 Interpreting Bar Graphs

Venture capital is money invested in new, often speculative, business enterprises. The amount of venture capital invested in companies has risen during the decade of the 1990s, as shown in the graph in Figure 1. Use the graph to determine the following.

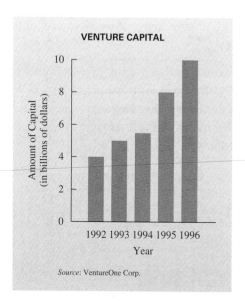

Figure 1

(a) What amount of venture capital was provided in 1992?

Move horizontally from the top of the bar for 1992 to the scale on the left to see that about $4 billion was provided.

(b) In what year was that amount doubled?

Follow the line for $8 billion across to the right. The bar for 1995 just touches that line, so the amount provided in 1992 was doubled in 1995.

EXAMPLE 2   Interpreting Line Graphs

The line graph in Figure 2 shows the total number of deals that involved venture capital in the 1990s. Use the graph to estimate the number of deals in 1993 and in 1995.

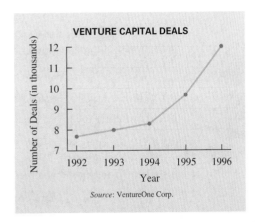

Figure 2

In 1993 the point lies on the line that corresponds to 8000, so 8000 venture capital deals were closed in 1993. The point for 1995 is about $\frac{2}{3}$, or .7, of the way between the lines for 9 and 10. Thus, we estimate 9700 deals were made in 1995.

We solved linear equations in one variable and explored their applications in Chapter 2. Now we want to extend those ideas to *linear equations in two variables.*

Linear Equation

A **linear equation in two variables** is an equation that can be put in the form

$$Ax + By = C,$$

where A, B, and C are real numbers and A and B are not both 0.

OBJECTIVE 2   Write a solution as an ordered pair. A solution of a linear equation in *two* variables requires *two* numbers, one for each variable. For example, the equation $y = 4x + 5$ is satisfied if x is replaced with 2 and y is replaced with 13, since

$$\mathbf{13} = 4(\mathbf{2}) + 5. \qquad \text{Let } x = 2; y = 13.$$

The pair of numbers $x = 2$ and $y = 13$ gives a solution of the equation $y = 4x + 5$. The phrase "$x = 2$ and $y = 13$" is abbreviated

x-value ⌐   ⌐ y-value

(2, 13)

Ordered pair

with the x-value, 2, and the y-value, 13, given as a pair of numbers written inside parentheses. *The x-value is always given first.* A pair of numbers such as (2, 13) is called an

ordered pair. As the name indicates, the order in which the numbers are written is important. The ordered pairs (**2**, **13**) and (**13**, **2**) are not the same. The second pair indicates that $x = 13$ and $y = 2$. (Of course, letters other than x and y may be used in the equation with the numbers.)

OBJECTIVE **3** Decide whether a given ordered pair is a solution of a given equation. An ordered pair that is a solution of an equation is said to *satisfy* the equation.

┌ **EXAMPLE** **3** Deciding Whether an Ordered Pair Satisfies an Equation

Decide whether the given ordered pair is a solution of the given equation.

(a) (3, 2); $2x + 3y = 12$

   To see whether (3, 2) is a solution of the equation $2x + 3y = 12$, we substitute 3 for x and 2 for y in the given equation.

$$2x + 3y = 12$$
$$2(3) + 3(2) = 12 \quad ? \qquad \text{Let } x = 3; \text{ let } y = 2.$$
$$6 + 6 = 12 \quad ?$$
$$12 = 12 \qquad\qquad \text{True}$$

This result is true, so (3, 2) satisfies $2x + 3y = 12$.

(b) (−2, −7); $m + 5n = 33$

$$(-2) + 5(-7) = 33 \quad ? \qquad \text{Let } m = -2; \text{ let } n = -7.$$
$$-2 + (-35) = 33 \quad ?$$
$$-37 = 33 \qquad\qquad \text{False}$$

This result is false, so (−2, −7) is *not* a solution of $m + 5n = 33$.

OBJECTIVE **4** Complete ordered pairs for a given equation. Choosing a number for one variable in a linear equation makes it possible to find the value of the other variable, as shown in the next example.

┌ **EXAMPLE** **4** Completing an Ordered Pair

Complete the ordered pair (7,   ) for the equation $y = 4x + 5$.

   In this ordered pair, $x = 7$. (Remember that x always comes first.) To find the corresponding value of y, replace x with 7 in the equation $y = 4x + 5$.

$$y = 4(7) + 5 = 28 + 5 = 33$$

The ordered pair is (7, 33).

   Ordered pairs often are displayed in a **table of values** as in the next example. The table may be written either vertically or horizontally.

EXAMPLE 5 Completing a Table of Values

Complete the given table of values for each equation. Then write the results as ordered pairs.

(a) $x - 2y = 8$

x	y
2	
10	
	0
	-2

To complete the first two ordered pairs, let $x = 2$ and $x = 10$, respectively.

	If	$x = 2,$			If	$x = 10,$
	then	$x - 2y = 8$			then	$x - 2y = 8$
	becomes	$2 - 2y = 8$			becomes	$10 - 2y = 8$
		$-2y = 6$				$-2y = -2$
		$y = -3.$				$y = 1.$

Now complete the last two ordered pairs by letting $y = 0$ and $y = -2$, respectively.

	If	$y = 0,$			If	$y = -2,$
	then	$x - 2y = 8$			then	$x - 2y = 8$
	becomes	$x - 2(0) = 8$			becomes	$x - 2(-2) = 8$
		$x - 0 = 8$				$x + 4 = 8$
		$x = 8.$				$x = 4.$

The completed table of values is as follows.

x	y
2	-3
10	1
8	0
4	-2

The corresponding ordered pairs are $(2, -3)$, $(10, 1)$, $(8, 0)$, and $(4, -2)$.

(b) $x = 5$

x	y
	-2
	6
	3

The given equation is $x = 5$. No matter which value of y might be chosen, the value of x is always the same, 5.

x	y
5	-2
5	6
5	3

The ordered pairs are $(5, -2)$, $(5, 6)$, and $(5, 3)$.

We can think of $x = 5$ in Example 5(b) as an equation in two variables by rewriting $x = 5$ as $x + 0y = 5$. This form of the equation shows that for any value of y, the value of x is 5. Similarly, $y = -2$ is the same as $0x + y = -2$.

Earlier in this book, we saw that linear equations in *one* variable had either one, zero, or an infinite number of real number solutions. Every linear equation in *two* variables has an infinite number of ordered pairs as solutions. Each choice of a number for one variable leads to a particular real number for the other variable.

To graph these solutions, represented as the ordered pairs (x, y), we need *two* number lines, one for each variable. These two number lines are drawn as shown in Figure 3. The horizontal number line is called the **x-axis.** The vertical line is called the **y-axis.** Together, the x-axis and y-axis form a **rectangular coordinate system.** It is also called the **Cartesian coordinate system,** in honor of René Descartes.

The coordinate system is divided into four regions, called **quadrants.** These quadrants are numbered counterclockwise, as shown in Figure 3. Points on the axes themselves are not in any quadrant. The point at which the x-axis and y-axis meet is called the **origin.** The origin, labeled 0 in Figure 3, is the point corresponding to $(0, 0)$.

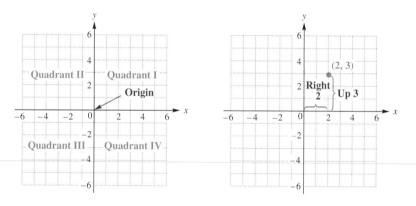

Figure 3   Figure 4

CONNECTIONS

The coordinate system that we use to plot points is credited to René Descartes (1596–1650), a French mathematician of the seventeenth century. It has been said that he developed the coordinate system while lying in bed, watching an insect move across the ceiling. He realized that he could locate the position of the insect at any given time by finding its distance from each of two perpendicular walls.

OBJECTIVE 5 Plot ordered pairs. By referring to the two axes, every point on the plane can be associated with an ordered pair. The numbers in the ordered pair are called the **coordinates** of the point. For example, locate the point associated with the ordered pair $(2, 3)$ by starting at the origin. Since the x-coordinate is 2, go 2 units to the right along the x-axis. Then, since the y-coordinate is 3, turn and go up 3 units on a line parallel to the y-axis. This is called **plotting** the point $(2, 3)$. (See Figure 4.) From now on we refer to the point with x-coordinate 2 and y-coordinate 3 as the point $(2, 3)$.

EXPLORE

Graphing Straight Lines (continued)

Equation	To Graph	Example
$Ax + By = 0$	Graph goes through (0, 0). Get additional points that lie on the graph by choosing any value of x or y, except 0.	(graph showing line $x = 2y$ through origin)
$Ax + By = C$ but not of the types above	Find any two points the line goes through. A good choice is to find the intercepts: let $x = 0$, and find the corresponding value of y; then let $y = 0$, and find x. As a check, get a third point by choosing a value of x or y that has not yet been used.	(graph showing line $3x - 2y = 6$ through points (2, 0) and (0, −3))

OBJECTIVE 5 Define a function. In the previous section, we saw several examples of relationships between two variables where each input value x produced an output value y. If each input x produces just *one* output y, we call the relationship a **function.** When a set of input values and a set of output values are related by a linear equation, $Ax + By = C$, each x-value corresponds to just one y-value, so linear equations define functions, unless $B = 0$. If $B = 0$, the equation becomes $Ax = C$ or $x = \frac{C}{A}$, where $\frac{C}{A}$ represents a real number. In this case, as we saw in Example 5, one input value x corresponds to an infinite number of output values y and the equation does not define a function.

┌ **EXAMPLE 6** Deciding Whether a Set of Ordered Pairs Defines a Function

Which of the following sets defines a function?

(a) $\{(1, -2), (2, 3), (0, 4), (4, 7)\}$
  The input 1 is paired with the output -2, the input 2 is paired with the output 3, and so on. Every input value corresponds to exactly one output value, so the set defines a function.

(b) $\{(2, -4), (1, -1), (0, 0), (1, 1), (2, 4)\}$
  Here, the input value 2 corresponds to the output values -4 and 4, so this set does not define a function. (What two output values correspond to the input value 1?)

(c) The ordered pairs that satisfy the equation $2x + 3y = 12$
  Because this is a linear equation with $B \neq 0$, it defines a function.

(d) The ordered pairs that satisfy the equation $5x = 10$
  In the context of two variables, when the coefficient of y is zero, as here, any value can be used for y. The only restriction imposed by the equation is that x must equal

2. Thus, $(2, 0)$, $(2, 5)$, $(2, -1)$, and any ordered pair with $x = 2$ satisfies this relationship. This is not a function, however, because $x = 2$ is paired with more than one real number.

(e) The ordered pairs that satisfy the equation $4y = 16$

Every ordered pair in this set has a y-value of 4. Since there is no restriction here on x, every real number is paired with $y = 4$. For example, $(-1, 4)$, $(2.5, 4)$, $(100, 4)$, and so on belong to this set, which defines a function, because every x corresponds to exactly one y, namely 4.

(f) The relationship between number of gallons needed to fill the gas tank and cost for a fill-up

This defines a function with ordered pairs of the form (number of gallons, cost). For a specific number of gallons, there is exactly one cost.

(g) The relationship between time spent traveling at a constant speed and number of miles traveled

Here the ordered pairs have the form (time, number of miles). A specific amount of time will correspond to exactly one number of miles, so this, too, defines a function.

Notice that in those sets that define functions, *no x-value appears more than once* (with a different *y*-value). The sets that were not functions had the same *x* as the input in at least two ordered pairs.

3.2 EXERCISES

In Exercises 1–6, match the information about the graphs with the linear equations in (a)–(e).

(a) $x = 5$ **(b)** $y = -3$ **(c)** $2x - 5y = 8$ **(d)** $x + 4y = 0$ **(e)** $3x + y = -4$

1. The graph of the equation has x-intercept $(4, 0)$.

2. The graph of the equation has y-intercept $(0, -4)$.

3. The graph of the equation goes through the origin.

4. The graph of the equation is a vertical line.

5. The graph of the equation is a horizontal line.

6. The graph of the equation goes through $(9, 2)$.

Complete the given ordered pairs using the given equation. Then graph each equation by plotting the points and drawing a line through them. See Examples 1 and 2.

7. $y = -x + 5$
$(0, \quad)$, $(\quad , 0)$, $(2, \quad)$

8. $y = x - 2$
$(0, \quad)$, $(\quad , 0)$, $(5, \quad)$

9. $y = \dfrac{2}{3}x + 1$
$(0, \quad)$, $(3, \quad)$, $(-3, \quad)$

10. $y = -\dfrac{3}{4}x + 2$
$(0, \quad)$, $(4, \quad)$, $(-4, \quad)$

11. $3x = -y - 6$
$(0, \quad)$, $(\quad , 0)$, $\left(-\dfrac{1}{3}, \quad \right)$

12. $x = 2y + 3$
$(\quad , 0)$, $(0, \quad)$, $\left(\quad , \dfrac{1}{2}\right)$

Find the x-intercept and the y-intercept for the graph of each equation. See Example 2.

13. $2x - 3y = 24$ **14.** $-3x + 8y = 48$ **15.** $x + 6y = 0$ **16.** $3x - y = 0$

17. What is the equation of the x-axis?

18. What is the equation of the y-axis?

19. A student attempted to graph $4x + 5y = 0$ by finding intercepts. She first let $x = 0$ and found y; then she let $y = 0$ and found x. In both cases, the resulting point was $(0, 0)$. She knew that she needed at least two different points to graph the line, but was unsure what to do next since finding intercepts gave her only one point. How would you explain to her what to do next?

20. Write a paragraph summarizing how to graph a linear equation in two variables.

Graph each linear equation. See Examples 1–5.

EXERCISES

21. $x = y + 2$	**22.** $x = -y + 6$	**23.** $x - y = 4$	**24.** $x - y = 5$
25. $2x + y = 6$	**26.** $-3x + y = -6$	**27.** $3x + 7y = 14$	**28.** $6x - 5y = 18$
29. $y - 2x = 0$	**30.** $y + 3x = 0$	**31.** $y = -6x$	**32.** $y = 4x$
33. $y + 1 = 0$	**34.** $y - 3 = 0$	**35.** $x = -2$	**36.** $x = 4$

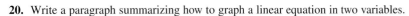

TECHNOLOGY INSIGHTS (EXERCISES 37–42)

In each exercise below, a calculator-generated graph of a linear equation in one variable with one side equal to 0 is shown. Accompanying the graph is the equation itself, where y is expressed in terms of x on the left side. Solve the equation using the methods of Section 2.2, and show that the solution you get is the same as the x-intercept (labeled "zero") on the calculator screen.

37. $8 - 2(3x - 4) - 2x = 0$

38. $5(2x - 1) - 4(2x + 1) - 7 = 0$

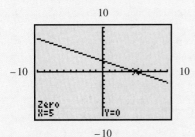

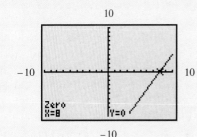

39. $.6x - .1x - x + 2.5 = 0$

40. $-\dfrac{2}{7}x + 2x - \dfrac{1}{2}x - \dfrac{17}{2} = 0$

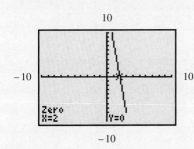

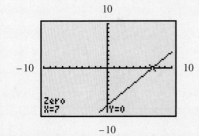

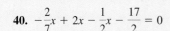

41. Use the results of Exercises 37–40 to explain how the x-intercept of the graph of an equation in two variables corresponds to the solution of an equation in one variable.

42. A horizontal line has no x-intercept. If you try to solve $5x - (3x + 2x) + 4 = 0$, you get no solution. What would the graph of $y = 5x - (3x + 2x) + 4$ look like on a graphing calculator?

Solve each problem.

43. The number of master's degrees earned in business management increased during the years 1971–1994 as shown in the figure. If $x = 0$ represents 1970, $x = 5$ represents 1975, $x = 10$ represents 1980, and so on, the number of master's degrees can be approximated by

$$y = 2.81x + 24.1,$$

where y is in thousands. (This is a *linear model* for the data.)

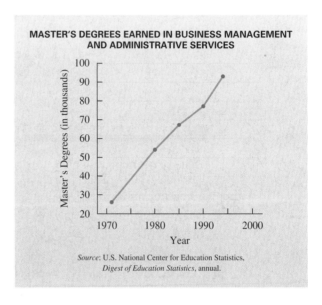

MASTER'S DEGREES EARNED IN BUSINESS MANAGEMENT AND ADMINISTRATIVE SERVICES

Source: U.S. National Center for Education Statistics, *Digest of Education Statistics*, annual.

(a) Use the equation to approximate the number of such degrees in the years 1980, 1985, 1990, and 1994.

(b) Estimate the y-values from the graph for the same years.

(c) Are the approximations using the equation close to the values you read from the graph?

44. Sporting goods sales (in billions of dollars) from 1988–1996 are approximated by the equation

$$y = 1.625x + 40.75,$$

where $x = 0$ corresponds to 1986, $x = 2$ corresponds to 1988, and so on. Sales for even-numbered years in that time period are plotted in the accompanying figure, which also shows the graph of the linear equation. As the graph indicates, the actual sales in 1992 (when the U.S. economy was depressed) of about $47 billion were less than the sales approximated by the equation.

(a) Use the *equation* to approximate the sales in each of the even-numbered years.

(b) Does this equation define a function? Why or why not?

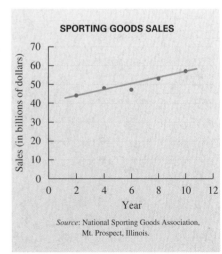

SPORTING GOODS SALES

Source: National Sporting Goods Association, Mt. Prospect, Illinois.

45. The height y of a woman (in centimeters) is a function of the length of her radius bone x (from the wrist to the elbow) and is defined by $y = 73.5 + 3.9x$. Estimate the heights of women with radius bones of the following lengths.

(a) 23 centimeters

(b) 25 centimeters

(c) 20 centimeters

(d) Graph $y = 73.5 + 3.9x$.

46. As a rough estimate, the weight of a man taller than about 60 inches is a linear function approximated by $y = 5.5x - 220$, where x is the height of the person in inches, and y is the weight in pounds. Estimate the weights of men whose heights are as follows.

(a) 62 inches   (b) 64 inches

(c) 68 inches   (d) 72 inches

(e) Graph $y = 5.5x - 220$.

47. The graph shows that the value of a certain automobile over its first five years is a function of the year. Use the graph to estimate the depreciation (loss in value) during the following years.

(a) First   (b) Second   (c) Fifth

(d) What is the total depreciation over the 5-year period?

48. The demand for an item is a function of its price. As price goes up, demand goes down. On the other hand, when price goes down, demand goes up. Suppose the demand for a certain Beanie Baby is 1000 when its price is $30 and 8000 when it costs $15.

(a) Let x be the price and y be the demand for the Beanie Baby. Graph the two given pairs of prices and demands.

(b) Assume the relationship is linear. Draw a line through the two points from part (a). From your graph estimate the demand if the price drops to $10.

(c) Use the graph to estimate the price if the demand is 4000.

Decide whether each set of ordered pairs defines a function. See Example 6.

49. $\{(0, 5), (2, 3), (4, 1), (6, -1), (8, -3)\}$

50. $\{(1, 3), (2, 3), (3, 3), (4, 3)\}$

51. The ordered pairs that satisfy $-x + 2y = 9$

52. The ordered pairs that satisfy $y = 4$

53. The ordered pairs that satisfy $x = 8$

54. The ordered pairs that satisfy $5x + y = 7$

55. The relationship in Exercise 43 between year and number of master's degrees

56. The relationship in Exercise 44 between year and sporting goods sales

3.3 The Slope of a Line

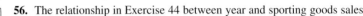

OBJECTIVES

1 Find the slope of a line given two points.

2 Find the slope from the equation of a line.

3 Use the slope to determine whether two lines are parallel, perpendicular, or neither.

When two variables are related in such a way that the value of one depends on the value of the other, we can form ordered pairs of corresponding numbers. For example, if $x + y = 7$, then when $x = 2$, $y = 5$, and when $x = -3$, $y = 10$. We write these pairs as $(2, 5)$ and $(-3, 10)$, respectively, with the understanding that the first number in the ordered pair represents x and the second number represents y. To indicate two nonspecific ordered pairs that satisfy a particular equation relating x and y, we use *subscript notation*. We write the pairs as (x_1, y_1) and (x_2, y_2). (Read x_1 as "x-sub-one" and x_2 as "x-sub-two.")

We can graph a straight line if at least two different points on the line are known. A line also can be graphed by using just one point on the line if the "steepness" of the line is known.

OBJECTIVE 1 Find the slope of a line given two points. One way to measure the steepness of a line is to compare the vertical change in the line (the rise) to the horizontal change (the run) while moving along the line from one fixed point to another. This measure of steepness is called the *slope* of the line.

Figure 15 shows a line with the points (x_1, y_1) and (x_2, y_2). As we move along the line from the point (x_1, y_1) to the point (x_2, y_2) y changes by $y_2 - y_1$ units. This is the vertical change. Similarly, x changes by $x_2 - x_1$ units, the horizontal change. The ratio of the change in y to the change in x gives the slope of the line. We usually denote slope with the letter m. The slope of a line is defined as follows.

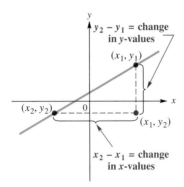

Figure 15

Slope Formula

The **slope** of the line through the points (x_1, y_1) and (x_2, y_2) is

$$m = \frac{\text{change in } y}{\text{change in } x} = \frac{y_2 - y_1}{x_2 - x_1} \quad \text{if } x_1 \neq x_2.$$

The slope of a line tells how fast y changes for each unit of change in x; that is, the slope gives the ratio of the change in y to the change in x. The change in y is called the **rise,** and the change in x is called the **run.**

CONNECTIONS

The idea of slope is used in many everyday situations. For example, because $10\% = \frac{1}{10}$, a highway with a 10% grade (or slope) rises one meter for every 10 horizontal meters. The highway sign shown below is used to warn of a downgrade ahead that may be long or steep. Architects specify the pitch of a roof using slope; a $\frac{5}{12}$ roof means that the roof rises 5 feet for every 12 feet in the horizontal direction. The slope of a stairwell also indicates the ratio of the vertical rise to the horizontal run. The slope of the stairs in the figure is $\frac{8}{14}$.

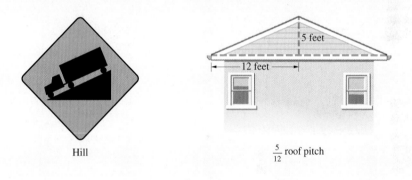

Hill

$\frac{5}{12}$ roof pitch

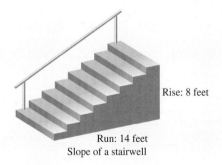

Slope of a stairwell

FOR DISCUSSION OR WRITING
Describe some other everyday examples of slope.

┌─ **E X A M P L E 1** **Finding the Slope of a Line**

Find the slope of each of the following lines.

(a) The line through $(-4, 7)$ and $(1, -2)$

Use the definition of slope. Let $(-4, 7) = (x_2, y_2)$ and $(1, -2) = (x_1, y_1)$. Then

$$\text{slope} = \frac{\text{change in } y}{\text{change in } x}$$

$$m = \frac{y_2 - y_1}{x_2 - x_1}$$

$$= \frac{7 - (-2)}{-4 - 1} = \frac{9}{-5} = -\frac{9}{5}.$$

As Figure 16 shows, this line has a slope of $-\frac{9}{5}$ (which can also be written as $\frac{-9}{5}$ or $\frac{9}{-5}$). One way of interpreting this is that the line drops vertically 9 units for a horizontal change of 5 units to the right.

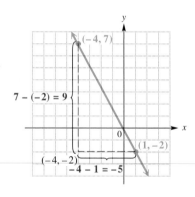

Figure 16

(b) The line through $(12, -5)$ and $(-9, -2)$

$$m = \frac{-5 - (-2)}{12 - (-9)} = \frac{-3}{21} = -\frac{1}{7}$$

The same slope is found by subtracting in reverse order.

$$\frac{-2 - (-5)}{-9 - 12} = \frac{3}{-21} = -\frac{1}{7}$$

 It makes no difference which point is (x_1, y_1) or (x_2, y_2); however, it is important to be consistent. Start with the x- and y-values of one point (either one) and subtract the corresponding values of the other point.

In Example 1(a) the slope is negative and the corresponding line in Figure 16 falls from left to right. As Figure 17(a) shows, this is generally true of lines with negative slopes. Lines with positive slopes go up (rise) from left to right, as shown in Figure 17(b).

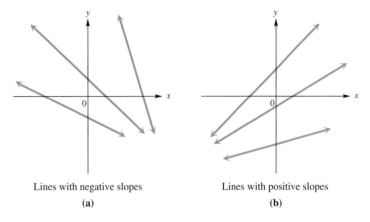

Lines with negative slopes   Lines with positive slopes

(a)   (b)

Figure 17

AUDIO

Positive and Negative Slopes

A line with positive slope rises from left to right.

A line with negative slope falls from left to right.

E X A M P L E   2   Finding the Slope of a Horizontal Line

Find the slope of the line through $(-8, 4)$ and $(2, 4)$.

Use the definition of slope.

$$m = \frac{4 - 4}{-8 - 2} = \frac{0}{-10} = 0$$

As shown in Figure 18, the line through these two points is horizontal, with equation $y = 4$. *All horizontal lines have a slope of 0,* since the difference in y-values is always 0.

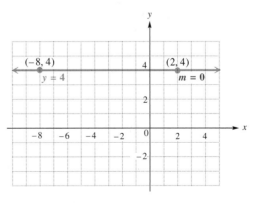

Figure 18

┌─
E X A M P L E   3   Finding the Slope of a Vertical Line

Find the slope of the line through $(6, 2)$ and $(6, -9)$.

$$m = \frac{2 - (-9)}{6 - 6} = \frac{11}{0} \qquad \text{Undefined}$$

Since division by 0 is undefined, the slope is undefined. The graph in Figure 19 shows that the line through these two points is vertical, with equation $x = 6$. All points on a vertical line have the same x-value, so *the slope of any vertical line is undefined.*

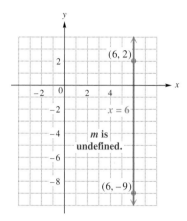

Figure 19
─┘

— AUDIO

Slopes of Horizontal and Vertical Lines

Horizontal lines, with equations of the form $y = k$, **have slope 0.**

Vertical lines, with equations of the form $x = k$, **have undefined slope.**

O B J E C T I V E   2   Find the slope from the equation of a line.   The slope of a line also can be found directly from its equation. For example, the slope of the line

$$y = -3x + 5$$

can be found using any two points on the line. Get two points by choosing two differ-ent values of x, say -2 and 4, and finding the corresponding y-values.

If $x = -2$:	If $x = 4$:
$y = -3(-2) + 5$	$y = -3(4) + 5$
$y = 6 + 5$	$y = -12 + 5$
$y = 11.$	$y = -7.$

The ordered pairs are $(-2, 11)$ and $(4, -7)$. Now use the slope formula to find the slope.

$$m = \frac{11 - (-7)}{-2 - 4} = \frac{18}{-6} = -3$$

The slope, -3, is the same number as the coefficient of x in the equation $y = -3x + 5$. It can be shown that this always happens, *as long as the equation is solved for y.* This fact is used to find the slope of a line from its equation.

Finding the Slope of a Line from its Equation

Step 1   Solve the equation for *y*.

Step 2   The slope is given by the coefficient of *x*.

E X A M P L E   4   Finding Slope from an Equation

Find the slope of each of the following lines.

(a) $2x - 5y = 4$

Solve the equation for *y*.

$$2x - 5y = 4$$
$$-5y = -2x + 4 \qquad \text{Subtract } 2x \text{ from each side.}$$
$$y = \frac{2}{5}x - \frac{4}{5} \qquad \text{Divide each side by } -5.$$

The slope is given by the coefficient of *x*, so the slope is $m = \dfrac{2}{5}$.

(b) $8x + 4y = 1$

Solve the equation for *y*.

$$8x + 4y = 1$$
$$4y = -8x + 1 \qquad \text{Subtract } 8x \text{ from each side.}$$
$$y = -2x + \frac{1}{4} \qquad \text{Divide each side by } 4.$$

The slope of this line is given by the coefficient of *x*, -2.

OBJECTIVE 3   Use the slope to determine whether two lines are parallel, perpendicular, or neither. Two lines in a plane that never intersect are **parallel.** We use slopes to tell whether two lines are parallel. For example, Figure 20 shows the graph of $x + 2y = 4$ and the graph of $x + 2y = -6$. These lines appear to be parallel. Solve for *y* to find that both $x + 2y = 4$ and $x + 2y = -6$ have a slope of $-\frac{1}{2}$. Nonvertical parallel lines always have equal slopes.

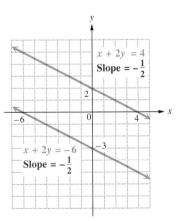

Figure 20

Figure 21 shows the graph of $x + 2y = 4$ and the graph of $2x - y = 6$. These lines appear to be **perpendicular** (meet at a 90° angle). Solving for y shows that the slope of $x + 2y = 4$ is $-\frac{1}{2}$, while the slope of $2x - y = 6$ is 2. The product of $-\frac{1}{2}$ and 2 is

$$-\frac{1}{2}(2) = -1.$$

This is true in general; the product of the slopes of two perpendicular lines, neither of which is vertical, is always -1. This means that the slopes of perpendicular lines are negative reciprocals: if one slope is the nonzero number a, the other is $-\frac{1}{a}$.

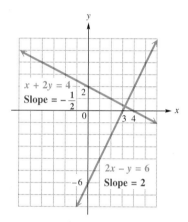

Figure 21

Parallel and Perpendicular Lines

Two nonvertical lines with the same slope are parallel; two perpendicular lines, neither of which is vertical, have slopes that are negative reciprocals of each other.

┌ **E X A M P L E 5**   Deciding Whether Two Lines Are Parallel or Perpendicular

Decide whether the lines are *parallel, perpendicular,* or *neither.*

(a) $x + 2y = 7$
$-2x + y = 3$

Find the slope of each line by first solving each equation for y.

$x + 2y = 7$	$-2x + y = 3$
$2y = -x + 7$	$y = 2x + 3$
$y = -\frac{1}{2}x + \frac{7}{2}$	
Slope: $-\frac{1}{2}$	Slope: 2

Since the slopes are not equal, the lines are not parallel. Check the product of the slopes: $-\frac{1}{2}(2) = -1.$ The two lines are perpendicular because the product of their slopes is -1, indicating that the slopes are negative reciprocals.

Solve.

10. $V = \dfrac{1}{3}\pi r^2 h$ *for* h

11. $6 - 3(1 + a) = 2(a + 5) - 2$

12. $-(m - 1) = 3 - 2m$

13. $\dfrac{y - 2}{3} = \dfrac{2y + 1}{5}$

14. $-5z \geq 4z - 18$

15. $2 < -6(z + 1) < 10$

Solve each problem.

16. Kimshana Lavoris earned $200 working part time during July, $375 during August, and $325 during September. If her average income for the four months from July through October must be at least $300, what possible amounts could she earn in October?

17. Mount Mayon in the Philippines is the most perfectly shaped conical volcano in the world. Its base is a perfect circle with a 39-mile circumference and it has a height of 8000 feet. (One mile is 5280 feet.) Find the radius of the circular base to the nearest mile. (*Hint:* This problem has some unneeded information.)

18. How much of a 20% chemical solution must be mixed with 30 liters of a 60% solution to get a 50% mixture?

19. The winning times in seconds for the women's 1000-meter speed skating event in the Winter Olympics for the years from 1960 to 1998 can be closely approximated by the equation

$$y = -.4685x + 95.07,$$

where x is the number of years since 1960. That is, $x = 5$ represents 1965, $x = 10$ represents 1970, and so on. Complete the table of ordered pairs for this linear equation. Round the y-values to the nearest hundredth of a second. (*Source: The Universal Almanac*, 1997, John W. Wright, General Editor.)

x	y
12	
28	
36	

20. Baby boomers are expected to inherit $10.4 trillion from their parents over the next 45 years, an average of $50,000 each. The pie chart shows how they plan to spend their inheritance. How much of the $50,000 is expected to go toward the purchase of a home? How much to retirement?

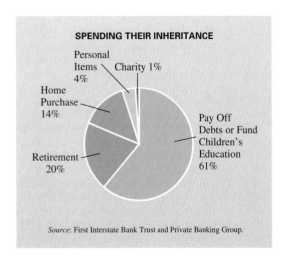

SPENDING THEIR INHERITANCE

Personal Items 4%
Charity 1%
Home Purchase 14%
Pay Off Debts or Fund Children's Education 61%
Retirement 20%

Source: First Interstate Bank Trust and Private Banking Group.

Consider the linear equation $3x + 2y = 12$. *Find the following.*

21. The x- and y-intercepts   **22.** The graph   **23.** The slope

24. Are the lines with equations $x + 5y = -6$ and $y = 5x - 8$ parallel, perpendicular, or neither?

25. California's exports to Pacific Rim nations, with the exception of China, fell during the first half of 1998, as shown in the bar graph. Is the set whose ordered pairs represent the data from the graph as (importer, % change) a function? Give the ordered pairs with the smallest change and the largest change. How is absolute value used in your answer?

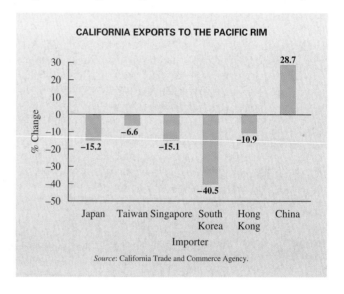

CALIFORNIA EXPORTS TO THE PACIFIC RIM

Source: California Trade and Commerce Agency.

Polynomials and Exponents

The number of passengers traveling by air has increased rapidly in the last decade. (See Exercise 46 in Section 4.1.) From 1985 to 1995, scheduled air carriers enjoyed an increase in net profits of $1,514,000,000. Surprisingly, in spite of more passengers traveling in planes that are consistently full, consumer complaints against U.S. airlines have generally decreased, as shown in the figure. There is one year, however, when all three graphs indicate an increase in each category from one year to the next, rather than a decrease. In which year did this occur?

Aeronautics

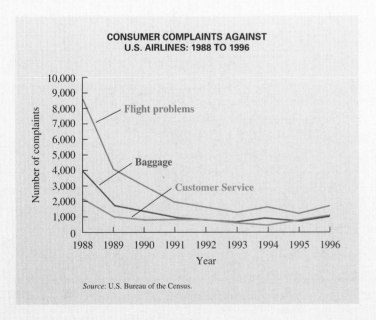

The graphs in the figure can each be approximated by a *polynomial function.* Polynomials are one of the topics studied in this chapter. Throughout the chapter we will see other examples of polynomials that describe information about the aeronautics industry.

223

4.1 Addition and Subtraction of Polynomials; Graphing Simple Polynomials

OBJECTIVES

1. Identify terms and coefficients.
2. Add like terms.
3. Know the vocabulary for polynomials.
4. Evaluate polynomials.
5. Add and subtract polynomials.
6. Graph equations defined by polynomials with degree 2.

OBJECTIVE 1 Identify terms and coefficients. In Chapter 1 we saw that in an expression such as

$$4x^3 + 6x^2 + 5x + 8,$$

the quantities $4x^3$, $6x^2$, $5x$, and 8 are called *terms*. As mentioned earlier, in the term $4x^3$, the number 4 is called the *numerical coefficient*, or simply the *coefficient*, of x^3. In the same way, 6 is the coefficient of x^2 in the term $6x^2$, 5 is the coefficient of x in the term $5x$, and 8 is the coefficient in the term 8. A constant term, like 8 in the expression above, can be thought of as $8x^0$, where

$$x^0 \text{ is defined to equal } 1.$$

We explain the reason for this definition later in this chapter.

EXAMPLE 1 Identifying Coefficients

Name the (numerical) coefficient of each term in these expressions.

(a) $4x^3$

The coefficient is 4.

(b) $x - 6x^4$

The coefficient of x is 1 because $x = 1 \cdot x$. The coefficient of x^4 is -6 since $x - 6x^4$ can be written as the sum $x + (-6x^4)$.

(c) $5 - v^3$

The coefficient of the term 5 is 5 because $5 = 5v^0$. By writing $5 - v^3$ as a sum, $5 + (-v^3)$, or $5 + (-1v^3)$, the coefficient of v^3 can be identified as -1.

OBJECTIVE 2 Add like terms. Recall from Section 1.8 that *like terms* have exactly the same combination of variables with the same exponents on the variables. Only the coefficients may differ. Examples of like terms are

$$19m^5 \quad \text{and} \quad 14m^5,$$
$$6y^9, \quad -37y^9, \quad \text{and} \quad y^9,$$
$$3pq \quad \text{and} \quad -2pq,$$
$$2xy^2 \quad \text{and} \quad -xy^2.$$

Using the distributive property, we add like terms by adding their coefficients.

VIDEO

EXAMPLE 2 Adding Like Terms

Simplify each expression by adding like terms.

(a) $-4x^3 + 6x^3 = (-4 + 6)x^3 = 2x^3$ Distributive property

(b) $9x^6 - 14x^6 + x^6 = (9 - 14 + 1)x^6 = -4x^6$

(c) $12m^2 + 5m + 4m^2 = (12 + 4)m^2 + 5m = 16m^2 + 5m$

(d) $3x^2y + 4x^2y - x^2y = (3 + 4 - 1)x^2y = 6x^2y$

In Example 2(c), we cannot combine $16m^2$ and $5m$. These two terms are unlike because the exponents on the variables are different. *Unlike terms* have different variables or different exponents on the same variables.

OBJECTIVE **3** Know the vocabulary for polynomials. A **polynomial in x** is a term or the sum of a finite number of terms of the form ax^n, for any real number a and any whole number n. For example,

$$16x^8 - 7x^6 + 5x^4 - 3x^2 + 4$$

is a polynomial in x (the 4 can be written as $4x^0$). This polynomial is written in **descending powers** of the variable, since the exponents on x decrease from left to right. On the other hand,

$$2x^3 - x^2 + \frac{4}{x}$$

is not a polynomial in x, since a variable appears in a denominator. Of course, we could define *polynomial* using any variable and not just x, as in Example 2(c). In fact, polynomials may have terms with *more* than one variable, as in Example 2(d).

The **degree of a term** is the sum of the exponents on the variables. For example, $3x^4$ has degree 4, while $6x^{17}$ has degree 17. The term $5x$ has degree 1, -7 has degree 0 (since -7 can be written as $-7x^0$), and $2x^2y$ has degree $2 + 1 = 3$ (y has an exponent of 1.) The **degree of a polynomial** is the highest degree of any nonzero term of the polynomial. For example, $3x^4 - 5x^2 + 6$ is of degree 4, the polynomial $5x + 7$ is of degree 1, 3 (or $3x^0$) is of degree 0, and $x^2y + xy - 5xy^2$ is of degree 3.

Three types of polynomials are very common and are given special names. A polynomial with exactly three terms is called a **trinomial.** (*Tri-* means "three," as in *tri*angle.) Examples are

$$9m^3 - 4m^2 + 6, \qquad 19y^2 + 8y + 5, \qquad \text{and} \qquad -3m^5n^2 + 2n^3 - m^4.$$

A polynomial with exactly two terms is called a **binomial.** (*Bi-* means "two," as in *bi*cycle.) Examples are

$$-9x^4 + 9x^3, \qquad 8m^2 + 6m, \qquad \text{and} \qquad 3m^5n^2 - 9m^2n^4.$$

A polynomial with only one term is called a **monomial.** (*Mon(o)-* means "one," as in *mono*rail.) Examples are

$$9m, \qquad -6y^5, \qquad a^2b^2, \qquad \text{and} \qquad 6.$$

EXAMPLE 3 Classifying Polynomials

For each polynomial, first simplify if possible by combining like terms. Then give the degree and tell whether it is a monomial, a binomial, a trinomial, or none of these.

(a) $2x^3 + 5$
  The polynomial cannot be simplified. The degree is 3. The polynomial is a binomial.

(b) $4xy - 5xy + 2xy$
  Add like terms to simplify: $4xy - 5xy + 2xy = xy$, which is a monomial of degree 2.

OBJECTIVE **4** Evaluate polynomials. A polynomial usually represents different numbers for different values of the variable, as shown in the next example.

┌───

E X A M P L E   4   Evaluating a Polynomial

Find the value of $3x^4 + 5x^3 - 4x - 4$ when $x = -2$ and when $x = 3$.
    First, substitute -2 for x.

$$3x^4 + 5x^3 - 4x - 4 = 3(-2)^4 + 5(-2)^3 - 4(-2) - 4$$
$$= 3 \cdot 16 + 5 \cdot (-8) + 8 - 4$$
$$= 48 - 40 + 8 - 4$$
$$= 12$$

Next, replace x with 3.

$$3x^4 + 5x^3 - 4x - 4 = 3(3)^4 + 5(3)^3 - 4(3) - 4$$
$$= 3 \cdot 81 + 5 \cdot 27 - 12 - 4$$
$$= 362$$

───

CAUTION
Notice the use of parentheses around the numbers that are substituted for the variable in Example 4. This is particularly important when substituting a negative number for a variable that is raised to a power, so that the sign of the product is correct.

───

CONNECTIONS

In Section 3.2 we introduced the idea of a function: for every input x, there is one output y. Polynomials often provide a way of defining functions that approximate data collected over a period of time. For example, according to the U.S. National Aeronautics and Space Administration (NASA), the budget in millions of dollars for space station research for 1996–2001 can be approximated by the polynomial equation

$$y = -10.25x^2 - 126.04x + 5730.21,$$

where $x = 0$ represents 1996, $x = 1$ represents 1997, and so on, up to $x = 5$ representing 2001. The actual budget for 1998 was 5327 million dollars; an input of $x = 2$ (for 1998) gives approximately $y = 5437$. Considering the magnitude of the numbers, this is a very good approximation.

FOR DISCUSSION OR WRITING
Use the given polynomial equation to approximate the budget in other years between 1996 and 2001. Compare to the actual figures given here.

Year	Budget (in millions of dollars)
1996	5710
1997	5675
1998	5327
1999	5306
2000	5077
2001	4832

OBJECTIVE 5   Add and subtract polynomials. Polynomials may be added, subtracted, multiplied, and divided.

Adding Polynomials

To add two polynomials, add like terms.

EXAMPLE 5   Adding Polynomials Vertically

Add $6x^3 - 4x^2 + 3$ and $-2x^3 + 7x^2 - 5$.

Write like terms in columns.

$$\begin{array}{r} 6x^3 - 4x^2 + 3 \\ -2x^3 + 7x^2 - 5 \\ \hline \end{array}$$

Now add, column by column.

$$\begin{array}{ccc} 6x^3 & -4x^2 & 3 \\ -2x^3 & 7x^2 & -5 \\ \hline 4x^3 & 3x^2 & -2 \end{array}$$

Add the three sums together.

$$4x^3 + 3x^2 + (-2) = 4x^3 + 3x^2 - 2$$

The polynomials in Example 5 also can be added horizontally, as shown in the next example.

EXAMPLE 6   Adding Polynomials Horizontally

Add $6x^3 - 4x^2 + 3$ and $-2x^3 + 7x^2 - 5$.

Write the sum as

$$(6x^3 - 4x^2 + 3) + (-2x^3 + 7x^2 - 5).$$

Use the associative and commutative properties to rewrite this sum with the parentheses removed and with the subtractions changed to additions of inverses.

$$6x^3 + (-4x^2) + 3 + (-2x^3) + 7x^2 + (-5)$$

Place like terms together.

$$6x^3 + (-2x^3) + (-4x^2) + 7x^2 + 3 + (-5)$$

Combine like terms to get

$$4x^3 + 3x^2 + (-2), \qquad \text{or simply} \qquad 4x^3 + 3x^2 - 2,$$

the same answer found in Example 5.

Earlier, we defined the difference $x - y$ as $x + (-y)$. (We find the difference $x - y$ by adding x and the opposite of y.) For example,

$$7 - 2 = 7 + (-2) = 5 \qquad \text{and} \qquad -8 - (-2) = -8 + 2 = -6.$$

A similar method is used to subtract polynomials.

Subtracting Polynomials

To subtract two polynomials, change all the signs on the second polynomial and add the result to the first polynomial.

EXAMPLE 7 Subtracting Polynomials

(a) Perform the subtraction $(5x - 2) - (3x - 8)$.
By the definition of subtraction,

$$(5x - 2) - (3x - 8) = (5x - 2) + [-(3x - 8)].$$

As shown in Chapter 1, the distributive property gives

$$-(3x - 8) = -1(3x - 8) = -3x + 8,$$

so

$$(5x - 2) - (3x - 8) = (5x - 2) + (-3x + 8) = 2x + 6.$$

(b) Subtract $6x^3 - 4x^2 + 2$ from $11x^3 + 2x^2 - 8$.
Write the problem.

$$(11x^3 + 2x^2 - 8) - (6x^3 - 4x^2 + 2)$$

Change all the signs in the second polynomial and add the two polynomials.

$$(11x^3 + 2x^2 - 8) + (-6x^3 + 4x^2 - 2) = 5x^3 + 6x^2 - 10$$

To check a subtraction problem, use the fact that if $a - b = c$, then $a = b + c$. For example, $6 - 2 = 4$, so we check by writing $6 = 2 + 4$, which is correct. Check the polynomial subtraction above by adding $6x^3 - 4x^2 + 2$ and $5x^3 + 6x^2 - 10$. Since the sum is $11x^3 + 2x^2 - 8$, the subtraction was performed correctly.

Subtraction also can be done in columns (vertically). We will use vertical subtraction in Section 4.6 when we study polynomial division.

EXAMPLE 8 Subtracting Polynomials Vertically

Use the method of subtracting by columns to find

$$(14y^3 - 6y^2 + 2y - 5) - (2y^3 - 7y^2 - 4y + 6).$$

Arrange like terms in columns.

$$\begin{array}{r} 14y^3 - 6y^2 + 2y - 5 \\ 2y^3 - 7y^2 - 4y + 6 \end{array}$$

Change all signs in the second row, and then add.

$$\begin{array}{r} 14y^3 - 6y^2 + 2y - 5 \\ -2y^3 + 7y^2 + 4y - 6 \\ \hline 12y^3 + y^2 + 6y - 11 \end{array}$$ Change all signs.

                                                                                                Add.

Either the horizontal or the vertical method may be used to add and subtract polynomials.

Polynomials in more than one variable are added and subtracted by combining like terms, just as with single variable polynomials.

51. Add.

$$\frac{2}{3}x^2 + \frac{1}{5}x + \frac{1}{6}$$
$$\frac{1}{2}x^2 - \frac{1}{3}x + \frac{2}{3}$$

52. Add.

$$\frac{4}{7}y^2 - \frac{1}{5}y + \frac{7}{9}$$
$$\frac{1}{3}y^2 - \frac{1}{3}y + \frac{2}{5}$$

53. Add.

$$9m^3 - 5m^2 + 4m - 8$$
$$-3m^3 + 6m^2 + 8m - 6$$

54. Add.

$$12r^5 + 11r^4 - 7r^3 - 2r^2$$
$$-8r^5 + 10r^4 + 3r^3 + 2r^2$$

55. Subtract.

$$12m^3 - 8m^2 + 6m + 7$$
$$-3m^3 + 5m^2 - 2m - 4$$

56. Subtract.

$$5a^4 - 3a^3 + 2a^2 - a + 6$$
$$-6a^4 + a^3 - a^2 + a - 1$$

57. After reading Examples 5–8, explain whether you have a preference regarding horizontal or vertical addition and subtraction of polynomials.

58. Write a paragraph explaining how to add and subtract polynomials. Give an example using addition.

Perform the indicated operations. See Examples 6–8.

59. $(8m^2 - 7m) - (3m^2 + 7m - 6)$

60. $(x^2 + x) - (3x^2 + 2x - 1)$

61. $(16x^3 - x^2 + 3x) + (-12x^3 + 3x^2 + 2x)$

62. $(-2b^6 + 3b^4 - b^2) + (b^6 + 2b^4 + 2b^2)$

63. $(7y^4 + 3y^2 + 2y) - (18y^4 - 5y^2 + y)$

64. $(8t^5 + 3t^3 + 5t) - (19t^5 - 6t^3 + t)$

65. $(9a^4 - 3a^2 + 2) + (4a^4 - 4a^2 + 2) + (-12a^4 + 6a^2 - 3)$

66. $(4m^2 - 3m + 2) + (5m^2 + 13m - 4) - (16m^2 + 4m - 3)$

67. $[(8m^2 + 4m - 7) - (2m^2 - 5m + 2)] - (m^2 + m + 1)$

68. $[(9b^3 - 4b^2 + 3b + 2) - (-2b^3 - 3b^2 + b)] - (8b^3 + 6b + 4)$

69. $[(3x^2 - 2x + 7) - (4x^2 + 2x - 3)] - [(9x^2 + 4x - 6) + (-4x^2 + 4x + 4)]$

70. $[(6t^2 - 3t + 1) - (12t^2 + 2t - 6)] - [(4t^2 - 3t - 8) + (-6t^2 + 10t - 12)]$

EXERCISES

71. Without actually performing the operations, determine mentally the coefficient of x^2 in the simplified form of $(-4x^2 + 2x - 3) - (-2x^2 + x - 1) + (-8x^2 + 3x - 4)$.

72. Without actually performing the operations, determine mentally the coefficient of x in the simplified form of $(-8x^2 - 3x + 2) - (4x^2 - 3x + 8) - (-2x^2 - x + 7)$.

Add or subtract as indicated. See Example 9.

73. $(6b + 3c) + (-2b - 8c)$

74. $(-5t + 13s) + (8t - 3s)$

75. $(4x + 2xy - 3) - (-2x + 3xy + 4)$

76. $(8ab + 2a - 3b) - (6ab - 2a + 3b)$

77. $(5x^2y - 2xy + 9xy^2) - (8x^2y + 13xy + 12xy^2)$

78. $(16t^3s^2 + 8t^2s^3 + 9ts^4) - (-24t^3s^2 + 3t^2s^3 - 18ts^4)$

For Exercises 79–82, use the formulas found on the inside covers.

Find the perimeter of each rectangle.

79.

$4x^2 + 3x + 1$

$x + 2$

80.

$5y^2 + 3y + 8$

$y + 4$

Find (a) a polynomial representing the perimeter of each triangle and (b) the measures of the angles of the triangle.

81.

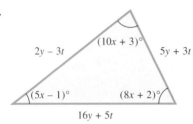

82.

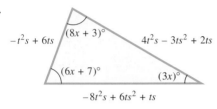

The concepts required to work Exercises 83–86 have been covered, but the usual wording of the problem has been changed. Perform the indicated operations.

83. Subtract $9x^2 - 6x + 5$ from $3x^2 - 2$.

84. Find the difference when $9x^4 + 3x^2 + 5$ is subtracted from $8x^4 - 2x^3 + x - 1$.

85. Find the difference between the sum of $5x^2 + 2x - 3$ and $x^2 - 8x + 2$ and the sum of $7x^2 - 3x + 6$ and $-x^2 + 4x - 6$.

86. Subtract the sum of $9t^3 - 3t + 8$ and $t^2 - 8t + 4$ from the sum of $12t + 8$ and $t^2 - 10t + 3$.

Graph each of the following by completing the table of values. See Example 10.

87. $y = x^2 - 4$

x	y
-2	
-1	
0	
1	
2	

88. $y = x^2 - 6$

x	y
-2	
-1	
0	
1	
2	

89. $y = 2x^2 - 1$

x	y
-2	
-1	
0	
1	
2	

90. $y = 2x^2 + 2$

x	y
-2	
-1	
0	
1	
2	

91. $y = 4 - x^2$

x	y
-2	
-1	
0	
1	
2	

92. $y = 2 - x^2$

x	y
-2	
-1	
0	
1	
2	

93. $y = (x + 3)^2$

x	-5	-4	-3	-2	-1
y					

94. $y = (x - 4)^2$

x	2	3	4	5	6
y					

4.2 The Product Rule and Power Rules for Exponents

OBJECTIVE 1 Identify bases and exponents. Recall from Section 1.2 that in the expression 5^2, the number 5 is the *base* and 2 is the *exponent*. The expression 5^2 is called an *exponential expression*. Usually we do not write the exponent when it is 1; however, sometimes it is convenient to do so. In general, for any quantity a, $a^1 = a$.

3   Use the rule $(a^m)^n = a^{mn}$.

4   Use the rule $(ab)^m = a^m b^m$.

5   Use the rule $\left(\dfrac{a}{b}\right)^m = \dfrac{a^m}{b^m}$.

E X A M P L E 1   Determining the Base and Exponent in an Exponential Expression

Evaluate each exponential expression. Name the base and the exponent.

		Base	Exponent
(a) $5^4 = 5 \cdot 5 \cdot 5 \cdot 5 = 625$		5	4
(b) $-5^4 = -1 \cdot 5^4 = -1 \cdot (5 \cdot 5 \cdot 5 \cdot 5) = -625$		5	4
(c) $(-5)^4 = (-5)(-5)(-5)(-5) = 625$		-5	4

Note the differences between parts (b) and (c) of Example 1. In -5^4 the lack of parentheses shows that the exponent 4 refers only to the base 5, and not -5; in $(-5)^4$ the parentheses show that the exponent 4 refers to the base -5. In summary, $-a^n$ and $(-a)^n$ are not necessarily the same.

Expression	Base	Exponent	Example
$-a^n$	a	n	$-3^2 = -(3 \cdot 3) = -9$
$(-a)^n$	$-a$	n	$(-3)^2 = (-3)(-3) = 9$

O B J E C T I V E 2   Use the product rule for exponents. By the definition of exponents,

$$2^4 \cdot 2^3 = \overbrace{(2 \cdot 2 \cdot 2 \cdot 2)}^{4 \text{ factors}}\overbrace{(2 \cdot 2 \cdot 2)}^{3 \text{ factors}}$$
$$= \underbrace{2 \cdot 2 \cdot 2 \cdot 2 \cdot 2 \cdot 2 \cdot 2}_{4 + 3 = 7 \text{ factors}}$$
$$= 2^7.$$

Also,

$$6^2 \cdot 6^3 = (6 \cdot 6)(6 \cdot 6 \cdot 6)$$
$$= 6 \cdot 6 \cdot 6 \cdot 6 \cdot 6$$
$$= 6^5.$$

Generalizing from these examples, $2^4 \cdot 2^3 = 2^{4+3} = 2^7$ and $6^2 \cdot 6^3 = 6^{2+3} = 6^5$, suggests the **product rule for exponents.**

Product Rule for Exponents

For any positive integers m and n, $a^m \cdot a^n = a^{m+n}$.
(Keep the same base; add the exponents.)
Example: $6^2 \cdot 6^5 = 6^{2+5} = 6^7$

E X A M P L E 2   Using the Product Rule

Use the product rule for exponents to find each result when possible.

(a) $6^3 \cdot 6^5 = 6^{3+5} = 6^8$ by the product rule.

(b) $(-4)^7(-4)^2 = (-4)^{7+2} = (-4)^9$

(c) $x^2 \cdot x = x^2 \cdot x^1 = x^{2+1} = x^3$

(d) $m^4 m^3 m^5 = m^{4+3+5} = m^{12}$

(e) $2^3 \cdot 3^2$

The product rule does not apply to the product $2^3 \cdot 3^2$, since the bases are different.

$$2^3 \cdot 3^2 = 8 \cdot 9 = 72$$

(f) $2^3 + 2^4$

The product rule does not apply to $2^3 + 2^4$, since this is a *sum*, not a *product*.

$$2^3 + 2^4 = 8 + 16 = 24$$

(g) $(2x^3)(3x^7)$

Since $2x^3$ means $2 \cdot x^3$ and $3x^7$ means $3 \cdot x^7$, we can use the associative and commutative properties to get

$$(2x^3)(3x^7) = (2 \cdot 3) \cdot (x^3 \cdot x^7) = 6x^{10}.$$

 Be sure that you understand the difference between *adding* and *multiplying* exponential expressions. For example,

$$8x^3 + 5x^3 = 13x^3, \qquad \text{but} \qquad (8x^3)(5x^3) = 8 \cdot 5x^{3+3} = 40x^6.$$

OBJECTIVE 3 **Use the rule** $(a^m)^n = a^{mn}$. We simplify an expression such as $(8^3)^2$ with the product rule for exponents.

$$(8^3)^2 = (8^3)(8^3) = 8^{3+3} = 8^6$$

The exponents in $(8^3)^2$ are multiplied to give the exponent in 8^6. As another example,

$$(5^2)^3 = 5^2 \cdot 5^2 \cdot 5^2 = 5^{2+2+2} = 5^6,$$

and $2 \cdot 3 = 6$. These examples suggest **power rule (a) for exponents.**

Power Rule (a) for Exponents

For any positive integers m and n, $\qquad (a^m)^n = a^{mn}$.
(Raise a power to a power by multiplying exponents.)
Example: $(3^2)^4 = 3^{2 \cdot 4} = 3^8$

EXAMPLE 3 Using Power Rule (a)

Use power rule (a) for exponents to simplify each expression.

(a) $(2^5)^3 = 2^{5 \cdot 3} = 2^{15}$ 　　　　　　 **(b)** $(5^7)^2 = 5^{7(2)} = 5^{14}$

(c) $(x^2)^5 = x^{2(5)} = x^{10}$ 　　　　　　 **(d)** $(n^3)^2 = n^{3(2)} = n^6$

OBJECTIVE 4 **Use the rule** $(ab)^m = a^m b^m$. The properties studied in Chapter 1 can be used to develop two more rules for exponents. Using the definition of an exponential expression and the commutative and associative properties, we can rewrite the expression $(4 \cdot 8)^3$ as follows.

$$(4 \cdot 8)^3 = (4 \cdot 8)(4 \cdot 8)(4 \cdot 8) \qquad \text{Definition of exponent}$$
$$= (4 \cdot 4 \cdot 4) \cdot (8 \cdot 8 \cdot 8) \qquad \text{Commutative and associative properties}$$
$$= 4^3 \cdot 8^3 \qquad \text{Definition of exponent}$$

This example suggests **power rule (b) for exponents.**

Power Rule (b) for Exponents

For any positive integer m,　　$(ab)^m = a^m b^m$.
(Raise a product to a power by raising each factor to the power.)
Example: $(2p)^5 = 2^5 p^5$

EXAMPLE 4   Using Power Rule (b)

Use power rule (b) for exponents to simplify each expression.

(a) $(3xy)^2 = 3^2 x^2 y^2 = 9x^2 y^2$

(b) $5(pq)^2 = 5(p^2 q^2)$ 　　Power rule (b)
$$= 5p^2 q^2 \qquad \text{Multiply.}$$

(c) $3(2m^2 p^3)^4 = 3[2^4 (m^2)^4 (p^3)^4]$ 　　Power rule (b)
$$= 3 \cdot 2^4 m^8 p^{12} \qquad \text{Power rule (a)}$$
$$= 48 m^8 p^{12} \qquad \text{Multiply.}$$

(d) $(-5^6)^3 = (-1 \cdot 5^6)^3 = (-1)^3 \cdot (5^6)^3 = -1 \cdot 5^{18} = -5^{18}$

OBJECTIVE 5 Use the rule $\left(\dfrac{a}{b}\right)^m = \dfrac{a^m}{b^m}$. Since the quotient $\dfrac{a}{b}$ can be written as $a\left(\dfrac{1}{b}\right)$, we can use power rule (b), together with some of the properties of real numbers, to get **power rule (c) for exponents.**

Power Rule (c) for Exponents

For any positive integer m,

$$\left(\frac{a}{b}\right)^m = \frac{a^m}{b^m} \quad (b \neq 0).$$

(Raise a quotient to a power by raising both numerator and denominator to the power.)
Example: $\left(\dfrac{5}{3}\right)^2 = \dfrac{5^2}{3^2}$

EXAMPLE 5   Using Power Rule (c)

Use power rule (c) for exponents to simplify each expression.

(a) $\left(\dfrac{2}{3}\right)^5 = \dfrac{2^5}{3^5}$ 　　　　　　　　　　　**(b)** $\left(\dfrac{m}{n}\right)^3 = \dfrac{m^3}{n^3}, \quad (n \neq 0)$

We list the rules for exponents discussed in this section below. These rules are basic to the study of algebra and should be *memorized*.

Rules for Exponents

For positive integers m and n:

		Examples
Product rule	$a^m \cdot a^n = a^{m+n}$	$6^2 \cdot 6^5 = 6^{2+5} = 6^7$
Power rules (a)	$(a^m)^n = a^{mn}$	$(3^2)^4 = 3^{2 \cdot 4} = 3^8$
(b)	$(ab)^m = a^m b^m$	$(2p)^5 = 2^5 p^5$
(c)	$\left(\dfrac{a}{b}\right)^m = \dfrac{a^m}{b^m} \quad (b \neq 0)$	$\left(\dfrac{5}{3}\right)^2 = \dfrac{5^2}{3^2}$

As shown in the next example, more than one rule may be needed to simplify an expression with exponents.

EXAMPLE 6 Using Combinations of Rules

Use the rules for exponents to simplify each expression.

(a) $\left(\dfrac{2}{3}\right)^2 \cdot 2^3 = \dfrac{2^2}{3^2} \cdot \dfrac{2^3}{1}$ Power rule (c)

$\qquad\quad = \dfrac{2^2 \cdot 2^3}{3^2 \cdot 1}$ Multiply the fractions.

$\qquad\quad = \dfrac{2^5}{3^2}$ Product rule

(b) $(5x)^3(5x)^4 = (5x)^7$ Product rule

$\qquad\qquad\quad = 5^7 x^7$ Power rule (b)

(c) $(2x^2y^3)^4(3xy^2)^3 = 2^4(x^2)^4(y^3)^4 \cdot 3^3 x^3(y^2)^3$ Power rule (b)

$\qquad\qquad\qquad = 2^4 \cdot 3^3 x^8 y^{12} x^3 y^6$ Power rule (a)

$\qquad\qquad\qquad = 16 \cdot 27 x^{11} y^{18}$ Product rule

$\qquad\qquad\qquad = 432 x^{11} y^{18}$

(d) $(-x^3y)^2(-x^5y^4)^3$

Think of the negative sign in each factor as -1.

$(-1 \cdot x^3y)^2 (-1 \cdot x^5y^4)^3$

$\qquad = (-1)^2(x^3)^2(y)^2 \cdot (-1)^3(x^5)^3(y^4)^3$ Power rule (b)

$\qquad = (-1)^2(x^6)(y^2)(-1)^3(x^{15})(y^{12})$ Power rule (a)

$\qquad = (-1)^5(x^{21})(y^{14})$ Product rule

$\qquad = -x^{21}y^{14}$

4.2 EXERCISES

Decide whether each statement is true or false.

1. $3^3 = 9$

2. $(-2)^4 = 2^4$

3. $(a^2)^3 = a^5$

4. $\left(\dfrac{1}{4}\right)^2 = \dfrac{1}{4^2}$

Write each expression using exponents.

5. $w \cdot w \cdot w \cdot w \cdot w \cdot w$

6. $t \cdot t \cdot t \cdot t \cdot t \cdot t \cdot t$

7. $\dfrac{1}{4 \cdot 4 \cdot 4 \cdot 4}$

8. $\dfrac{1}{3 \cdot 3 \cdot 3}$

9. $(-7x)(-7x)(-7x)(-7x)$

10. $(-8p)(-8p)$

11. $\left(\dfrac{1}{2}\right)\left(\dfrac{1}{2}\right)\left(\dfrac{1}{2}\right)\left(\dfrac{1}{2}\right)\left(\dfrac{1}{2}\right)\left(\dfrac{1}{2}\right)$

12. $\left(-\dfrac{1}{4}\right)\left(-\dfrac{1}{4}\right)\left(-\dfrac{1}{4}\right)\left(-\dfrac{1}{4}\right)\left(-\dfrac{1}{4}\right)$

13. Explain how the expressions $(-3)^4$ and -3^4 are different.

14. Explain how the expressions $(5x)^3$ and $5x^3$ are different.

Identify the base and the exponent for each exponential expression. In Exercises 15–18, also evaluate each expression. See Example 1.

15. 3^5

16. 2^7

17. $(-3)^5$

18. $(-2)^7$

19. $(-6x)^4$

20. $(-8x)^4$

21. $-6x^4$

22. $-8x^4$

23. Explain why the product rule does not apply to the expression $5^2 + 5^3$. Then evaluate the expression by finding the individual powers and adding the results.

24. Repeat Exercise 23 for the expression $(-4)^3 + (-4)^4$.

Use the product rule to simplify each expression. Write the answer in exponential form. See Example 2.

25. $5^2 \cdot 5^6$

26. $3^6 \cdot 3^7$

27. $4^2 \cdot 4^7 \cdot 4^3$

28. $5^3 \cdot 5^8 \cdot 5^2$

29. $(-7)^3(-7)^6$

30. $(-9)^8(-9)^5$

31. $t^3 \cdot t^8 \cdot t^{13}$

32. $n^5 \cdot n^6 \cdot n^9$

33. $(-8r^4)(7r^3)$

34. $(10a^7)(-4a^3)$

35. $(-6p^5)(-7p^5)$

36. $(-5w^8)(-9w^8)$

37. Explain why the product rule does not apply to the expression $3^2 \cdot 4^3$. Then evaluate the expression by finding the individual powers and multiplying the results.

38. Repeat Exercise 37 for the expression $(-3)^3 \cdot (-2)^5$.

TECHNOLOGY INSIGHTS (EXERCISES 39–42)

The graphing calculator screen shown here on the left reinforces the fact that $(-3)^4$ and -3^4 are not equal, since the first is 81 and the second is -81. The screen on the right shows how a calculator displays whether a statement is true or false: If the calculator returns a 1, the statement is true; it returns a 0 for a false statement.

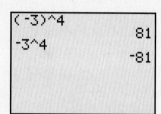

Evaluate each expression in the statement, and then tell whether the calculator would return a 1 or a 0.

39.

```
(-2)^5=-2^5
```

40.

```
(-2)^6=-2^6
```

41.

```
(-3)^6= -3^6
```

42.

```
(-3)^5=-3^5
```

Use the power rules for exponents to simplify each expression. Write the answer in exponential form. See Examples 3–5.

43. $(4^3)^2$

44. $(8^3)^6$

45. $(t^4)^5$

46. $(y^6)^5$

47. $(7r)^3$

48. $(11x)^4$

49. $(5xy)^5$

50. $(9pq)^6$

51. $(-5^2)^6$

52. $(-9^4)^8$

53. $(-8^3)^5$

54. $(-7^5)^7$

55. $8(qr)^3$

56. $4(vw)^5$

57. $\left(\dfrac{1}{2}\right)^3$

58. $\left(\dfrac{1}{3}\right)^5$

59. $\left(\dfrac{a}{b}\right)^3 \quad (b \neq 0)$

60. $\left(\dfrac{r}{t}\right)^4 \quad (t \neq 0)$

61. $\left(\dfrac{9}{5}\right)^8$

62. $\left(\dfrac{12}{7}\right)^3$

Use a combination of the rules of exponents introduced in this section to simplify each expression. See Example 6.

63. $\left(\dfrac{5}{2}\right)^3 \cdot \left(\dfrac{5}{2}\right)^2$

64. $\left(\dfrac{3}{4}\right)^5 \cdot \left(\dfrac{3}{4}\right)^6$

65. $\left(\dfrac{9}{8}\right)^3 \cdot 9^2$

66. $\left(\dfrac{8}{5}\right)^4 \cdot 8^3$

67. $(2x)^9(2x)^3$

68. $(6y)^5(6y)^8$

69. $(-6p)^4(-6p)$

70. $(-13q)^3(-13q)$

71. $(6x^2y^3)^5$

72. $(5r^5t^6)^7$

73. $(x^2)^3(x^3)^5$

74. $(y^4)^5(y^3)^5$

75. $(2w^2x^3y)^2(x^4y)^5$

76. $(3x^4y^2z)^3(yz^4)^5$

77. $(-r^4s)^2(-r^2s^3)^5$

78. $(-ts^6)^4(-t^3s^5)^3$

79. $\left(\dfrac{5a^2b^5}{c^6}\right)^3 \quad (c \neq 0)$

80. $\left(\dfrac{6x^3y^9}{z^5}\right)^4 \quad (z \neq 0)$

81. A student tried to simplify $(10^2)^3$ as 1000^6. Is this correct? If not, how is it simplified using the product rule for exponents?

82. Explain why $(3x^2y^3)^4$ is *not* equivalent to $(3 \cdot 4)x^8y^{12}$.

Find the area of each figure.* (Leave π in the answer for Exercise 86.) Use the formulas found on the inside covers.

83.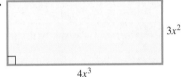

84.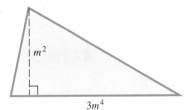

85.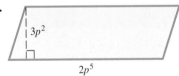

86.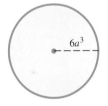

Find the volume of each figure. Use the formulas found on the inside covers.

87. **88.**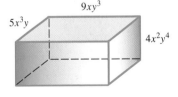

89. Assume a is a positive number greater than 1. Arrange the following terms in order from smallest to largest: $-(-a)^3, \ -a^3, \ (-a)^4, \ -a^4$. Explain how you decided on the order.

90. Describe a rule to tell whether an exponential expression with a negative base is positive or negative.

---

*The small square in the figures for Exercises 83–85 indicates a right angle (90°).

In Chapter 2 we used the formula for simple interest, $I = prt$, which deals with interest paid only on the principal. With **compound interest,** interest is paid on the principal and the interest earned earlier. The formula for compound interest, which involves an exponential expression, is

$$A = P(1 + r)^n.$$

Here A is the amount accumulated from a principal of P dollars left untouched for n years with an annual interest rate r (expressed as a decimal).

Use this formula and a calculator to find A to the nearest cent in Exercises 91–94.

91. $P = \$250, r = .04, n = 5$ **92.** $P = \$400, r = .04, n = 3$

93. $P = \$1500, r = .035, n = 6$ **94.** $P = \$2000, r = .025, n = 4$

EXERCISES

4.3 Multiplication of Polynomials

OBJECTIVES

1. Multiply a monomial and a polynomial.

2. Multiply two polynomials.

3. Multiply binomials by the FOIL method.

OBJECTIVE **1** Multiply a monomial and a polynomial. As shown in the previous section, the product of two monomials is found by using the rules for exponents and the commutative and associative properties. For example,

$$(-8m^6)(-9m^4) = -8(-9)(m^6)(m^4) = 72m^{6+4} = 72m^{10}.$$

CAUTION Do not confuse *addition* of terms with *multiplication* of terms. For example,

$$7q^5 + 2q^5 = 9q^5, \quad \text{but} \quad (7q^5)(2q^5) = 7 \cdot 2q^{5+5} = 14q^{10}.$$

To find the product of a monomial and a polynomial with more than one term, we use the distributive property with the method just shown.

EXAMPLE 1 Multiplying a Monomial and a Polynomial

Use the distributive property to find each product.
(a) $4x^2(3x + 5)$

$$4x^2(3x + 5) = (4x^2)(3x) + (4x^2)(5) \quad \text{Distributive property}$$
$$= 12x^3 + 20x^2 \quad \text{Multiply monomials.}$$

(b) $-8m^3(4m^3 + 3m^2 + 2m - 1)$

$$-8m^3(4m^3 + 3m^2 + 2m - 1)$$
$$= (-8m^3)(4m^3) + (-8m^3)(3m^2) \quad \text{Distributive property}$$
$$+ (-8m^3)(2m) + (-8m^3)(-1)$$
$$= -32m^6 - 24m^5 - 16m^4 + 8m^3 \quad \text{Multiply monomials.}$$

VIDEO

OBJECTIVE **2** Multiply two polynomials. We can use the distributive property repeatedly to find the product of any two polynomials. For example, to find the product of the polynomials $x + 1$ and $x - 4$, think of $x - 4$ as a single quantity and use the distributive property as follows.

$$(x + 1)(x - 4) = x(x - 4) + 1(x - 4)$$

Now use the distributive property twice to find $x(x-4)$ and $1(x-4)$.

$$x(x-4) + 1(x-4) = x(x) + x(-4) + 1(x) + 1(-4)$$
$$= x^2 - 4x + x - 4$$
$$= x^2 - 3x - 4$$

(We could have treated $x + 1$ as the single quantity instead to get $(x+1)x + (x+1)(-4)$ in the first step. Verify that this approach gives the same final result, $x^2 - 3x - 4$.)

We give a rule for multiplying any two polynomials below.

Multiplying Polynomials

To multiply two polynomials, multiply each term of the second polynomial by each term of the first polynomial and add the products.

EXAMPLE 2   Multiplying Two Polynomials

Multiply $(m^2 + 5)(4m^3 - 2m^2 + 4m)$.

Multiply each term of the second polynomial by each term of the first.

$$(m^2 + 5)(4m^3 - 2m^2 + 4m)$$
$$= m^2(4m^3) - m^2(2m^2) + m^2(4m) + 5(4m^3) - 5(2m^2) + 5(4m)$$
$$= 4m^5 - 2m^4 + 4m^3 + 20m^3 - 10m^2 + 20m$$

Now combine like terms.

$$= 4m^5 - 2m^4 + 24m^3 - 10m^2 + 20m$$

When at least one of the factors in a product of polynomials has three or more terms, the multiplication can be simplified by writing one polynomial above the other vertically.

EXAMPLE 3   Multiplying Vertically

Multiply $(x^3 + 2x^2 + 4x + 1)(3x + 5)$ using the vertical method.

Write the polynomials as follows.

$$\begin{array}{r} x^3 + 2x^2 + 4x + 1 \\ 3x + 5 \\ \hline \end{array}$$

It is not necessary to line up terms in columns, because any terms may be multiplied (not just like terms). Begin by multiplying each of the terms in the top row by 5.

Step 1

$$\begin{array}{r} x^3 + \ 2x^2 + \ 4x + 1 \\ 3x + 5 \\ \hline 5x^3 + 10x^2 + 20x + 5 \end{array}$$ $5(x^3 + 2x^2 + 4x + 1)$

Notice how this process is similar to multiplication of whole numbers. Now multiply each term in the top row by $3x$. Be careful to place like terms in columns, since the final step will involve addition (as in multiplying two whole numbers).

Step 2

$$\begin{array}{r} x^3 + \ 2x^2 + \ 4x + 1 \\ 3x + 5 \\ \hline 5x^3 + 10x^2 + 20x + 5 \\ 3x^4 + 6x^3 + 12x^2 + \ 3x \end{array}$$ $3x(x^3 + 2x^2 + 4x + 1)$

Step 3   Add like terms.

$$x^3 + \ 2x^2 + \ 4x + 1$$
$$3x + 5$$
$$\overline{5x^3 + 10x^2 + 20x + 5}$$
$$\underline{3x^4 + \ 6x^3 + 12x^2 + \ 3x}$$
$$\overline{3x^4 + 11x^3 + 22x^2 + 23x + 5}$$

The product is $3x^4 + 11x^3 + 22x^2 + 23x + 5$.

OBJECTIVE 3 Multiply binomials by the FOIL method. In algebra, many of the polynomials to be multiplied are both binomials (with just two terms). For these products a shortcut that eliminates the need to write out all the steps is used. To develop this shortcut, multiply $x + 3$ and $x + 5$ using the distributive property.

$$(x + 3)(x + 5) = x(x + 5) + 3(x + 5)$$
$$= x(x) + x(5) + 3(x) + 3(5)$$
$$= x^2 + 5x + 3x + 15$$
$$= x^2 + 8x + 15$$

The first term in the second line, $x(x)$, is the product of the first terms of the two binomials.

$$(x + 3)(x + 5) \qquad \text{Multiply the first terms: } x(x).$$

The term $x(5)$ is the product of the first term of the first binomial and the last term of the second binomial. This is the **outer product.**

$$(x + 3)(x + 5) \qquad \text{Multiply the outer terms: } x(5).$$

The term $3(x)$ is the product of the last term of the first binomial and the first term of the second binomial. The product of these middle terms is the **inner product.**

$$(x + 3)(x + 5) \qquad \text{Multiply the inner terms: } 3(x).$$

Finally, $3(5)$ is the product of the last terms of the two binomials.

$$(x + 3)(x + 5) \qquad \text{Multiply the last terms: } 3(5).$$

The inner product and the outer product should be added mentally, so that the three terms of the answer can be written without extra steps as

$$(x + 3)(x + 5) = x^2 + 8x + 15.$$

A summary of these steps is given below. This procedure is sometimes called the **FOIL method,** which comes from the abbreviation for *First, Outer, Inner, Last.*

Multiplying Binomials by the FOIL Method

Step 1 **Multiply the first terms.** Multiply the two first terms of the binomials to get the first term of the answer.

Step 2 **Find the outer and inner products.** Find the outer product and the inner product and add them (mentally if possible) to get the middle term of the answer.

Step 3 **Multiply the last terms.** Multiply the two last terms of the binomials to get the last term of the answer.

┌─ **E X A M P L E 4** Using the FOIL Method

Find the product $(x + 8)(x - 6)$ by the FOIL method.

Step 1 F Multiply the *first* terms: $x(x) = x^2$.

Step 2 O Find the product of the *outer* terms: $x(-6) = -6x$.

 I Find the product of the *inner* terms: $8(x) = 8x$.

 Add the outer and inner products mentally: $-6x + 8x = 2x$.

Step 3 L Multiply the *last* terms: $8(-6) = -48$.

The product $(x + 8)(x - 6)$ is the sum of the terms found in the three steps above, so

$$(x + 8)(x - 6) = x^2 - 6x + 8x - 48 = x^2 + 2x - 48.$$

As a shortcut, this product can be found in the following manner.

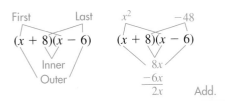

It is not possible to add the inner and outer products of the FOIL method if unlike terms result, as shown in the next example.

┌─ **E X A M P L E 5** Using the FOIL Method

Multiply $(9x - 2)(3y + 1)$.

First	$(9x - 2)(3y + 1)$	$27xy$
Outer	$(9x - 2)(3y + 1)$	$9x$ ←
Inner	$(9x - 2)(3y + 1)$	$-6y$ ←
Last	$(9x - 2)(3y + 1)$	-2

Unlike terms

$$\overset{F\qquad O\quad I\quad\; L}{(9x - 2)(3y + 1) = 27xy + 9x - 6y - 2}$$

┌───
EXAMPLE 6 Using the FOIL Method

Find the following products.

$$\qquad\qquad\qquad\qquad\quad\; F \qquad\quad O \qquad\quad I \qquad\quad L$$

(a) $(2k + 5y)(k + 3y) = (2k)(k) + (2k)(3y) + (5y)(k) + (5y)(3y)$

$$= 2k^2 + 6ky + 5ky + 15y^2$$

$$= 2k^2 + 11ky + 15y^2 \qquad\qquad \text{Combine like terms.}$$

(b) $(7p + 2q)(3p - q) = 21p^2 - pq - 2q^2 \qquad$ FOIL

(c) $2x^2(x - 3)(3x + 4) = 2x^2(3x^2 - 5x - 12) \qquad$ FOIL

$$= 6x^4 - 10x^3 - 24x^2 \qquad \text{Distributive property}$$
───┘

4.3 EXERCISES

1. Match each product in Column I with the correct monomial in Column II.

I	II
(a) $(5x^3)(6x^5)$	**A.** $125x^{15}$
(b) $(-5x^5)(6x^3)$	**B.** $30x^8$
(c) $(5x^5)^3$	**C.** $-216x^9$
(d) $(-6x^3)^3$	**D.** $-30x^8$

2. Match each product in Column I with the correct polynomial in Column II.

I	II
(a) $(x - 5)(x + 3)$	**A.** $x^2 + 8x + 15$
(b) $(x + 5)(x + 3)$	**B.** $x^2 - 8x + 15$
(c) $(x - 5)(x - 3)$	**C.** $x^2 - 2x - 15$
(d) $(x + 5)(x - 3)$	**D.** $x^2 + 2x - 15$

Find each product. See Example 1.

3. $(-5a^9)(-8a^5)$ 　　　　**4.** $(-3m^6)(-5m^4)$ 　　　　**5.** $-2m(3m + 2)$

6. $-5p(6 + 3p)$ 　　　　**7.** $3p(8 - 6p + 12p^3)$ 　　　　**8.** $4x(3 + 2x + 5x^3)$

9. $-8z(2z + 3z^2 + 3z^3)$ 　　　　**10.** $-7y(3 + 5y^2 - 2y^3)$

11. $7x^2y(2x^3y^2 + 3xy - 4y)$ 　　　　**12.** $9xy^3(-3x^2y^4 + 6xy - 2x)$

EXERCISES

Find each product. See Examples 2 and 3.

13. $(6x + 1)(2x^2 + 4x + 1)$ 　　　　**14.** $(9y - 2)(8y^2 - 6y + 1)$

15. $(4m + 3)(5m^3 - 4m^2 + m - 5)$ 　　　　**16.** $(y + 4)(3y^3 - 2y^2 + y + 3)$

17. $(2x - 1)(3x^5 - 2x^3 + x^2 - 2x + 3)$ 　　　　**18.** $(2a + 3)(a^4 - a^3 + a^2 - a + 1)$

19. $(5x^2 + 2x + 1)(x^2 - 3x + 5)$ 　　　　**20.** $(2m^2 + m - 3)(m^2 - 4m + 5)$

Find each product. Use the FOIL method. See Examples 4–6.

21. $(n - 2)(n + 3)$ 　　　　**22.** $(r - 6)(r + 8)$ 　　　　**23.** $(x + 6)(x - 6)$

24. $(y + 9)(y - 9)$ 　　　　**25.** $(4r + 1)(2r - 3)$ 　　　　**26.** $(5x + 2)(2x - 7)$

27. $(3x + 2)(3x - 2)$ 　　　　**28.** $(7x + 3)(7x - 3)$ 　　　　**29.** $(3q + 1)(3q + 1)$

EXERCISES

30. $(4w + 7)(4w + 7)$ 　　　　**31.** $(3x + y)(x - 2y)$ 　　　　**32.** $(5p + m)(2p - 3m)$

33. $(-3t + 4)(t + 6)$ 　　　　**34.** $(-5x + 9)(x - 2)$

35. $3y^3(2y + 3)(y - 5)$ 　　　　**36.** $5t^4(t + 3)(3t - 1)$

37. Find a polynomial that represents the area of this square.

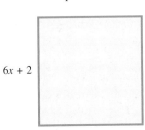

$6x + 2$

38. Find a polynomial that represents the area of this rectangle.

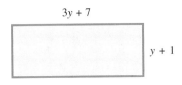

$3y + 7$

$y + 1$

39. Perform the following multiplications:

$$(x + 4)(x - 4); \quad (y + 2)(y - 2); \quad (r + 7)(r - 7).$$

Observe your answers, and explain the pattern that can be found.

40. Repeat Exercise 39 for the following:

$$(x + 4)(x + 4); \quad (y - 2)(y - 2); \quad (r + 7)(r + 7).$$

Find each product.

EXERCISES

41. $\left(3p + \dfrac{5}{4}q\right)\left(2p - \dfrac{5}{3}q\right)$

42. $\left(-x + \dfrac{2}{3}y\right)\left(3x - \dfrac{3}{4}y\right)$

43. $(m^3 - 4)(2m^3 + 3)$

44. $(4a^2 + b^2)(a^2 - 2b^2)$

45. $(2k^3 + h^2)(k^2 - 3h^2)$

46. $(4x^3 - 5y^4)(x^2 + y)$

47. $3p^3(2p^2 + 5p)(p^3 + 2p + 1)$

48. $5k^2(k^2 - k + 4)(k^3 - 3)$

49. $-2x^5(3x^2 + 2x - 5)(4x + 2)$

50. $-4x^3(3x^4 + 2x^2 - x)(-2x + 1)$

Find a polynomial that represents the area of each shaded region. In Exercises 53 and 54 leave π in your answer. Use the formulas found on the inside covers.

51.

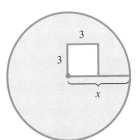

$x + 7$

x

$x + 7$

x

52.

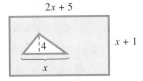

$2x + 5$

$x + 1$

4

x

53.

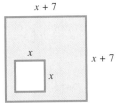

3

3

x

54.

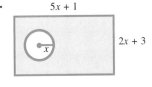

$5x + 1$

$2x + 3$

x

RELATING CONCEPTS (EXERCISES 55–62)

Work Exercises 55–62 in order. *Refer to the figure as necessary.*

$3x + 6$

10

55. Find a polynomial that represents the area of the rectangle.

RELATING CONCEPTS (EXERCISES 55-62) (CONTINUED)

56. Suppose you know that the area of the rectangle is 600 square yards. Use this information and the polynomial from Exercise 55 to write an equation that allows you to solve for x.

57. Solve for x.

58. What are the dimensions of the rectangle (assume units are all in yards)?

59. Suppose the rectangle represents a lawn and it costs $3.50 per square yard to lay sod on the lawn. How much will it cost to sod the entire lawn?

60. Use the result of Exercise 58 to find the perimeter of the lawn.

61. Again, suppose the rectangle represents a lawn and it costs $9.00 per yard to fence the lawn. How much will it cost to fence the lawn?

62. (a) Suppose that it costs k dollars per square yard to sod the lawn. Determine a polynomial in the variables x and k that represents the cost to sod the entire lawn.

(b) Suppose that it costs r dollars per yard to fence the lawn. Determine a polynomial in the variables x and r that represents the cost to fence the lawn.

Did you make the connection that sodding requires knowing the area, while fencing requires knowing the perimeter?

63. Explain the FOIL method of multiplying two binomials. Give an example.

64. Why does the FOIL method not apply to the product of a binomial and a trinomial? Give an example.

4.4 Special Products

OBJECTIVES

1. Square binomials.
2. Find the product of the sum and difference of two terms.
3. Find higher powers of binomials.

In this section, we develop patterns for certain binomial products that occur frequently.

OBJECTIVE 1 Square binomials. The square of a binomial can be found quickly by using the method shown in Example 1.

EXAMPLE 1 Squaring a Binomial

Find $(m + 3)^2$.

Squaring $m + 3$ by the FOIL method gives

$$(m + 3)(m + 3) = m^2 + 3m + 3m + 9 = m^2 + 6m + 9.$$

The result has the square of both the first and the last terms of the binomial:

$$m^2 = m^2 \qquad \text{and} \qquad 3^2 = 9.$$

The middle term is twice the product of the two terms of the binomial, since both the outer and inner products are $(m)(3)$ and

$$(m)(3) + (m)(3) = 2(m)(3) = 6m.$$

This example suggests the following rule.

Square of a Binomial

The square of a binomial is a trinomial consisting of the square of the first term, plus twice the product of the two terms, plus the square of the last term of the binomial. For x and y,

$$(x + y)^2 = x^2 + 2xy + y^2$$

and

$$(x - y)^2 = x^2 - 2xy + y^2.$$

EXAMPLE 2   Squaring Binomials

Use the rule to find each product.

(a) $(5z - 1)^2 = (5z)^2 - 2(5z)(1) + 1^2 = 25z^2 - 10z + 1$
Recall that $(5z)^2 = 5^2z^2 = 25z^2$.

(b) $(3b + 5r)^2 = (3b)^2 + 2(3b)(5r) + (5r)^2 = 9b^2 + 30br + 25r^2$

(c) $(2a - 9x)^2 = 4a^2 - 36ax + 81x^2$

(d) $\left(4m + \dfrac{1}{2}\right)^2 = (4m)^2 + 2(4m)\left(\dfrac{1}{2}\right) + \left(\dfrac{1}{2}\right)^2 = 16m^2 + 4m + \dfrac{1}{4}$

(e) $t(a + 2b)^2 = t(a^2 + 4ab + 4b^2)$ Square the binomial.
$= a^2t + 4abt + 4b^2t$ Distributive property

 A common error when squaring a binomial is forgetting the middle term of the product. In general, $(x + y)^2 \neq x^2 + y^2$ and $(x - y)^2 \neq x^2 - y^2$.

OBJECTIVE 2   Find the product of the sum and difference of two terms. Binomial products of the form $(x + y)(x - y)$ also occur frequently. In these products, one binomial is the sum of two terms, and the other is the difference of the same two terms. As an example, the product of $a + 2$ and $a - 2$ is

$$(a + 2)(a - 2) = a^2 - 2a + 2a - 4 = a^2 - 4.$$

Using the FOIL method, the product of $x + y$ and $x - y$ is the difference of two squares.

Product of the Sum and Difference of Two Terms

The product of the sum and difference of the two terms x and y is

$$(x + y)(x - y) = x^2 - y^2.$$

 The product $(x + y)(x - y)$ cannot be written as $(x + y)^2$ or as $(x - y)^2$, since one factor involves addition and the other involves subtraction.

┌─
E X A M P L E   3   Finding the Product of the Sum and Difference of Two Terms

Find each product.

(a) $(x + 4)(x - 4)$

Use the pattern for the sum and difference of two terms.

$$(x + 4)(x - 4) = x^2 - 4^2 = x^2 - 16$$

(b) $(3 - w)(3 + w)$

By the commutative property, this product is the same as $(3 + w)(3 - w)$.

$$(3 - w)(3 + w) = (3 + w)(3 - w) = 3^2 - w^2 = 9 - w^2$$

(c) $(a - b)(a + b) = a^2 - b^2$
─┘

┌─
E X A M P L E   4   Finding the Product of the Sum and Difference of Two Terms

Find each product.

(a) $(5m + 3)(5m - 3)$

Use the rule for the product of the sum and difference of two terms.

$$(5m + 3)(5m - 3) = (5m)^2 - 3^2 = 25m^2 - 9$$

(b) $(4x + y)(4x - y) = (4x)^2 - y^2 = 16x^2 - y^2$

(c) $\left(z - \dfrac{1}{4}\right)\left(z + \dfrac{1}{4}\right) = z^2 - \dfrac{1}{16}$

(d) $r(x^2 + y)(x^2 - y) = r(x^4 - y^2)$
$$= rx^4 - ry^2$$
─┘

The product formulas of this section will be important later, particularly in Chapters 5 and 6. Therefore, it is important to memorize these formulas and practice using them.

OBJECTIVE **3**   Find higher powers of binomials. The methods used in the previous section and this section can be combined to find higher powers of binomials.

┌─
E X A M P L E   5   Finding Higher Powers of Binomials

Find each product.

(a) $(x + 5)^3$

$$(x + 5)^3 = (x + 5)^2(x + 5)$$ $a^3 = a^2 \cdot a$
$$= (x^2 + 10x + 25)(x + 5)$$ Square the binomial.
$$= x^3 + 10x^2 + 25x + 5x^2 + 50x + 125$$ Multiply polynomials.
$$= x^3 + 15x^2 + 75x + 125$$ Combine like terms.

(b) $(2y - 3)^4$

$$(2y - 3)^4 = (2y - 3)^2(2y - 3)^2$$ $a^4 = a^2 \cdot a^2$
$$= (4y^2 - 12y + 9)(4y^2 - 12y + 9)$$ Square each binomial.
$$= 16y^4 - 48y^3 + 36y^2 - 48y^3 + 144y^2$$ Multiply polynomials.
$$\quad - 108y + 36y^2 - 108y + 81$$
$$= 16y^4 - 96y^3 + 216y^2 - 216y + 81$$ Combine like terms.
─┘

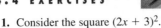

4.4 EXERCISES

1. Consider the square $(2x + 3)^2$.
   (a) What is the square of the first term, $(2x)^2$?
   (b) What is twice the product of the two terms, $2(2x)(3)$?
   (c) What is the square of the last term, 3^2?
   (d) Write the final product, which is a trinomial, using your results in parts (a)–(c).

2. Explain in your own words how to square a binomial. Give an example.

Find each square. See Examples 1 and 2.

3. $(p + 2)^2$ 4. $(r + 5)^2$ 5. $(a - c)^2$

6. $(p - y)^2$ 7. $(4x - 3)^2$ 8. $(5y + 2)^2$

9. $(8t + 7s)^2$ 10. $(7z - 3w)^2$ 11. $\left(5x + \dfrac{2}{5}y\right)^2$

12. $\left(6m - \dfrac{4}{5}n\right)^2$ 13. $x(2x + 5)^2$ 14. $t(3t - 1)^2$

15. $-(4r - 2)^2$ 16. $-(3y - 8)^2$

17. Consider the product $(7x + 3y)(7x - 3y)$.
   (a) What is the product of the first terms, $(7x)(7x)$?
   (b) Multiply the outer terms, $(7x)(-3y)$. Then multiply the inner terms, $(3y)(7x)$. Add the results. What is this sum?
   (c) What is the product of the last terms, $(3y)(-3y)$?
   (d) Write the complete product using your answers in parts (a) and (c). Why is the sum found in part (b) omitted here?

18. Explain in your own words how to find the product of the sum and the difference of two terms. Give an example.

Find each product. See Examples 3 and 4.

19. $(q + 2)(q - 2)$ 20. $(x + 8)(x - 8)$

21. $(2w + 5)(2w - 5)$ 22. $(3z + 8)(3z - 8)$

23. $(10x + 3y)(10x - 3y)$ 24. $(13r + 2z)(13r - 2z)$

25. $(2x^2 - 5)(2x^2 + 5)$ 26. $(9y^2 - 2)(9y^2 + 2)$

27. $\left(7x + \dfrac{3}{7}\right)\left(7x - \dfrac{3}{7}\right)$ 28. $\left(9y + \dfrac{2}{3}\right)\left(9y - \dfrac{2}{3}\right)$

29. $p(3p + 7)(3p - 7)$ 30. $q(5q - 1)(5q + 1)$

RELATING CONCEPTS (EXERCISES 31–40)

Special products can be illustrated by using areas of rectangles. Use the figure, and **work Exercises 31–36 in order** *to justify the special product* $(a + b)^2 = a^2 + 2ab + b^2$.

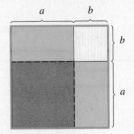

31. Express the area of the large square as the square of a binomial.

32. Give the monomial that represents the area of the red square.

33. Give the monomial that represents the sum of the areas of the blue rectangles.

34. Give the monomial that represents the area of the yellow square.

35. What is the sum of the monomials you obtained in Exercises 32–34?

36. Explain why the binomial square you found in Exercise 31 must equal the polynomial you found in Exercise 35.

To understand how the special product $(a + b)^2 = a^2 + 2ab + b^2$ can be applied to a purely numerical problem, **work Exercises 37–40 in order.**

37. Evaluate 35^2 using either traditional paper-and-pencil methods or a calculator.

38. The number 35 can be written as $30 + 5$. Therefore, $35^2 = (30 + 5)^2$. Use the special product for squaring a binomial with $a = 30$ and $b = 5$ to write an expression for $(30 + 5)^2$. Do not simplify at this time.

39. Use the rules for order of operations to simplify the expression you found in Exercise 38.

40. How do the answers in Exercises 37 and 39 compare?

Did you make the connections among geometry, algebra, and arithmetic in these exercises?

EXERCISES

The special product

$$(a + b)(a - b) = a^2 - b^2$$

can be used to perform some multiplication problems. For example,

$$51 \times 49 = (50 + 1)(50 - 1) \qquad 102 \times 98 = (100 + 2)(100 - 2)$$
$$= 50^2 - 1^2 = 2500 - 1 \qquad = 100^2 - 2^2$$
$$= 2499 \qquad = 10{,}000 - 4$$
$$= 9996.$$

Once these patterns are recognized, multiplications of this type can be done mentally.

Use this method to calculate each product mentally.

41. 101×99 **42.** 103×97

43. 201×199 **44.** 301×299

45. $20\frac{1}{2} \times 19\frac{1}{2}$ **46.** $30\frac{1}{3} \times 29\frac{2}{3}$

Find each product. See Example 5.

47. $(m - 5)^3$ **48.** $(p + 3)^3$ **49.** $(2a + 1)^3$

50. $(3m - 1)^3$ **51.** $(3r - 2t)^4$ **52.** $(2z + 5y)^4$

EXERCISES

53. Explain how the expressions $x^2 + y^2$ and $(x + y)^2$ differ.

54. Does $a^3 + b^3$ equal $(a + b)^3$? Explain your answer.

69. $\dfrac{(x^{-1}y^2z)^{-2}}{(x^{-3}y^3z)^{-1}}$ **70.** $\dfrac{(a^{-2}b^{-3}c^{-4})^{-5}}{(a^2b^3c^4)^5}$ **71.** $\left(\dfrac{xy^{-2}}{x^2y}\right)^{-3}$ **72.** $\left(\dfrac{wz^{-5}}{w^{-3}z}\right)^{-2}$

73. Consider the following typical student **error**:

$$\frac{16^3}{2^2} = \left(\frac{16}{2}\right)^{3-2} = 8^1 = 8.$$

Explain what the student did incorrectly, and then give the correct answer.

74. Consider the following typical student **error**:

$$-5^4 = (-5)^4 = 625.$$

Explain what the student did incorrectly, and then give the correct answer.

4.6 Division of Polynomials

OBJECTIVES

1 Divide a polynomial by a monomial.

2 Divide a polynomial by a polynomial.

OBJECTIVE 1 Divide a polynomial by a monomial. We add two fractions with a common denominator as follows.

$$\frac{a}{c} + \frac{b}{c} = \frac{a+b}{c}.$$

Looking at this statement in reverse gives us a rule for dividing a polynomial by a monomial.

Dividing a Polynomial by a Monomial

To divide a polynomial by a monomial, divide each term of the polynomial by the monomial:

$$\frac{a+b}{c} = \frac{a}{c} + \frac{b}{c} \quad (c \neq 0).$$

The parts of a division problem are named in the diagram.

$$\text{Dividend} \rightarrow \quad \frac{12x^2 + 6x}{6x} = 2x + 1 \quad \leftarrow \text{Quotient}$$
$$\text{Divisor} \rightarrow$$

EXAMPLE 1 Dividing a Polynomial by a Monomial

Divide $5m^5 - 10m^3$ by $5m^2$.

Use the rule above, with $+$ replaced by $-$. Then use the quotient rule for exponents.

$$\frac{5m^5 - 10m^3}{5m^2} = \frac{5m^5}{5m^2} - \frac{10m^3}{5m^2} = m^3 - 2m$$

Recall from arithmetic that division problems can be checked by multiplication:

$$\frac{63}{7} = 9 \quad \text{because} \quad 7 \cdot 9 = 63.$$

To check the polynomial quotient, multiply $m^3 - 2m$ by $5m^2$. Because

$$5m^2(m^3 - 2m) = 5m^5 - 10m^3,$$

the quotient is correct.

Since division by 0 is undefined, the quotient

$$\frac{5m^5 - 10m^3}{5m^2}$$

is undefined if $m = 0$. From now on we assume that no denominators are 0.

E X A M P L E 2   Dividing a Polynomial by a Monomial

Divide $\dfrac{16a^5 - 12a^4 + 8a^2}{4a^3}$.

Divide each term of $16a^5 - 12a^4 + 8a^2$ by $4a^3$.

$$\frac{16a^5 - 12a^4 + 8a^2}{4a^3} = \frac{16a^5}{4a^3} - \frac{12a^4}{4a^3} + \frac{8a^2}{4a^3}$$

$$= 4a^2 - 3a + \frac{2}{a}$$

The result is not a polynomial because of the expression $\frac{2}{a}$, which has a variable in the denominator. While the sum, difference, and product of two polynomials are always polynomials, the quotient of two polynomials may not be.

Again, check by multiplying.

$$4a^3\left(4a^2 - 3a + \frac{2}{a}\right) = 4a^3(4a^2) - 4a^3(3a) + 4a^3\left(\frac{2}{a}\right)$$

$$= 16a^5 - 12a^4 + 8a^2$$

E X A M P L E 3   Dividing a Polynomial by a Monomial with a Negative Coefficient

Divide $-7x^3 + 12x^4 - 4x$ by $-4x$.

The polynomial should be written in descending powers before dividing. Write it as $12x^4 - 7x^3 - 4x$; then divide by $-4x$.

$$\frac{12x^4 - 7x^3 - 4x}{-4x} = \frac{12x^4}{-4x} - \frac{7x^3}{-4x} + \frac{-4x}{-4x}$$

$$= -3x^3 + \frac{7x^2}{4} + 1 = -3x^3 + \frac{7}{4}x^2 + 1$$

Check by multiplying.

In Example 3, notice the quotient $\dfrac{-4x}{-4x} = 1$. It is a common error to leave

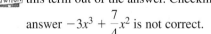 this term out of the answer. Checking by multiplication will show that the

answer $-3x^3 + \dfrac{7}{4}x^2$ is not correct.

OBJECTIVE **2** Divide a polynomial by a polynomial. We use a method of "long division" to divide a polynomial by a polynomial (other than a monomial). This method is similar to the method of long division used for two whole numbers. For comparison, the division of whole numbers is shown alongside the division of polynomials. Both polynomials must be written with descending powers before beginning the division process.

Step 1   Divide 27 into 6696.

Divide $2x + 3$ into $8x^3 - 4x^2 - 14x + 15$.

$$27\overline{)6696} \qquad 2x + 3\overline{)8x^3 - 4x^2 - 14x + 15}$$

Step 2   27 divides into 66 **2** times; $2 \cdot 27 = $ **54** .

$2x$ divides into $8x^3$ **$4x^2$** times; $4x^2(2x + 3) = $ **$8x^3 + 12x^2$** .

$$\begin{array}{r} 2 \\ 27\overline{)6696} \\ \rightarrow 54 \end{array} \qquad \begin{array}{r} 4x^2 \\ 2x + 3\overline{)8x^3 - 4x^2 - 14x + 15} \\ 8x^3 + 12x^2 \leftarrow \end{array}$$

Step 3   Subtract: $66 - 54 = 12$; then bring down the next digit.

Subtract: $-4x^2 - 12x^2 = -16x^2$; then bring down the next term.

$$\begin{array}{r} 2 \\ 27\overline{)6696} \\ 54\downarrow \\ \hline 129 \end{array} \qquad \begin{array}{r} 4x^2 \\ 2x + 3\overline{)8x^3 - 4x^2 - 14x + 15} \\ 8x^3 + 12x^2 \quad \downarrow \\ \hline -16x^2 - 14x \end{array}$$

(To subtract two polynomials, change the sign of the second and then add.)

Step 4   27 divides into 129 **4** times; $4 \cdot 27 = $ **108** .

$2x$ divides into $-16x^2$ **$-8x$** times; $-8x(2x + 3) = $ **$-16x^2 - 24x$** .

$$\begin{array}{r} 24 \\ 27\overline{)6696} \\ 54 \\ \hline 129 \\ \rightarrow 108 \end{array} \qquad \begin{array}{r} 4x^2 - 8x \\ 2x + 3\overline{)8x^3 - 4x^2 - 14x + 15} \\ 8x^3 + 12x^2 \\ \hline -16x^2 - 14x \\ -16x^2 - 24x \leftarrow \end{array}$$

Step 5   Subtract: $129 - 108 = 21$; then bring down the next digit.

Subtract: $-14x - (-24x) = 10x$; then bring down the next term.

$$\begin{array}{r} 24 \\ 27\overline{)6696} \\ 54 \\ \hline 129 \\ 108\downarrow \\ \hline 216 \end{array} \qquad \begin{array}{r} 4x^2 - 8x \\ 2x + 3\overline{)8x^3 - 4x^2 - 14x + 15} \\ 8x^3 + 12x^2 \\ \hline -16x^2 - 14x \\ -16x^2 - 24x \quad \downarrow \\ \hline 10x + 15 \end{array}$$

Step 6   27 divides into 216 **8** times; $2x$ divides into $10x$ **5** times;
       $8 \cdot 27 = \boxed{216}$. $5(2x + 3) = \boxed{10x + 15}$.

$$\begin{array}{r} 248 \\ 27\overline{)6696} \\ 54 \\ \hline 129 \\ 108 \\ \hline 216 \\ \underline{216} \end{array}$$

$$\begin{array}{r} 4x^2 - 8x + 5 \\ 2x + 3\overline{)8x^3 - 4x^2 - 14x + 15} \\ \underline{8x^3 + 12x^2} \\ -16x^2 - 14x \\ \underline{-16x^2 - 24x} \\ 10x + 15 \\ \underline{10x + 15} \end{array}$$

6696 divided by 27 is 248. $8x^3 - 4x^2 - 14x + 15$ divided by
There is no remainder. $2x + 3$ is $4x^2 - 8x + 5$. There is no
                                  remainder.

Step 7   Check by multiplying.   Check by multiplying.
       $27 \cdot 248 = 6696$ $(2x + 3)(4x^2 - 8x + 5)$
                                         $= 8x^3 - 4x^2 - 14x + 15$

Notice that at each step in the polynomial division process, the *first* term was divided into the *first* term.

E X A M P L E   4   Dividing a Polynomial by a Polynomial

Divide $5x + 4x^3 - 8 - 4x^2$ by $2x - 1$.

Both polynomials must be written in descending powers of x. Rewrite the first polynomial as $4x^3 - 4x^2 + 5x - 8$. Then begin the division process.

$$\begin{array}{r} 2x^2 - x + 2 \\ 2x - 1\overline{)4x^3 - 4x^2 + 5x - 8} \\ \underline{4x^3 - 2x^2} \\ -2x^2 + 5x \\ \underline{-2x^2 + x} \\ 4x - 8 \\ \underline{4x - 2} \\ -6 \end{array}$$

Step 1 $2x$ divides into $4x^3$ **($2x^2$)** times;
       $2x^2(2x - 1) = 4x^3 - 2x^2$.

Step 2   Subtract; bring down the next term.

Step 3 $2x$ divides into $-2x^2$ **($-x$)** times;
       $-x(2x - 1) = -2x^2 + x$.

Step 4   Subtract; bring down the next term.

Step 5 $2x$ divides into $4x$ **2** times;
       $2(2x - 1) = 4x - 2$.

Step 6   Subtract. The remainder is -6.

In a division problem like this one, the division process stops when the degree of the remainder is less than the degree of the divisor. Thus, $2x - 1$ divides into $4x^3 - 4x^2 + 5x - 8$ with a quotient of $2x^2 - x + 2$ and a remainder of -6. Write the remainder as a fraction with $2x - 1$ as the denominator. The result is not a polynomial because of the remainder.

$$\frac{4x^3 - 4x^2 + 5x - 8}{2x - 1} = 2x^2 - x + 2 + \frac{-6}{2x - 1}$$

Step 7 Check by multiplying.

$$(2x - 1)\left(2x^2 - x + 2 + \frac{-6}{2x - 1}\right)$$

$$= (2x - 1)(2x^2) + (2x - 1)(-x) + (2x - 1)(2) + (2x - 1)\left(\frac{-6}{2x - 1}\right)$$

$$= 4x^3 - 2x^2 - 2x^2 + x + 4x - 2 - 6$$

$$= 4x^3 - 4x^2 + 5x - 8$$

EXAMPLE 5 Dividing into a Polynomial with Missing Terms

Divide $x^3 + 2x - 3$ by $x - 1$.

Here the polynomial $x^3 + 2x - 3$ is missing the x^2 term. When terms are missing, use 0 as the coefficient for the missing terms; in this case, $0x^2$ is a "placeholder."

$$x^3 + 2x - 3 = x^3 + 0x^2 + 2x - 3$$

Now divide.

$$
\begin{array}{r}
x^2 + x + 3 \\
x - 1{\overline{\smash{\big)}\,x^3 + 0x^2 + 2x - 3}} \\
\underline{x^3 - x^2} \\
x^2 + 2x \\
\underline{x^2 - x} \\
3x - 3 \\
\underline{3x - 3}
\end{array}
$$

The remainder is 0. The quotient is $x^2 + x + 3$. Check by multiplying.

$$(x^2 + x + 3)(x - 1) = x^3 + 2x - 3$$

EXAMPLE 6 Dividing by a Polynomial with Missing Terms

Divide $x^4 + 2x^3 + 2x^2 - x - 1$ by $x^2 + 1$.

Since $x^2 + 1$ has a missing x term, write it as $x^2 + 0x + 1$. Then proceed through the division process.

$$
\begin{array}{r}
x^2 + 2x + 1 \\
x^2 + 0x + 1{\overline{\smash{\big)}\,x^4 + 2x^3 + 2x^2 - x - 1}} \\
\underline{x^4 + 0x^3 + x^2} \\
2x^3 + x^2 - x \\
\underline{2x^3 + 0x^2 + 2x} \\
x^2 - 3x - 1 \\
\underline{x^2 + 0x + 1} \\
-3x - 2 \qquad \leftarrow\text{Remainder}
\end{array}
$$

When the result of subtracting $(-3x - 2$, in this case) is a polynomial of smaller degree than the divisor $(x^2 + 0x + 1)$, that polynomial is the remainder. Write the result as

$$x^2 + 2x + 1 + \frac{-3x - 2}{x^2 + 1}.$$

CONNECTIONS

In Section 4.1, we found the value of a polynomial in x for a given value of x by substituting that number for x. Surprisingly, we can accomplish the same thing by division. Suppose we want to find the value of $2x^3 - 4x^2 + 3x - 5$ for $x = -3$. Instead of substituting -3 for x in the polynomial, we divide the polynomial by $x - (-3) = x + 3$. The remainder will give the value of the polynomial for $x = -3$. In general, when a polynomial P is divided by $x - r$, the remainder is equal to P evaluated at $x = r$.

FOR DISCUSSION OR WRITING

1. Evaluate $2x^3 - 4x^2 + 3x - 5$ for $x = -3$.

2. Divide $2x^3 - 4x^2 + 3x - 5$ by $x + 3$. Give the remainder.

3. Compare the answers to Exercises 1 and 2. What do you notice?

4. Choose another polynomial and evaluate it both ways at some value of the variable. Do the answers agree?

4.6 EXERCISES

Fill in each blank with the correct response.

1. In the statement $\dfrac{6x^2 + 8}{2} = 3x^2 + 4$, _____ is the dividend, _____ is the divisor, and _____ is the quotient.

2. The expression $\dfrac{3x + 12}{x}$ is undefined if $x =$ _____ .

3. To check the division shown in Exercise 1, multiply _____ by _____ and show that the product is _____ .

4. The expression $5x^2 - 3x + 6 + \dfrac{2}{x}$ _____ a polynomial.
   (is/is not)

5. Explain why the division problem $\dfrac{16m^3 - 12m^2}{4m}$ can be performed using the methods of this section, while the division problem $\dfrac{4m}{16m^3 - 12m^2}$ cannot.

6. Suppose that a polynomial in the variable x has degree 5 and it is divided by a monomial in the variable x having degree 3. What is the degree of the quotient?

Perform each division. See Examples 1–3.

7. $\dfrac{60x^4 - 20x^2 + 10x}{2x}$

8. $\dfrac{120x^6 - 60x^3 + 80x^2}{2x}$

9. $\dfrac{20m^5 - 10m^4 + 5m^2}{5m^2}$

10. $\dfrac{12t^5 - 6t^3 + 6t^2}{6t^2}$

11. $\dfrac{8t^5 - 4t^3 + 4t^2}{2t}$

12. $\dfrac{8r^4 - 4r^3 + 6r^2}{2r}$

13. $\dfrac{4a^5 - 4a^2 + 8}{4a}$

14. $\dfrac{5t^8 + 5t^7 + 15}{5t}$

Divide each polynomial by $3x^2$. See Examples 1–3.

15. $12x^5 - 9x^4 + 6x^3$

16. $24x^6 - 12x^5 + 30x^4$

17. $3x^2 + 15x^3 - 27x^4$

18. $3x^2 - 18x^4 + 30x^5$

19. $36x + 24x^2 + 6x^3$

20. $9x - 12x^2 + 9x^3$

21. $4x^4 + 3x^3 + 2x$

22. $5x^4 - 6x^3 + 8x$

Perform each division. See Examples 1–3.

23. $\dfrac{-27r^4 + 36r^3 - 6r^2 - 26r + 2}{-3r}$

24. $\dfrac{-8k^4 + 12k^3 + 2k^2 - 7k + 3}{-2k}$

25. $\dfrac{2m^5 - 6m^4 + 8m^2}{-2m^3}$

26. $\dfrac{6r^5 - 8r^4 + 10r^2}{-2r^4}$

27. $(20a^4 - 15a^5 + 25a^3) \div (5a^4)$

28. $(16y^5 - 8y^2 + 12y) \div (4y^2)$

29. $(120x^{11} - 60x^{10} + 140x^9 - 100x^8) \div (10x^{12})$

30. $(120x^{12} - 84x^9 + 60x^8 - 36x^7) \div (12x^9)$

31. The quotient in Exercise 21 is $\dfrac{4x^2}{3} + x + \dfrac{2}{3x}$. Notice how the third term is written with x in the denominator. Would $\dfrac{2}{3}x$ be an acceptable form for this term? Why or why not?

32. Refer to the quotient given in Example 2 in this section. Write it as an equivalent expression using negative exponents as necessary.

Use the appropriate formula, found on the inside covers, to answer each question.

33. The area of the rectangle is given by the polynomial $15x^3 + 12x^2 - 9x + 3$. What is the polynomial that expresses the length?

34. The area of the triangle is given by the polynomial $24m^3 + 48m^2 + 12m$. What is the polynomial that expresses the length of the base?

35. The quotient of a certain polynomial and $-7m^2$ is $9m^2 + 3m + 5 - \dfrac{2}{m}$. Find the polynomial.

36. Suppose that a polynomial of degree n is divided by a monomial of degree m to get a *polynomial* quotient.
(a) How do m and n compare in value?
(b) What is the expression that gives the degree of the quotient?

▚ RELATING CONCEPTS (EXERCISES 37–40)

Our system of numeration is called a decimal system. It is based on powers of ten. In a whole number such as 2846, each digit is understood to represent the number of powers of ten for its place value. The 2 represents two thousands (2×10^3), the 8 represents eight hundreds (8×10^2), the 4 represents four tens (4×10^1), and the 6 represents six ones (or units) (6×10^0). In expanded form we write

$$2846 = (2 \times 10^3) + (8 \times 10^2) + (4 \times 10^1) + (6 \times 10^0).$$

*Keeping this information in mind, **work Exercises 37–40 in order.***

37. Divide 2846 by 2, using paper-and-pencil methods: $2\overline{)2846}$.

38. Write your answer in Exercise 37 in expanded form.

39. Use the methods of this section to divide the polynomial $2x^3 + 8x^2 + 4x + 6$ by 2.

40. Compare your answers in Exercises 38 and 39. How are they similar? How are they different? For what value of x does the answer in Exercise 39 equal the answer in Exercise 38?

Did you make the connection between division of whole numbers and division of polynomials?

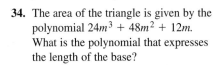

Perform each division. See Example 4.

41. $\dfrac{x^2 - x - 6}{x - 3}$

42. $\dfrac{m^2 - 2m - 24}{m - 6}$

43. $\dfrac{2y^2 + 9y - 35}{y + 7}$

44. $\dfrac{2y^2 + 9y + 7}{y + 1}$

45. $\dfrac{p^2 + 2p + 20}{p + 6}$

46. $\dfrac{x^2 + 11x + 16}{x + 8}$

47. $(r^2 - 8r + 15) \div (r - 3)$

48. $(t^2 + 2t - 35) \div (t - 5)$

49. $\dfrac{12m^2 - 20m + 3}{2m - 3}$

50. $\dfrac{12y^2 + 20y + 7}{2y + 1}$

51. $\dfrac{4a^2 - 22a + 32}{2a + 3}$

52. $\dfrac{9w^2 + 6w + 10}{3w - 2}$

53. $\dfrac{8x^3 - 10x^2 - x + 3}{2x + 1}$

54. $\dfrac{12t^3 - 11t^2 + 9t + 18}{4t + 3}$

55. $\dfrac{8k^4 - 12k^3 - 2k^2 + 7k - 6}{2k - 3}$

56. $\dfrac{27r^4 - 36r^3 - 6r^2 + 26r - 24}{3r - 4}$

57. $\dfrac{5y^4 + 5y^3 + 2y^2 - y - 3}{y + 1}$

58. $\dfrac{2r^3 - 5r^2 - 6r + 15}{r - 3}$

59. $\dfrac{3k^3 - 4k^2 - 6k + 10}{k - 2}$

60. $\dfrac{5z^3 - z^2 + 10z + 2}{z + 2}$

61. $\dfrac{6p^4 - 16p^3 + 15p^2 - 5p + 10}{3p + 1}$

62. $\dfrac{6r^4 - 11r^3 - r^2 + 16r - 8}{2r - 3}$

Perform each division. See Examples 5 and 6.

EXERCISES

63. $\dfrac{5 - 2r^2 + r^4}{r^2 - 1}$

64. $\dfrac{4t^2 + t^4 + 7}{t^2 + 1}$

65. $\dfrac{y^3 + 1}{y + 1}$

66. $\dfrac{y^3 - 1}{y - 1}$

67. $\dfrac{a^4 - 1}{a^2 - 1}$

68. $\dfrac{a^4 - 1}{a^2 + 1}$

69. $\dfrac{x^4 - 4x^3 + 5x^2 - 3x + 2}{x^2 + 3}$

70. $\dfrac{3t^4 + 5t^3 - 8t^2 - 13t + 2}{t^2 - 5}$

71. $\dfrac{2x^5 + 9x^4 + 8x^3 + 10x^2 + 14x + 5}{2x^2 + 3x + 1}$

72. $\dfrac{4t^5 - 11t^4 - 6t^3 + 5t^2 - t + 3}{4t^2 + t - 3}$

73. $(3a^2 - 11a + 17) \div (2a + 6)$

74. $(4x^2 + 11x - 8) \div (3x + 6)$

75. Suppose that one of your classmates asks you the following question: "How do I know when to stop the division process in a problem like the one in Exercise 69?" How would you respond?

76. Suppose that someone asks you if the following division problem is correct:

$$(6x^3 + 4x^2 - 3x + 9) \div (2x - 3) = 4x^2 + 9x - 3.$$

Tell how, by looking only at the *first term* of the quotient, you immediately know that the problem has been worked incorrectly.

Find a polynomial that describes each quantity required. Use the formulas found on the inside covers.

77. Give the length of the rectangle.

$5x + 2$

The area is $5x^3 + 7x^2 - 13x - 6$ square units.

78. Find the measure of the base of the parallelogram.

$x - 1$

The area is $2x^3 + 2x^2 - 3x - 1$ square units.

26. In 1995 and 1996, there were the same number of aircraft accidents involving passenger fatalities throughout worldwide scheduled air services. In 1994, there were 2 more than in each of the other years. How many accidents were there in each of these years if there was a total of 68 altogether? (*Source:* International Civil Aviation Organization.)

Solve each inequality.

27. $-8x \le -80$ **28.** $-2(x + 4) > 3x + 6$ **29.** $-3 \le 2x + 5 < 9$

Solve the problem.

30. One side of a triangle is twice as long as a second side. The third side of the triangle is 17 feet long. The perimeter of the triangle cannot be more than 50 feet. Find the longest possible values for the other two sides of the triangle.

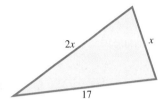

31. Complete the table of values for $-2x + 4y = 8$.

x	0		2		4
y		0		1	

32. Graph $y = -3x + 6$.

33. Graph $y = (x + 4)^2$, using the x-values $-6, -5, -4, -3,$ and -2 to obtain a set of points.

Perform the indicated operations.

34. $(7x^3 - 12x^2 - 3x + 8) + (6x^2 + 4) - (-4x^3 + 8x^2 - 2x - 2)$

35. $6x^5(3x^2 - 9x + 10)$ **36.** $(7x + 4)(9x + 3)$

37. $(5x + 8)^2$ **38.** $\dfrac{14x^3 - 21x^2 + 7x}{7x}$

39. $\dfrac{y^3 - 3y^2 + 8y - 6}{y - 1}$

Evaluate each expression.

40. $4^{-1} + 3^0$ **41.** $2^{-4} \cdot 2^5$ **42.** $\dfrac{8^{-5} \cdot 8^7}{8^2}$

43. Write with positive exponents only: $\dfrac{(a^{-3}b^2)^2}{(2a^{-4}b^{-3})^{-1}}$.

44. Write in scientific notation: 34,500.

45. Write without exponents: 5.36×10^{-7}.

5

Factoring and Applications

In the early days of the American frontier, it took four horses 75 days to haul one wagonload of goods a thousand miles. By the late 1800s, roads, canals, and railways connected "island" communities, greatly expediting the movement of people and goods. Whereas a stagecoach traveled 50 miles in a day, trains now covered 50 miles in an *hour*— over 700 miles per day. The vehicle that really changed twentieth-century American travel, however, was the automobile. Today emerging technologies are further transforming modern transportation systems. Just one example is the global network of computers moving millions of bits of data per second.

Much of this emerging technology depends on mathematics. Understanding mathematics makes it possible to model data using functions. In Chapter 4, we used polynomials of degree 2 to model data. Now, in this chapter, we introduce a method for solving polynomial equations of degree 2 called *quadratic equations,* and we give further examples of them as models. For instance, average fuel economy trends in miles per gallon for the automotive industry are closely approximated by the quadratic equation

$$y = -.04x^2 + .93x + 21,$$

where x represents the year. The years are coded so that $x = 0$ represents 1978, $x = 2$ represents 1980, $x = 4$ represents 1982, and so on. This equation was developed from the data in the table.

Fuel Economy

Year	1978	1980	1982	1984	1986	1988	1990	1992	1994	1996
Miles per Gallon	19.9	23.1	25.1	25.0	25.9	26.0	25.4	25.1	24.7	24.9

Source: National Highway Traffic Safety Administration.

Substituting 0 for x in the equation gives $y = 21$, the approximate average miles per gallon in 1978. For 1996, substituting $1996 - 1978 = 18$ for x in the equation gives $y = 24.78$ as the approximate average miles per gallon. Which of these is a closer approximation to the data in the table? We will return to this transportation model in Section 5.6, Example 4.

5.1 The Greatest Common Factor; Factoring by Grouping

OBJECTIVES

1 Find the greatest common factor of a list of terms.

2 Factor out the greatest common factor.

3 Factor by grouping.

Recall from Chapter 1 that to **factor** means to write a quantity as a product. That is, factoring is the opposite of multiplying. For example,

Multiplying	*Factoring*
$6 \cdot 2 = 12$,	$12 = 6 \cdot 2$.

Factors Product   Product Factors

Other factored forms of 12 are

$$(-6)(-2), \quad 3 \cdot 4, \quad (-3)(-4), \quad 12 \cdot 1, \quad \text{and} \quad (-12)(-1).$$

More than two factors may be used, so another factored form of 12 is $2 \cdot 2 \cdot 3$. The positive integer factors of 12 are

$$1, 2, 3, 4, 6, 12.$$

OBJECTIVE 1 Find the greatest common factor of a list of terms. An integer that is a factor of two or more integers is called a **common factor** of those integers. For example, 6 is a common factor of 18 and 24 since 6 is a factor of both 18 and 24. Other common factors of 18 and 24 are 1, 2, and 3. The **greatest common factor** of a list of integers is the largest common factor of those integers. Thus, 6 is the greatest common factor of 18 and 24, since it is the largest of the common factors of these numbers.

 Factors of a number are also divisors of the number. The greatest common factor is actually the same as the greatest common divisor.

CONNECTIONS

There are many rules for deciding what numbers divide into a given number. Here are some especially useful divisibility rules for small numbers. It is surprising how many people do not know them.

A Whole Number Divisible by:	Must Have the Following Property:
2	Ends in 0, 2, 4, 6, or 8
3	Sum of its digits is divisible by 3
4	Last two digits form a number divisible by 4
5	Ends in 0 or 5
6	Divisible by both 2 and 3
8	Last three digits form a number divisible by 8
9	Sum of its digits is divisible by 9
10	Ends in 0

Recall from Chapter 1 that a prime number has only itself and 1 as factors. In Section 1.1 we factored numbers into prime factors. This is the first step in finding the greatest common factor of a list of numbers. We find the greatest common factor (GCF) of a list of numbers as follows.

AUDIO

Finding the Greatest Common Factor (GCF)

Step 1 **Factor.** Write each number in prime factored form.

Step 2 **List common factors.** List each prime number that is a factor of every number in the list.

Step 3 **Choose smallest exponents.** Use as exponents on the common prime factors the *smallest* exponent from the prime factored forms. (If a prime does not appear in one of the prime factored forms, it cannot appear in the greatest common factor.)

Step 4 **Multiply.** Multiply the primes from Step 3. If there are no primes left after Step 3, the greatest common factor is 1.

┌ **E X A M P L E   1**   Finding the Greatest Common Factor for Numbers

Find the greatest common factor for each list of numbers.

(a) 30, 45

First write each number in prime factored form.

$$30 = 2 \cdot 3 \cdot \mathbf{5}$$
$$45 = \mathbf{3} \cdot 3 \cdot \mathbf{5}$$

Now, take each prime the *least* number of times it appears in all the factored forms. There is no 2 in the prime factored form of 45, so there will be no 2 in the greatest common factor. The least number of times 3 appears in all the factored forms is 1, and the least number of times 5 appears is also 1. From this, the GCF is

$$3^1 \cdot 5^1 = 15.$$

(b) 72, 120, 432

Find the prime factored form of each number.

$$72 = \mathbf{2} \cdot 2 \cdot 2 \cdot \mathbf{3} \cdot 3$$
$$120 = \mathbf{2} \cdot 2 \cdot 2 \cdot \mathbf{3} \cdot 5$$
$$432 = \mathbf{2} \cdot 2 \cdot 2 \cdot 2 \cdot \mathbf{3} \cdot 3 \cdot 3$$

The least number of times 2 appears in all the factored forms is 3, and the least number of times 3 appears is 1. There is no 5 in the prime factored form of either 72 or 432, so the GCF is

$$2^3 \cdot 3 = 24.$$

(c) 10, 11, 14

Write the prime factored form of each number.

$$10 = 2 \cdot 5$$
$$11 = 11$$
$$14 = 2 \cdot 7$$

There are no primes common to all three numbers, so the GCF is 1.

The greatest common factor can also be found for a list of variable terms. For example, the terms x^4, x^5, x^6, and x^7 have x^4 as the greatest common factor because each of these terms can be written with x^4 as a factor.

$$x^4 = 1 \cdot x^4, \qquad x^5 = x \cdot x^4, \qquad x^6 = x^2 \cdot x^4, \qquad x^7 = x^3 \cdot x^4$$

 The exponent on a variable in the GCF is the *smallest* exponent that appears in the factors.

E X A M P L E 2   Finding the Greatest Common Factor for Variable Terms

Find the greatest common factor for each list of terms.

(a) $21m^7$, $-18m^6$, $45m^8$, $-24m^5$

$$21m^7 = 3 \cdot 7 \cdot m^7$$
$$-18m^6 = -1 \cdot 2 \cdot 3^2 \cdot m^6$$
$$45m^8 = 3^2 \cdot 5 \cdot m^8$$
$$-24m^5 = -1 \cdot 2^3 \cdot 3 \cdot m^5$$

First, 3 is the greatest common factor of the coefficients 21, -18, 45, and -24. The smallest exponent on m is 5, so the GCF of the terms is $3m^5$.

(b) $x^4 y^2$, $x^7 y^5$, $x^3 y^7$, y^{15}

$$x^4 y^2 = x^4 \cdot y^2$$
$$x^7 y^5 = x^7 \cdot y^5$$
$$x^3 y^7 = x^3 \cdot y^7$$
$$y^{15} = y^{15}$$

There is no x in the last term, y^{15}, so x will not appear in the greatest common factor. There is a y in each term, however, and 2 is the smallest exponent on y. The GCF is y^2.

OBJECTIVE 2 Factor out the greatest common factor. We use the idea of a greatest common factor to write a polynomial (a sum) in factored form as a product. For example, the polynomial

$$3m + 12$$

has two terms, $3m$ and 12. The greatest common factor for these two terms is 3. We can write $3m + 12$ so that each term is a product with 3 as one factor.

$$3m + 12 = 3 \cdot m + 3 \cdot 4$$

Now use the distributive property.

$$3m + 12 = 3 \cdot m + 3 \cdot 4 = 3(m + 4)$$

The factored form of $3m + 12$ is $3(m + 4)$. This process is called **factoring out the greatest common factor.**

The polynomial $3m + 12$ is *not* in factored form when written as

$$3 \cdot m + 3 \cdot 4.$$

 The *terms* are factored, but the polynomial is not. The factored form of $3m + 12$ is the *product*

$$3(m + 4).$$

EXAMPLE 3 Factoring Out the Greatest Common Factor

Factor out the greatest common factor.

(a) $20m^5 + 10m^4 + 15m^3$

The GCF for the terms of this polynomial is $5m^3$.

$$20m^5 + 10m^4 + 15m^3 = (5m^3)(4m^2) + (5m^3)(2m) + (5m^3)3$$
$$= 5m^3(4m^2 + 2m + 3)$$

Check by multiplying $5m^3$ and $4m^2 + 2m + 3$. You should get the original polynomial.

(b) $x^5 + x^3 = (x^3)x^2 + (x^3)1 = x^3(x^2 + 1)$

(c) $20m^7p^2 - 36m^3p^4 = 4m^3p^2(5m^4 - 9p^2)$

(d) $a(a + 3) + 4(a + 3)$

The binomial $a + 3$ is the greatest common factor here.

$$a(a + 3) + 4(a + 3) = (a + 3)(a + 4)$$

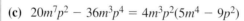

Be sure to include the 1 in a problem like Example 3(b). Always check that the factored form can be multiplied out to give the original polynomial.

OBJECTIVE 3 Factor by grouping. Common factors are used in **factoring by grouping,** as explained in the next example.

EXAMPLE 4 Factoring by Grouping

Factor by grouping.

(a) $2x + 6 + ax + 3a$

The first two terms have a common factor of 2, and the last two terms have a common factor of a.

$$2x + 6 + ax + 3a = 2(x + 3) + a(x + 3)$$

The expression is still not in factored form because it is the *sum* of two terms. Now, however, $x + 3$ is a common factor and can be factored out.

$$2x + 6 + ax + 3a = 2(x + 3) + a(x + 3) = (x + 3)(2 + a)$$

The final result is in factored form because it is a *product*. Note that the goal in factoring by grouping is to get a common factor, $x + 3$ here, so that the last step is possible.

Same

(b) $m^2 + 6m + 2m + 12 = m(m + 6) + 2(m + 6)$
$$= (m + 6)(m + 2)$$

(c) $6xy - 21x - 8y + 28 = 3x(2y - 7) - 4(2y - 7) = (2y - 7)(3x - 4)$

Must be same

Since the quantities in parentheses in the second step must be the same, it was necessary here to factor out -4 rather than 4.

 Use negative signs carefully when grouping, as in Example 4(c). Otherwise, sign errors may result.

Use these steps when factoring four terms by grouping.

Factoring by Grouping

Step 1 **Group terms.** Collect the terms into two groups so that each group has a common factor.

Step 2 **Factor within groups.** Factor out the greatest common factor from each group.

Step 3 **Factor the entire polynomial.** Factor a common binomial factor from the results of Step 2.

Step 4 **If necessary, rearrange terms.** If Step 2 does not result in a common binomial factor, try a different grouping.

EXAMPLE 5 Rearranging Terms Before Factoring by Grouping

Factor by grouping.

(a) $10x^2 - 12y^2 + 15xy - 8xy$

Factoring out the common factor of 2 from the first two terms and the common factor of xy terms from the last two terms gives

$$10x^2 - 12y^2 + 15xy - 8xy = 2(5x^2 - 6y^2) + xy(15 - 8).$$

This did not lead to a common factor, so we try rearranging the terms. There is usually more than one way to do this. Let's try

$$10x^2 - 8xy - 12y^2 + 15xy,$$

grouping the first two terms and the last two terms as follows.

$$10x^2 - 8xy - 12y^2 + 15xy = 2x(5x - 4y) + 3y(-4y + 5x)$$
$$= 2x(5x - 4y) + 3y(5x - 4y)$$
$$= (5x - 4y)(2x + 3y)$$

(b) $2xy + 12 - 3y - 8x$

We need to rearrange these terms to get two groups that each have a common factor. Trial and error suggests the following grouping.

$$2xy + 12 - 3y - 8x = (2xy - 3y) + (-8x + 12)$$
$$= y(2x - 3) - 4(2x - 3) \qquad \text{Factor each group.}$$
$$= (2x - 3)(y - 4) \qquad \text{Factor out the common binomial factor.}$$

5.1 EXERCISES

1. Is 3 the greatest common factor of 18, 24, and 42? If not, what is?

2. Is pq the greatest common factor of pq^2, p^2, and p^2q^2? If not, what is?

3. Factoring is the opposite of what operation?

4. How can you check your answer when you factor a polynomial?

5. Give an example of three numbers whose greatest common factor is 5.

6. Explain how to find the greatest common factor of a list of terms. Use examples.

Find the greatest common factor for each list of terms. See Examples 1 and 2.

7. $16y$, 24

8. $18w$, 27

9. $30x^3$, $40x^6$, $50x^7$

10. $60z^4$, $70z^8$, $90z^9$

11. $12m^3n^2$, $18m^5n^4$, $36m^8n^3$

12. $25p^5r^7$, $30p^7r^8$, $50p^5r^3$

13. $-x^4y^3$, $-xy^2$

14. $-a^4b^5$, $-a^3b$

15. $42ab^3$, $-36a$, $90b$, $-48ab$

16. $45c^3d$, $75c$, $90d$, $-105cd$

An expression is factored when it is written as a product, not a sum. Which of the following are not factored?

17. $2k^2(5k)$

18. $2k^2(5k + 1)$

19. $2k^2 + (5k + 1)$

20. $(2k^2 + 1)(5k + 1)$

21. Is $-xy$ a common factor of $-x^4y^3$ and $-xy^2$? If so, what is the other factor that when multiplied by $-xy$ gives $-x^4y^3$?

22. Is $-a^5b^2$ a common factor of $-a^4b^5$ and $-a^3b$?

Complete each factoring.

23. $12 = 6(\quad)$

24. $18 = 9(\quad)$

25. $3x^2 = 3x(\quad)$

26. $8x^3 = 8x(\quad)$

27. $9m^4 = 3m^2(\quad)$

28. $12p^5 = 6p^3(\quad)$

29. $-8z^9 = -4z^5(\quad)$

30. $-15k^{11} = -5k^8(\quad)$

31. $6m^4n^5 = 3m^3n(\quad)$

32. $27a^3b^2 = 9a^2b(\quad)$

33. $-14x^4y^3 = 2xy(\quad)$

34. $-16m^3n^3 = 4mn^2(\quad)$

Factor out the greatest common factor. See Example 3.

35. $12y - 24$

36. $18p + 36$

37. $10a^2 - 20a$

38. $15x^3 - 30x^2$

70. $2x^2 - 7.2x + 6.3 = 0$

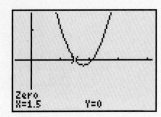

71. $2x^2 + 7.2x + 5.5 = 0$

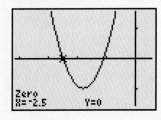

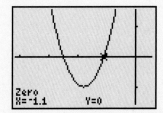

72. $4x^2 - x - 33 = 0$

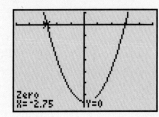

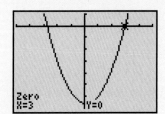

EXERCISES

5.6 Applications of Quadratic Equations

OBJECTIVES

1. Solve problems about geometric figures.

2. Solve problems using the Pythagorean formula.

3. Solve problems using quadratic models.

We can now use factoring to solve quadratic equations that arise from applied problems.

OBJECTIVE 1 Solve problems about geometric figures. Most problems in this section require one of the formulas given on the inside covers. We still follow the six-step problem-solving method from Section 2.3 and continue the work with formulas and geometric problems begun in Section 2.4.

VIDEO

┌ **EXAMPLE 1** Solving an Area Problem

The Goldsteins are planning to add a rectangular porch to their house. The width of the porch will be 4 feet less than its length, and they want it to have an area of 96 square feet. Find the length and width of the porch.

Step 1 Let x = the length of the porch;

Step 2 $x - 4$ = the width (the width is 4 less than the length). See Figure 1.

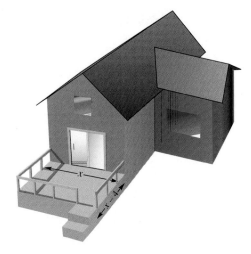

Figure 1

Step 3   The area of a rectangle is given by the formula

$$\text{area} = \text{length} \times \text{width}.$$
$$A = LW$$

Step 4

$96 = x(x - 4)$	Let $A = 96$, $L = x$, $W = x - 4$.
$96 = x^2 - 4x$	Distributive property
$0 = x^2 - 4x - 96$	Subtract 96 from both sides.
$0 = (x - 12)(x + 8)$	Factor.
$x - 12 = 0$ or $x + 8 = 0$	Zero-factor property
$x = 12$ or $x = -8$	Solve.

Step 5   The solutions of the equations are 12 and -8. Since a rectangle cannot have a negative length, discard the solution -8. Then 12 feet is the length of the porch and $12 - 4 = 8$ feet is the width.

Step 6   As a check, note that the width is 4 less than the length and the area is $8 \cdot 12 = 96$ square feet, as required.

When solving an applied problem, remember Step 6. Always check solutions against physical facts.

The next application involves *perimeter,* the distance around a figure, as well as area.

E X A M P L E   2   Solving an Area and Perimeter Problem

The length of a rectangular rug is 4 feet more than the width. The area of the rug is numerically 1 more than the perimeter. See Figure 2. Find the length and width of the rug.

$x + 4$

x

Figure 2

Let x = the width of the rug.

Then $x + 4$ = the length of the rug.

The area is the product of the length and width, so

$$A = LW.$$

Substituting $x + 4$ for the length and x for the width gives

$$A = (x + 4)x.$$

Now substitute into the formula for perimeter.

$$P = 2L + 2W$$
$$P = 2(x + 4) + 2x$$

According to the information given in the problem, the area is numerically 1 more than the perimeter. Write this as an equation.

The area	is	1	more than	the perimeter.
↓	↓	↓	↓	↓

$$(x + 4)x \; = \; 1 \; + \; 2(x + 4) + 2x$$

Simplify and solve this equation.

$$x^2 + 4x = 1 + 2x + 8 + 2x \quad \text{Distributive property}$$
$$x^2 + 4x = 9 + 4x \quad \text{Combine terms.}$$
$$x^2 = 9 \quad \text{Subtract } 4x \text{ from both sides.}$$
$$x^2 - 9 = 0 \quad \text{Subtract 9 from both sides.}$$
$$(x + 3)(x - 3) = 0 \quad \text{Factor.}$$
$$x + 3 = 0 \quad \text{or} \quad x - 3 = 0 \quad \text{Zero-factor property}$$
$$x = -3 \quad \text{or} \quad x = 3$$

A rectangle cannot have a negative width, so ignore -3. The only valid solution is 3, so the width is 3 feet and the length is $3 + 4 = 7$ feet. Check to see that the area is numerically 1 more than the perimeter. The rug is 3 feet wide and 7 feet long.

OBJECTIVE 2 Solve problems using the Pythagorean formula. The next example requires the **Pythagorean formula** from geometry.

Pythagorean Formula

If a right triangle (a triangle with a 90° angle) has longest side of length c and two other sides of lengths a and b, then

$$a^2 + b^2 = c^2.$$

(See the figure.) The longest side, the **hypotenuse,** is opposite the right angle. The two shorter sides are the **legs** of the triangle.

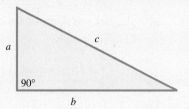

┌─ **EXAMPLE 3** Using the Pythagorean Formula

Ed and Mark leave their office with Ed traveling north and Mark traveling east. When Mark is 1 mile farther than Ed from the office, the distance between them is 2 miles more than Ed's distance from the office. Find their distances from the office and the distance between them.

Let x represent Ed's distance from the office.

Then $x + 1$ represents Mark's distance from the office,

and $x + 2$ represents the distance between them.

Place these on a right triangle as in Figure 3.

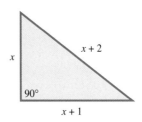

Figure 3

Substitute into the Pythagorean formula,

$$a^2 + b^2 = c^2$$
$$x^2 + (x + 1)^2 = (x + 2)^2.$$

Since $(x + 1)^2 = x^2 + 2x + 1$, and since $(x + 2)^2 = x^2 + 4x + 4$, the equation becomes

$$x^2 + x^2 + 2x + 1 = x^2 + 4x + 4.$$

$x^2 - 2x - 3 = 0$	Standard form
$(x - 3)(x + 1) = 0$	Factor.
$x - 3 = 0$ or $x + 1 = 0$	Zero-factor property
$x = 3$ or $x = -1$	Solve.

Since -1 cannot be the length of a side of a triangle, 3 is the only possible answer. Therefore, Ed is 3 miles north of the office, Mark is $3 + 1 = 4$ miles east of the office, and they are $3 + 2 = 5$ miles apart. Check to see that $3^2 + 4^2 = 5^2$ is true. ■

When solving a problem involving the Pythagorean formula, be sure that the expressions for the sides are properly placed.

$$\text{leg}^2 + \text{leg}^2 = \text{hypotenuse}^2$$

(The hypotenuse is opposite the right angle.)

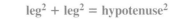

When a carpenter builds a floor for a rectangular room, it is essential that the corners of the floor are at right angles; otherwise, problems will occur when the walls are constructed, when flooring is laid, and so on. To check that the floor is "squared off," the carpenter can use the *converse* of the Pythagorean formula: If $a^2 + b^2 = c^2$, then the angle opposite side c is a right angle.

FOR DISCUSSION OR WRITING

Suppose a carpenter is building an 8-foot by 12-foot room. After the floor is built, the carpenter finds that the length of the diagonal of the floor is 14 feet, 8 inches. Is the floor "squared off" properly? If not, what should the diagonal measure?

OBJECTIVE **3** Solve problems using quadratic models. In the next example, we use the quadratic model for fuel economy from the beginning of this chapter.

EXAMPLE 4 Using a Quadratic Equation That Models Gasoline Mileage

Earlier we gave the quadratic expression $-.04x^2 + .93x + 21$, that models the automotive industry fuel economy trend. If we set the expression equal to y, then y is the average miles per gallon for the industry in year x. Recall, x is coded so that $x = 0$ represents 1978, $x = 2$ represents 1980, and so on. The equation

$$y = -.04x^2 + .93x + 21$$

was developed from the following data.

Fuel Economy

Year	Miles per Gallon
1978	19.9
1980	23.1
1982	25.1
1984	25.0
1986	25.9
1988	26.0
1990	25.4
1992	25.1
1994	24.7
1996	24.9

Source: National Highway Traffic Safety Administration.

(a) From the table, in what year did the number of miles per gallon appear to peak?

Look down the second column for the largest number, 26.0 miles per gallon. Reading across to the first column, we see that this mileage corresponds to 1988.

(b) Use the equation to find the miles per gallon in 1992.

The miles per gallon in year x is given by y. In 1992, $x = 1992 - 1978 = 14$. Substitute 14 for x in the equation $y = -.04x^2 + .93x + 21$.

$$y = -.04(14)^2 + .93(14) + 21 = 26.18 \qquad \text{Use a calculator.}$$

From the table, the actual data for 1992 is 25.1, so this approximation is a little high.

(c) Repeat part (b) for 1986.

For 1986, $x = 1986 - 1978 = 8$, and

$$y = -.04(8)^2 + .93(8) + 21 = 25.88. \qquad \text{Let } x = 8.$$

How does this approximation compare to the actual data in the table?

(d) Which of the results in parts (b) and (c) is the better approximation of the data in the table?

The approximation for 1986 is much closer than the approximation for 1992.

5.6 EXERCISES

Complete these statements which review the six-step method for solving applied problems first introduced in Chapter 2.

1. Read the problem carefully, choose _____ to represent _____ and write it down.

2. Write down a mathematical _____ for any other unknown quantities. Draw a _____ if it would be helpful.

3. Translate the problem into _____ .

4. Solve the _____ and answer the _____ in the problem. _____ your answer.

In Exercises 5–8, a figure and a corresponding geometric formula are given. Complete each problem using the problem-solving steps from Chapter 2.

(a) Write an equation using the formula and the given information. (Step 3)
(b) Solve the equation, giving only the solution(s) that make sense in the problem. (Step 4)
(c) Use the solution(s) to find the indicated dimensions of the figure. (Step 5)
(d) Check your solution. (Step 6)

5. Area of a rectangle: $A = LW$
The area of this rectangle is 80 square units. Find its length and its width.

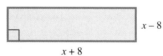

6. Area of a parallelogram: $A = bh$
The area of this parallelogram is 45 square units. Find its base and its height.

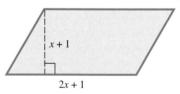

7. Area of a triangle: $A = \frac{1}{2}bh$
The area of this triangle is 60 square units. Find its base and its height.

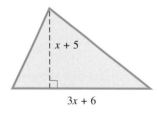

8. Volume of a rectangular Chinese box: $V = LWH$
The volume of this box is 192 cubic units. Find its length and its width.

Solve each problem. Check your answer to be sure that it is reasonable. Refer to the formulas found on the inside covers. See Examples 1 and 2.

9. The length of a VHS videocassette shell is 3 inches more than its width. The area of the rectangular top side of the shell is 28 square inches. Find the length and the width of the videocassette shell.

10. A plastic box that holds a standard audiocassette has a length 4 centimeters longer than its width. The area of the rectangular top of the box is 77 square centimeters. Find the length and the width of the box.

11. The dimensions of a certain IBM computer monitor screen are such that its length is 3 inches more than its width. If the length is increased by 1 inch while the width remains the same, the area is increased by 8 square inches. What are the dimensions of the screen?

12. The keyboard of the computer mentioned in Exercise 11 is 11 inches longer than it is wide. If both its length and width are increased by 2 inches, the area of the top of the keyboard is increased by 58 square inches. What are the length and the width of the keyboard?

13. The area of a triangle is 30 square inches. The base of the triangle measures 2 inches more than twice the height of the triangle. Find the measures of the base and the height.

14. A certain triangle has its base equal in measure to its height. The area of the triangle is 72 square meters. Find the equal base and height measure.

15. A ten-gallon aquarium holding African cichlids is 3 inches higher than it is wide. Its length is 21 inches, and its volume is 2730 cubic inches. What are the height and width of the aquarium?

16. Nana Nantambu wishes to build a box to hold her tools. It is to be 2 feet high, and the width is to be 3 feet less than its length. If its volume is to be 80 cubic feet, find the length and the width of the box.

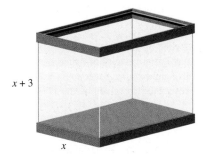

Use the Pythagorean formula to solve each problem. See Example 3.

17. The hypotenuse of a right triangle is 1 centimeter longer than the longer leg. The shorter leg is 7 centimeters shorter than the longer leg. Find the length of the longer leg of the triangle.

18. The longer leg of a right triangle is 1 meter longer than the shorter leg. The hypotenuse is 1 meter shorter than twice the shorter leg. Find the length of the shorter leg of the triangle.

EXERCISES

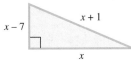

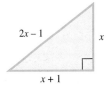

19. A ladder is resting against a wall. The top of the ladder touches the wall at a height of 15 feet. Find the distance from the wall to the bottom of the ladder if the length of the ladder is one foot more than twice its distance from the wall.

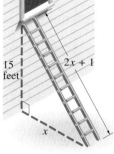

20. Two cars leave an intersection. One car travels north; the other travels east. When the car traveling north had gone 24 miles, the distance between the cars was four miles more than three times the distance traveled by the car heading east. Find the distance between the cars at that time.

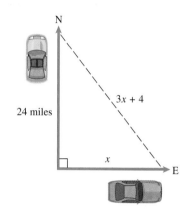

EXERCISES

21. A garden has the shape of a right triangle with one leg 2 meters longer than the other. The hypotenuse is two meters less than twice the length of the shorter leg. Find the length of the shorter leg.

22. The hypotenuse of a right triangle is 1 foot more than twice the length of the shorter leg. The longer leg is 1 foot less than twice the length of the shorter leg. Find the length of the shorter leg.

If an object is propelled upward from a height of s feet at an initial velocity of v feet per second, then its height h after t seconds (disregarding air resistance*) is given by the equation

$$h = -16t^2 + vt + s,$$

where h is in feet. For example, if the object is propelled from a height of 48 feet with an initial velocity of 32 feet per second, its height h is given by the equation $h = -16t^2 + 32t + 48$.

Use this information in Exercises 23–26.

23. After how many seconds is the height 64 feet? (*Hint:* Let $h = 64$ and solve.)

24. After how many seconds is the height 60 feet?

25. After how many seconds does the object hit the ground? (*Hint:* When the object hits the ground, $h = 0$.)

26. The quadratic equation from Exercise 25 has two solutions, yet only one of them is appropriate for answering the question. Why is this so?

EXERCISES

---

*From now on, in all problems of this sort involving an object propelled upward or dropped from a height, we give formulas that disregard air resistance.

55. A 9-inch by 12-inch picture is to be placed on a cardboard mat so that there is an equal border around the picture. The area of the finished mat and picture is to be 208 square inches. How wide will the border be?

Mat

56. A box is made from a 12-centimeter by 10-centimeter piece of cardboard by cutting equal-sized squares from each corner and folding up the sides. The area of the bottom of the box is to be 48 square centimeters. Find the length of a side of the cutout squares.

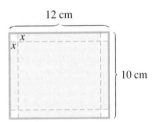

12 cm

10 cm

57. In 1994, Greyhound Lines, Inc. was near its second bankruptcy in three years. Since then, a new CEO, Craig R. Lentzsch, appears to have turned the company around. The bar graph and table of values show the operating income for 1994, when Lentzsch took over, 1995, and 1996. (*Source:* Company Reports, Rothschild Inc.)

Year	Operating Income (in millions of dollars)
1994	−65
1995	9.4
1996	37

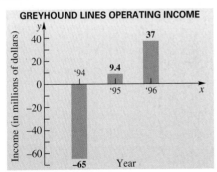

GREYHOUND LINES OPERATING INCOME

Using the data, we constructed the quadratic equation

$$y = -23.4x^2 + 285x - 831,$$

which gives the operating income y (in millions of dollars) in year x. We use $x = 4$ to represent 1994, $x = 5$ to represent 1995, and so on.

(a) Use the equation to predict the operating income in 1997.

(b) Use the equation to predict the operating income in 1998.

(c) Comment on the validity of the answers for parts (a) and (b).

[5.7] *Solve each inequality.*

58. $(q + 5)(q - 3) > 0$

59. $(2r - 1)(r + 4) \geq 0$

60. $m^2 - 5m + 6 \leq 0$

61. $2x^2 + 5x - 12 \geq 0$

62. $2p^2 + 5p - 12 < 0$

63. Suppose you know that the solution set of a quadratic inequality involving the $<$ symbol is $(-5, 7)$. If the symbol is changed to $>$, what is the solution set of the new inequality?

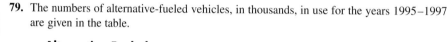

MIXED REVIEW EXERCISES

Factor completely.

64. $z^2 - 11zx + 10x^2$

65. $3k^2 + 11k + 10$

66. $15m^2 + 20mp - 12mp - 16p^2$

67. $y^4 - 625$

68. $6m^3 - 21m^2 - 45m$

69. $24ab^3c^2 - 56a^2bc^3 + 72a^2b^2c$

70. $25a^2 + 15ab + 9b^2$

71. $12x^2yz^3 + 12xy^2z - 30x^3y^2z^4$

72. $2a^5 - 8a^4 - 24a^3$

73. $12r^2 + 18rq - 10rq - 15q^2$

74. $1000a^3 + 27$

75. $49t^2 + 56t + 16$

Solve.

76. $t(t - 7) = 0$

77. $x(x + 3) = 10$

78. $4x^2 - x - 3 \le 0$

79. The numbers of alternative-fueled vehicles, in thousands, in use for the years 1995–1997 are given in the table.

Alternative-Fueled Vehicles

Year	Number (in thousands)
1995	333
1996	357
1997	386

Source: Energy Information Administration, *Alternatives to Traditional Fuels,* 1993.

Using statistical methods, we constructed the quadratic equation

$$y = 2.5x^2 - 453.5x + 20{,}850$$

to model the number of vehicles y in year x. Here we used $x = 95$ for 1995, $x = 96$ for 1996, and so on. Because only three years of data were used to determine the model, we must be particularly careful about using it to estimate for years before 1995 or after 1997.
(a) What prediction for 1998 is given by the equation?
(b) Why might the prediction for 1998 be unreliable?

80. The sum of two consecutive even integers is 34 less than their product. Find the integers.

81. The floor plan for a house is a rectangle with length 7 meters more than its width. The area is 170 square meters. Find the width and length of the house.

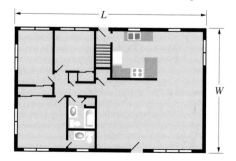

┌─
│ E X A M P L E   4   Writing in Lowest Terms
│ Write each rational expression in lowest terms.
│
│ **(a)** $\dfrac{3x - 12}{5x - 20}$
│
│ Begin by factoring both numerator and denominator. Then use the fundamental
│ property.
│
│ $$\frac{3x - 12}{5x - 20} = \frac{3(x - 4)}{5(x - 4)}$$
│
│ $$= \frac{3}{5}$$
│
│ The rational expression $\dfrac{3x - 12}{5x - 20}$ equals $\dfrac{3}{5}$ for all values of x, where $x \neq 4$ (since the
│ denominator of the original rational expression is 0 when x is 4).
│
│ **(b)** $\dfrac{m^2 + 2m - 8}{2m^2 - m - 6}$
│
│ $$\frac{m^2 + 2m - 8}{2m^2 - m - 6} = \frac{(m + 4)(m - 2)}{(2m + 3)(m - 2)} \qquad \textit{Factor.}$$
│
│ $$= \frac{m + 4}{2m + 3} \qquad \textit{Fundamental property}$$
│
│ Thus, $\dfrac{m^2 + 2m - 8}{2m^2 - m - 6} = \dfrac{m + 4}{2m + 3}$ for $m \neq -\dfrac{3}{2}$ and $m \neq 2$, since the denominator of the
│ original expression is 0 for these values of m.
└─

From now on, we will write statements of equality of rational expressions with the
understanding that they apply only to those real numbers that make neither denominator equal to 0.

One of the most common errors in algebra occurs when students attempt to
write rational expressions in lowest terms *before factoring*. The fundamental
property is applied only *after* the numerator and denominator are expressed in
factored form. For example, although x appears in both the numerator and
denominator in Example 4(a), and 12 and 20 have a common factor of 4, the
fundamental property cannot be used before factoring because $3x$, $5x$, 12, and
 20 are *terms*, not *factors*. Terms are *added* or *subtracted;* factors are *multiplied*
or *divided.* For example,

$$\frac{6 + 2}{3 + 2} = \frac{8}{5}, \quad \textbf{not} \quad \frac{6}{3} + \frac{2}{2} = 2 + 1 = 3.$$

Also, $\dfrac{2x + 3}{4x + 6} = \dfrac{2x + 3}{2(2x + 3)} = \dfrac{1}{2}$,   **but** $\dfrac{x^2 + 6}{x + 3} \neq x + 2.$

┌ **E X A M P L E 5** Writing in Lowest Terms (Factors Are Opposites)

Write $\dfrac{x - y}{y - x}$ in lowest terms.

At first glance, there does not seem to be any way in which $x - y$ and $y - x$ can be factored to get a common factor. This is not the case, however, since the numerator can be factored as

$$x - y = -1(-x + y) = -1(y - x).$$

Now use the fundamental property to simplify.

$$\frac{x - y}{y - x} = \frac{-1(y - x)}{1(y - x)} = \frac{-1}{1} = -1$$

Either the numerator or the denominator could have been factored in the first step.

In Example 5, the binomials in the numerator and denominator are opposites. A general rule for this situation follows.

A fraction in which the numerator and denominator show subtraction of the same terms but in opposite order is equal to -1. More generally, if a factor of the numerator is the opposite of a factor in the denominator, the quotient of those two factors is -1.

┌ **E X A M P L E 6** Writing in Lowest Terms (Factors Are Opposites)

Write each rational expression in lowest terms.

(a) $\dfrac{2 - m}{m - 2}$

Since $2 - m$ and $m - 2$ (or $-2 + m$) are opposites, we use the rule above.

$$\frac{2 - m}{m - 2} = -1$$

(b) $\dfrac{4x^2 - 9}{6 - 4x}$

Factor the numerator and denominator and use the rule above.

$$\frac{4x^2 - 9}{6 - 4x} = \frac{(2x + 3)(2x - 3)}{2(3 - 2x)}$$

$$= \frac{2x + 3}{2}(-1)$$

$$= -\frac{2x + 3}{2}$$

(c) $\dfrac{3 + r}{3 - r}$

The quantity $3 - r$ *is not* the opposite of $3 + r$. This rational expression cannot be simplified.

OBJECTIVE 4 Recognize equivalent forms of rational expressions. When working with rational expressions, it is important to be able to recognize equivalent forms of an expression. For example, the common fraction $-\frac{5}{6}$ can also be written as $\frac{-5}{6}$ and as $\frac{5}{-6}$. Look again at Example 6(b). The form of the answer given there is only one of several acceptable forms. The $-$ sign representing the -1 factor is in front of the fraction, on the same line as the fraction bar. The -1 factor may be placed in front of the fraction, in the numerator, or in the denominator. Some other acceptable forms of the answer are

$$\frac{-(2x + 3)}{2}, \qquad \frac{-2x - 3}{2}, \qquad \text{and} \qquad \frac{2x + 3}{-2}.$$

However, $\dfrac{-2x + 3}{2}$ is *not* an acceptable form, because the sign preceding 3 in the numerator should be $-$ rather than $+$. (The form $\dfrac{2x + 3}{-2}$, listed above, is seldom used.)

E X A M P L E 7 Writing Equivalent Forms of a Rational Expression

Write four equivalent forms of the rational expression $-\dfrac{3x + 2}{x - 6}$.

If we let the negative sign preceding the fraction apply to the numerator, we have the equivalent form

$$\frac{-(3x + 2)}{x - 6}.$$

By distributing the negative sign in this expression, we have another equivalent form,

$$\frac{-3x - 2}{x - 6}.$$

If we let the negative sign apply to the denominator of the fraction, we get

$$\frac{3x + 2}{-(x - 6)}$$

or, distributing once again,

$$\frac{3x + 2}{-x + 6}.$$

CAUTION In Example 7, it would be incorrect to distribute the negative sign to *both* the numerator *and* the denominator. This would lead to the *opposite* of the original expression.

CONNECTIONS

In Chapter 4 we used long division to find the quotient of two polynomials. For example, we found $(2x^2 + 5x - 12) \div (2x - 3)$ as follows:

$$
\begin{array}{r}
x + 4 \\
2x - 3\overline{\smash{)}2x^2 + 5x - 12} \\
\underline{2x^2 - 3x} \\
8x - 12 \\
\underline{8x - 12} \\
0
\end{array}
$$

The quotient is $x + 4$. We also get the same quotient by expressing the division problem as a rational expression (fraction) and writing this rational expression in lowest terms.

$$
\frac{2x^2 + 5x - 12}{2x - 3} = \frac{(2x - 3)(x + 4)}{2x - 3} = x + 4
$$

FOR DISCUSSION OR WRITING

What kind of division problem has a quotient that cannot be found by reducing a fraction to lowest terms? Try using rational expressions to solve the following division problems. Then use long division to compare.

1. $(3x^2 + 11x + 8) \div (x + 2)$

2. $(x^3 - 8) \div (x^2 + 2x + 4)$

6.1 EXERCISES

1. Fill in the blanks with the correct responses.

 (a) The rational expression $\dfrac{x + 5}{x - 3}$ is undefined when $x =$ _____, and is equal to 0 when $x =$ _____.

 (b) The rational expression $\dfrac{p - q}{q - p}$ is undefined when $p =$ _____, and in all other cases its simplified form is _____.

2. Make the correct choice for the blank.

 (a) $\dfrac{4 - r^2}{4 + r^2}$ _____ equal to -1.
  (is/is not)

 (b) $\dfrac{5 + 2x}{3 - x}$ and $\dfrac{-5 - 2x}{x - 3}$ _____ equivalent rational expressions.
  (are/are not)

3. Define *rational expression* in your own words, and give an example.

4. Give an example of a rational expression that is not in lowest terms, and then show the steps required to write it in lowest terms.

Find any values for which each rational expression is undefined. See Example 1.

5. $\dfrac{12}{5y}$ **6.** $\dfrac{-7}{3z}$ **7.** $\dfrac{4x^2}{3x + 5}$ **8.** $\dfrac{2x^3}{3x + 4}$

9. $\dfrac{5m + 2}{m^2 + m - 6}$ **10.** $\dfrac{2r - 5}{r^2 - 5r + 4}$ **11.** $\dfrac{3x - 1}{x^2 + 2}$ **12.** $\dfrac{4q + 2}{q^2 + 9}$

EXERCISES

*Find the numerical value of each rational expression when (**a**) x = 2 and (**b**) x = −3. See Example 2.*

13. $\dfrac{5x - 2}{4x}$

14. $\dfrac{3x + 1}{5x}$

15. $\dfrac{2x^2 - 4x}{3x}$

16. $\dfrac{4x^2 - 1}{5x}$

17. $\dfrac{(-3x)^2}{4x + 12}$

18. $\dfrac{(-2x)^3}{3x + 9}$

19. $\dfrac{5x + 2}{2x^2 + 11x + 12}$

20. $\dfrac{7 - 3x}{3x^2 - 7x + 2}$

21. If 2 is substituted for x in the rational expression $\dfrac{x - 2}{x^2 - 4}$, the result is $\dfrac{0}{0}$. An often-heard statement is "Any number divided by itself is 1." Does this mean that this expression is equal to 1 for $x = 2$? If not, explain.

22. For $x \neq 2$, the rational expression $\dfrac{2(x - 2)}{x - 2}$ is equal to 2. Can the same be said for $\dfrac{2x - 2}{x - 2}$? Explain.

Write each rational expression in lowest terms. See Examples 3 and 4.

23. $\dfrac{18r^3}{6r}$

24. $\dfrac{27p^2}{3p}$

25. $\dfrac{4(y - 2)}{10(y - 2)}$

26. $\dfrac{15(m - 1)}{9(m - 1)}$

27. $\dfrac{(x + 1)(x - 1)}{(x + 1)^2}$

28. $\dfrac{(t + 5)(t - 3)}{(t - 1)(t + 5)}$

29. $\dfrac{7m + 14}{5m + 10}$

30. $\dfrac{8z - 24}{4z - 12}$

31. $\dfrac{m^2 - n^2}{m + n}$

32. $\dfrac{a^2 - b^2}{a - b}$

33. $\dfrac{12m^2 - 3}{8m - 4}$

34. $\dfrac{20p^2 - 45}{6p - 9}$

35. $\dfrac{3m^2 - 3m}{5m - 5}$

36. $\dfrac{6t^2 - 6t}{2t - 2}$

37. $\dfrac{9r^2 - 4s^2}{9r + 6s}$

38. $\dfrac{16x^2 - 9y^2}{12x - 9y}$

39. $\dfrac{zw + 4z - 3w - 12}{zw + 4z + 5w + 20}$

40. $\dfrac{km + 4k + 4m + 16}{km + 4k + 5m + 20}$

41. $\dfrac{5k^2 - 13k - 6}{5k + 2}$

42. $\dfrac{7t^2 - 31t - 20}{7t + 4}$

43. $\dfrac{2x^2 - 3x - 5}{2x^2 - 7x + 5}$

44. $\dfrac{3x^2 + 8x + 4}{3x^2 - 4x - 4}$

Write each expression in lowest terms. See Examples 5 and 6.

45. $\dfrac{6 - t}{t - 6}$

46. $\dfrac{2 - k}{k - 2}$

47. $\dfrac{m^2 - 1}{1 - m}$

48. $\dfrac{a^2 - b^2}{b - a}$

49. $\dfrac{q^2 - 4q}{4q - q^2}$

50. $\dfrac{z^2 - 5z}{5z - z^2}$

51. $\dfrac{p + 6}{p - 6}$

52. $\dfrac{5 - x}{5 + x}$

53. The area of the rectangle is represented by $x^4 + 10x^2 + 21$. What is the width?

$\left(\textit{Hint: Use } W = \dfrac{A}{L}.\right)$

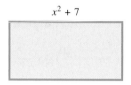

$x^2 + 7$

54. The volume of the box is represented by

$$(x^2 + 8x + 15)(x + 4).$$

Find the polynomial that represents the area of the bottom of the box.

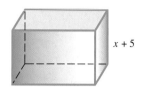

$x + 5$

Write four equivalent expressions for each of the following. See Example 7.

55. $-\dfrac{x + 4}{x - 3}$

56. $-\dfrac{x + 6}{x - 1}$

57. $-\dfrac{2x - 3}{x + 3}$

58. $-\dfrac{5x - 6}{x + 4}$

59. $\dfrac{-3x + 1}{5x - 6}$

60. $\dfrac{-2x - 9}{3x + 1}$

Write each expression in lowest terms.

61. $\dfrac{m^2 - n^2 - 4m - 4n}{2m - 2n - 8}$

62. $\dfrac{x^2 y + y + x^2 z + z}{xy + xz}$

63. $\dfrac{b^3 - a^3}{a^2 - b^2}$

64. $\dfrac{k^3 + 8}{k^2 - 4}$

65. $\dfrac{z^3 + 27}{z^3 - 3z^2 + 9z}$

66. $\dfrac{1 - 8r^3}{8r^2 + 4r + 2}$

TECHNOLOGY INSIGHTS (EXERCISES 67–72)

In the tables shown in Exercises 67–72, the expressions for Y_1 and Y_2 are equivalent except for the value for which Y_1 shows an error. In each case, Y_1 is defined by a rational expression. Predict the denominator of Y_1.

X	Y₁	Y₂
-5	-7	-7
-4	-6	-6
-3	-5	-5
-2	ERROR	-4
-1	-3	-3
0	-2	-2
1	-1	-1

Y₂◻X−2

67.

X	Y₁	Y₂
-6	-9	-9
-5	-8	-8
-4	-7	-7
-3	ERROR	-6
-2	-5	-5
-1	-4	-4
0	-3	-3

Y₂◻X−3

68.

X	Y₁	Y₂
-7	-11	-11
-6	-10	-10
-5	-9	-9
-4	ERROR	-8
-3	-7	-7
-2	-6	-6
-1	-5	-5

Y₂◻X−4

69.

X	Y₁	Y₂
-8	13	13
-7	12	12
-6	11	11
-5	ERROR	10
-4	9	9
-3	8	8
-2	7	7

Y₂◻5−X

70.

X	Y₁	Y₂
-9	15	15
-8	14	14
-7	13	13
-6	ERROR	12
-5	11	11
-4	10	10
-3	9	9

Y₂◻6−X

EXERCISES

71.

X	Y₁	Y₂
0	9	9
1	13	13
2	19	19
3	ERROR	27
4	37	37
5	49	49
6	63	63

$Y_2 \equiv X^2 + 3X + 9$

72.

X	Y₁	Y₂
1	21	21
2	28	28
3	37	37
4	ERROR	48
5	61	61
6	76	76
7	93	93

$Y_2 \equiv X^2 + 4X + 16$

6.2 Multiplication and Division of Rational Expressions

OBJECTIVES

1 Multiply rational expressions.

2 Divide rational expressions.

OBJECTIVE 1   Multiply rational expressions. The product of two fractions is found by multiplying the numerators and multiplying the denominators. Rational expressions are multiplied in the same way.

Multiplying Rational Expressions

The product of the rational expressions $\frac{P}{Q}$ and $\frac{R}{S}$ is

$$\frac{P}{Q} \cdot \frac{R}{S} = \frac{PR}{QS}.$$

The next example shows the multiplication of two common fractions and the multiplication of two rational expressions with variables so that you can compare the steps.

EXAMPLE 1   Multiplying Rational Expressions

Multiply. Write answers in lowest terms.

(a) $\dfrac{3}{10} \cdot \dfrac{5}{9}$ 　　　　　　　　　　**(b)** $\dfrac{6}{x} \cdot \dfrac{x^2}{12}$

Find the product of the numerators and the product of the denominators.

$$\frac{3}{10} \cdot \frac{5}{9} = \frac{3 \cdot 5}{10 \cdot 9} \qquad\qquad \frac{6}{x} \cdot \frac{x^2}{12} = \frac{6 \cdot x^2}{x \cdot 12}$$

Use the fundamental property to write each product in lowest terms.

$$\frac{3}{10} \cdot \frac{5}{9} = \frac{1 \cdot 3 \cdot 5}{2 \cdot 5 \cdot 3 \cdot 3} = \frac{1}{6} \qquad\qquad \frac{6}{x} \cdot \frac{x^2}{12} = \frac{6 \cdot x \cdot x}{2 \cdot 6 \cdot x} = \frac{x}{2}$$

Notice in the second step above that the products were left in factored form since common factors must be identified to write the product in lowest terms.

It is also possible to divide out common factors in the numerator and denominator *before* multiplying the rational expressions. For example,

$$\frac{6}{5} \cdot \frac{35}{22} = \frac{2 \cdot 3}{5} \cdot \frac{5 \cdot 7}{2 \cdot 11} \qquad \text{Identify common factors.}$$

$$= \frac{3 \cdot 7}{11} \qquad \text{Lowest terms}$$

$$= \frac{21}{11}. \qquad \text{Multiply in numerator.}$$

E X A M P L E 2   Multiplying Rational Expressions

Multiply. Write the product in lowest terms.

$$\frac{x + y}{2x} \cdot \frac{x^2}{(x + y)^2}$$

Use the definition of multiplication.

$$\frac{x + y}{2x} \cdot \frac{x^2}{(x + y)^2} = \frac{(x + y)x^2}{2x(x + y)^2} \qquad \begin{array}{l}\text{Multiply numerators; multiply}\\ \text{denominators.}\end{array}$$

$$= \frac{(x + y)x \cdot x}{2x(x + y)(x + y)} \qquad \text{Factor; identify common factors.}$$

$$= \frac{x}{2(x + y)} \cdot \frac{x(x + y)}{x(x + y)} \qquad \text{Definition of multiplication}$$

$$= \frac{x}{2(x + y)} \qquad \text{Lowest terms}$$

Notice the factor $\dfrac{x(x + y)}{x(x + y)}$ in the third line. Since it equals 1, the final product is $\dfrac{x}{2(x + y)}$.

E X A M P L E 3   Multiplying Rational Expressions

Multiply. Write the product in lowest terms.

$$\frac{x^2 + 3x}{x^2 - 3x - 4} \cdot \frac{x^2 - 5x + 4}{x^2 + 2x - 3}$$

First factor the numerators and denominators. Then use the fundamental property to write the product in lowest terms.

$$\frac{x^2 + 3x}{x^2 - 3x - 4} \cdot \frac{x^2 - 5x + 4}{x^2 + 2x - 3}$$

$$= \frac{x(x + 3)}{(x - 4)(x + 1)} \cdot \frac{(x - 4)(x - 1)}{(x + 3)(x - 1)} \qquad \text{Factor.}$$

$$= \frac{x(x + 3)(x - 4)(x - 1)}{(x - 4)(x + 1)(x + 3)(x - 1)} \qquad \text{Multiply numerators; multiply denominators.}$$

$$= \frac{x}{x + 1} \qquad \text{Lowest terms}$$

The quotients $\dfrac{x+3}{x+3}$, $\dfrac{x-4}{x-4}$, and $\dfrac{x-1}{x-1}$ are all equal to 1, justifying the final product

$\dfrac{x}{x+1}$.

OBJECTIVE 2 Divide rational expressions. To develop a method for dividing rational numbers and rational expressions, consider the following problem. Suppose that you have $\frac{7}{8}$ gallon of milk and you wish to find how many quarts you have. Since a quart is $\frac{1}{4}$ gallon, you must ask yourself, "How many $\frac{1}{4}$s are there in $\frac{7}{8}$?" This would be interpreted as

$$\frac{7}{8} \div \frac{1}{4} \quad \text{or} \quad \frac{\frac{7}{8}}{\frac{1}{4}}$$

since the fraction bar means division.

The fundamental property of rational expressions discussed earlier can be applied to rational number values of P, Q, and K. With $P = \frac{7}{8}$, $Q = \frac{1}{4}$, and $K = 4$,

$$\frac{P}{Q} = \frac{P \cdot K}{Q \cdot K} = \frac{\frac{7}{8} \cdot 4}{\frac{1}{4} \cdot 4} = \frac{\frac{7}{8} \cdot 4}{1} = \frac{7}{8} \cdot \frac{4}{1}.$$

So, to divide $\frac{7}{8}$ by $\frac{1}{4}$, we must multiply $\frac{7}{8}$ by the reciprocal of $\frac{1}{4}$, namely 4. Since $\left(\frac{7}{8}\right)(4) = \frac{7}{2}$, there are $\frac{7}{2}$ or $3\frac{1}{2}$ quarts in $\frac{7}{8}$ gallon.

The discussion above illustrates the rule for dividing common fractions. To divide $\frac{a}{b}$ by $\frac{c}{d}$, multiply $\frac{a}{b}$ by the reciprocal of $\frac{c}{d}$. Division of rational expressions is defined in the same way.

Dividing Rational Expressions

If $\frac{P}{Q}$ and $\frac{R}{S}$ are any two rational expressions, with $\frac{R}{S} \neq 0$, then

$$\frac{P}{Q} \div \frac{R}{S} = \frac{P}{Q} \cdot \frac{S}{R} = \frac{PS}{QR}.$$

The next example shows the division of two common fractions and the division of two rational expressions involving variables.

EXAMPLE 4   Dividing Rational Expressions

Divide. Write answers in lowest terms.

(a) $\dfrac{5}{8} \div \dfrac{7}{16}$

(b) $\dfrac{y}{y+3} \div \dfrac{4y}{y+5}$

Multiply the first expression and the reciprocal of the second.

$$\frac{5}{8} \div \frac{7}{16} = \frac{5}{8} \cdot \frac{16}{7} \quad \text{Reciprocal of } \tfrac{7}{16}$$

$$\frac{y}{y+3} \div \frac{4y}{y+5}$$

$$= \frac{5 \cdot 16}{8 \cdot 7}$$

$$= \frac{y}{y+3} \cdot \frac{y+5}{4y} \quad \begin{array}{l}\text{Reciprocal}\\ \text{of } \tfrac{4y}{y+5}\end{array}$$

$$= \frac{5 \cdot 8 \cdot 2}{8 \cdot 7}$$

$$= \frac{y(y+5)}{(y+3)(4y)}$$

$$= \frac{5 \cdot 2}{7}$$

$$= \frac{y+5}{4(y+3)} \quad \begin{array}{l}\text{Fundamental}\\ \text{property}\end{array}$$

$$= \frac{10}{7}$$

E X A M P L E   5   Dividing Rational Expressions

Divide. Write the quotient in lowest terms.

$$\frac{(3m)^2}{(2p)^3} \div \frac{6m^3}{16p^2}$$

Use the properties of exponents as necessary.

$$\frac{(3m)^2}{(2p)^3} \div \frac{6m^3}{16p^2} = \frac{(3m)^2}{(2p)^3} \cdot \frac{16p^2}{6m^3} \qquad \text{Multiply by reciprocal.}$$

$$= \frac{(3m)(3m)}{(2p)(2p)(2p)} \cdot \frac{16p^2}{6m^3} \qquad \text{Meaning of exponent}$$

$$= \frac{9 \cdot 16m^2p^2}{8 \cdot 6p^3m^3} \qquad \begin{array}{l}\text{Multiply numerators.}\\ \text{Multiply denominators.}\end{array}$$

$$= \frac{3}{mp} \qquad \text{Lowest terms}$$

E X A M P L E   6   Dividing Rational Expressions

Divide. Write the quotient in lowest terms.

$$\frac{x^2 - 4}{(x+3)(x-2)} \div \frac{(x+2)(x+3)}{2x}$$

$$= \frac{x^2 - 4}{(x+3)(x-2)} \cdot \frac{2x}{(x+2)(x+3)} \qquad \text{Use the definition of division.}$$

$$= \frac{(x+2)(x-2)}{(x+3)(x-2)} \cdot \frac{2x}{(x+2)(x+3)} \qquad \begin{array}{l}\text{Be sure numerators and}\\ \text{denominators are factored.}\end{array}$$

$$= \frac{(x+2)(x-2)(2x)}{(x+3)(x-2)(x+2)(x+3)} \qquad \begin{array}{l}\text{Multiply numerators and}\\ \text{multiply denominators.}\end{array}$$

$$= \frac{2x}{(x+3)^2} \qquad \begin{array}{l}\text{Use the fundamental property}\\ \text{to write in lowest terms.}\end{array}$$

In Example 6, only the numerator had to be factored. Remember that *all* numerators and denominators must be factored before the fundamental property can be applied.

E X A M P L E   7   Dividing Rational Expressions (Factors Are Opposites)

Divide. Write the quotient in lowest terms.

$$\frac{m^2 - 4}{m^2 - 1} \div \frac{2m^2 + 4m}{1 - m}$$

$$= \frac{m^2 - 4}{m^2 - 1} \cdot \frac{1 - m}{2m^2 + 4m} \qquad \text{Use the definition of division.}$$

$$= \frac{(m + 2)(m - 2)}{(m + 1)(m - 1)} \cdot \frac{1 - m}{2m(m + 2)} \qquad \begin{array}{l}\text{Factor; } 1 - m \text{ and } m - 1 \\ \text{differ only in sign.}\end{array}$$

$$= \frac{-1(m - 2)}{2m(m + 1)} \qquad \text{From Section 6.1, } \frac{1-m}{m-1} = -1.$$

$$= \frac{2 - m}{2m(m + 1)} \qquad \begin{array}{l}\text{Use the distributive property} \\ \text{in the numerator.}\end{array}$$

In summary, follow these steps to multiply and divide rational expressions.

Multiplying or Dividing Rational Expressions

Step 1 **Note the operation.** If the operation is division, use the definition of division to rewrite as multiplication.

Step 2 **Factor.** Factor all numerators and denominators completely.

Step 3 **Multiply.** Multiply numerators and multiply denominators.

Step 4 **Write in lowest terms.** Use the fundamental property to write the answer in lowest terms.

6.2 EXERCISES

1. Match each multiplication problem in Column I with the correct product in Column II.

I	II
(a) $\dfrac{5x^3}{10x^4} \cdot \dfrac{10x^7}{2x}$	**A.** $\dfrac{2}{5x^5}$
(b) $\dfrac{10x^4}{5x^3} \cdot \dfrac{10x^7}{2x}$	**B.** $\dfrac{5x^5}{2}$
(c) $\dfrac{5x^3}{10x^4} \cdot \dfrac{2x}{10x^7}$	**C.** $\dfrac{1}{10x^7}$
(d) $\dfrac{10x^4}{5x^3} \cdot \dfrac{2x}{10x^7}$	**D.** $10x^7$

2. Match each division problem in Column I with the correct quotient in Column II.

I   II

(a) $\dfrac{5x^3}{10x^4} \div \dfrac{10x^7}{2x}$ A. $\dfrac{5x^5}{2}$

(b) $\dfrac{10x^4}{5x^3} \div \dfrac{10x^7}{2x}$ B. $10x^7$

(c) $\dfrac{5x^3}{10x^4} \div \dfrac{2x}{10x^7}$ C. $\dfrac{2}{5x^5}$

(d) $\dfrac{10x^4}{5x^3} \div \dfrac{2x}{10x^7}$ D. $\dfrac{1}{10x^7}$

Multiply. Write each answer in lowest terms. See Examples 1 and 2.

3. $\dfrac{15a^2}{14} \cdot \dfrac{7}{5a}$

4. $\dfrac{27k^3}{9k} \cdot \dfrac{24}{9k^2}$

5. $\dfrac{12x^4}{18x^3} \cdot \dfrac{-8x^5}{4x^2}$

6. $\dfrac{12m^5}{-2m^2} \cdot \dfrac{6m^6}{28m^3}$

7. $\dfrac{2(c+d)}{3} \cdot \dfrac{18}{6(c+d)^2}$

8. $\dfrac{4(y-2)}{x} \cdot \dfrac{3x}{6(y-2)^2}$

Divide. Write each answer in lowest terms. See Examples 4 and 5.

9. $\dfrac{9z^4}{3z^5} \div \dfrac{3z^2}{5z^3}$

10. $\dfrac{35q^8}{9q^5} \div \dfrac{25q^6}{10q^5}$

11. $\dfrac{4t^4}{2t^5} \div \dfrac{(2t)^3}{-6}$

12. $\dfrac{-12a^6}{3a^2} \div \dfrac{(2a)^3}{27a}$

13. $\dfrac{3}{2y-6} \div \dfrac{6}{y-3}$

14. $\dfrac{4m+16}{10} \div \dfrac{3m+12}{18}$

RELATING CONCEPTS (EXERCISES 15–18)

We know that division by 0 is undefined. For example, $5 \div 0$ cannot be a real number because there is no real number that when multiplied by 0 gives 5 as a product. When dividing rational expressions, we know that no denominator can be 0 in either of the expressions in the problem, and that furthermore, the numerator of the divisor cannot be 0.

Work Exercises 15–18 in order, referring to the division problem

$$\frac{x-6}{x+4} \div \frac{x+7}{x+5}.$$

15. Why must we have the restriction $x \neq -4$?

16. Why must we have the restriction $x \neq -5$?

17. Why must we have the restriction $x \neq -7$?

18. Why is 6 allowed as a replacement for x even though 6 causes the numerator in the first fraction to be 0?

Did you make the connection that division by 0 is not allowed in both the individual fractions and the problem as a whole?

19. Explain in your own words how to multiply rational expressions. Illustrate with an example.

20. Explain in your own words how to divide rational expressions. Illustrate with an example.

Multiply or divide. Write each answer in lowest terms. See Examples 3, 6, and 7.

21. $\dfrac{5x - 15}{3x + 9} \cdot \dfrac{4x + 12}{6x - 18}$

22. $\dfrac{8r + 16}{24r - 24} \cdot \dfrac{6r - 6}{3r + 6}$

23. $\dfrac{2 - t}{8} \div \dfrac{t - 2}{6}$

24. $\dfrac{4}{m - 2} \div \dfrac{16}{2 - m}$

25. $\dfrac{27 - 3z}{4} \cdot \dfrac{12}{2z - 18}$

26. $\dfrac{5 - x}{5 + x} \cdot \dfrac{x + 5}{x - 5}$

27. $\dfrac{p^2 + 4p - 5}{p^2 + 7p + 10} \div \dfrac{p - 1}{p + 4}$

28. $\dfrac{z^2 - 3z + 2}{z^2 + 4z + 3} \div \dfrac{z - 1}{z + 1}$

29. $\dfrac{2k^2 - k - 1}{2k^2 + 5k + 3} \div \dfrac{4k^2 - 1}{2k^2 + k - 3}$

30. $\dfrac{2m^2 - 5m - 12}{m^2 + m - 20} \div \dfrac{4m^2 - 9}{m^2 + 4m - 5}$

31. $\dfrac{2k^2 + 3k - 2}{6k^2 - 7k + 2} \cdot \dfrac{4k^2 - 5k + 1}{k^2 + k - 2}$

32. $\dfrac{2m^2 - 5m - 12}{m^2 - 10m + 24} \div \dfrac{4m^2 - 9}{m^2 - 9m + 18}$

33. $\dfrac{m^2 + 2mp - 3p^2}{m^2 - 3mp + 2p^2} \div \dfrac{m^2 + 4mp + 3p^2}{m^2 + 2mp - 8p^2}$

34. $\dfrac{r^2 + rs - 12s^2}{r^2 - rs - 20s^2} \div \dfrac{r^2 - 2rs - 3s^2}{r^2 + rs - 30s^2}$

35. $\dfrac{m^2 + 3m + 2}{m^2 + 5m + 4} \cdot \dfrac{m^2 + 10m + 24}{m^2 + 5m + 6}$

36. $\dfrac{z^2 - z - 6}{z^2 - 2z - 8} \cdot \dfrac{z^2 + 7z + 12}{z^2 - 9}$

37. $\dfrac{y^2 + y - 2}{y^2 + 3y - 4} \div \dfrac{y + 2}{y + 3}$

38. $\dfrac{r^2 + r - 6}{r^2 + 4r - 12} \div \dfrac{r + 3}{r - 1}$

39. $\dfrac{2m^2 + 7m + 3}{m^2 - 9} \cdot \dfrac{m^2 - 3m}{2m^2 + 11m + 5}$

40. $\dfrac{m^2 + 2mp - 3p^2}{m^2 - 3mp + 2p^2} \div \dfrac{m^2 + 4mp + 3p^2}{m^2 + 2mp - 8p^2}$

41. $\dfrac{r^2 + rs - 12s^2}{r^2 - rs - 20s^2} \div \dfrac{r^2 - 2rs - 3s^2}{r^2 + rs - 30s^2}$

42. $\dfrac{(x + 1)^3(x + 4)}{x^2 + 5x + 4} \div \dfrac{x^2 + 2x + 1}{x^2 + 3x + 2}$

43. $\dfrac{(q - 3)^4(q + 2)}{q^2 + 3q + 2} \div \dfrac{q^2 - 6q + 9}{q^2 + 4q + 4}$

44. $\dfrac{(x + 4)^3(x - 3)}{x^2 - 9} \div \dfrac{x^2 + 8x + 16}{x^2 + 6x + 9}$

In working each exercise, remember how grouping symbols are used (Section 1.2), how to factor sums and differences of cubes (Section 5.4), and how to factor by grouping (Section 5.1).

45. $\dfrac{x + 5}{x + 10} \div \left(\dfrac{x^2 + 10x + 25}{x^2 + 10x} \cdot \dfrac{10x}{x^2 + 15x + 50} \right)$

46. $\dfrac{m - 8}{m - 4} \div \left(\dfrac{m^2 - 12m + 32}{8m} \cdot \dfrac{m^2 - 8m}{m^2 - 8m + 16} \right)$

47. $\dfrac{3a - 3b - a^2 + b^2}{4a^2 - 4ab + b^2} \cdot \dfrac{4a^2 - b^2}{2a^2 - ab - b^2}$

48. $\dfrac{4r^2 - t^2 + 10r - 5t}{2r^2 + rt + 5r} \cdot \dfrac{4r^3 + 4r^2t + rt^2}{2r + t}$

49. $\dfrac{-x^3 - y^3}{x^2 - 2xy + y^2} \div \dfrac{3y^2 - 3xy}{x^2 - y^2}$

50. $\dfrac{b^3 - 8a^3}{4a^3 + 4a^2b + ab^2} \div \dfrac{4a^2 + 2ab + b^2}{-a^3 - ab^3}$

51. If the rational expression $\dfrac{5x^2y^3}{2pq}$ represents the area of a rectangle and $\dfrac{2xy}{p}$ represents the length, what rational expression represents the width?

	Width

Length = $\dfrac{2xy}{p}$

The area is $\dfrac{5x^2y^3}{2pq}$.

52. If you are given the problem $\dfrac{4y + 12}{2y - 10} \div \dfrac{?}{y^2 - y - 20} = \dfrac{2(y + 4)}{y - 3}$, what must be the polynomial that is represented by the question mark?

6.3 The Least Common Denominator

OBJECTIVE **1** Find the least common denominator for a group of fractions. Just as with common fractions, adding or subtracting rational expressions (to be discussed in the next section) often requires finding a **least common denominator (LCD),** the least expression that all denominators divide into without a remainder. For example, the least common denominator for $\frac{2}{9}$ and $\frac{5}{12}$ is 36, since 36 is the smallest positive number that both 9 and 12 divide into evenly.

Least common denominators often can be found by inspection. For example, the LCD for $\frac{1}{6}$ and $\frac{2}{3m}$ is $6m$. In other cases, the LCD can be found by a procedure similar to that used in Chapter 5 for finding the greatest common factor.

Finding the Least Common Denominator

Step 1 **Factor.** Factor each denominator into prime factors.

Step 2 **List the factors.** List each different denominator factor the *greatest* number of times it appears in any of the denominators.

Step 3 **Multiply.** Multiply the denominator factors from Step 2 to get the LCD.

When each denominator is factored into prime factors, every prime factor must divide evenly into the LCD.

In Example 1, the LCD is found both for numerical denominators and algebraic denominators.

EXAMPLE 1 Finding the Least Common Denominator

Find the LCD for each pair of fractions.

(a) $\dfrac{1}{24}, \dfrac{7}{15}$ **(b)** $\dfrac{1}{8x}, \dfrac{3}{10x}$

Step 1 Write each denominator in factored form with numerical coefficients in prime factored form.

$$24 = 2 \cdot 2 \cdot 2 \cdot 3 \qquad\qquad 8x = 2 \cdot 2 \cdot 2 \cdot x$$
$$= 2^3 \cdot 3 \qquad\qquad\qquad = 2^3 \cdot x$$
$$15 = 3 \cdot 5 \qquad\qquad\qquad 10x = 2 \cdot 5 \cdot x$$

Step 2 We find the LCD by taking each different factor the *greatest* number of times it appears as a factor in any of the denominators.

The factor 2 appears three times in one product and not at all in the other, so the greatest number of times 2 appears is three. The greatest number of times both 3 and 5 appear is one.   Here 2 appears three times in one product and once in the other, so the greatest number of times 2 appears is three. The greatest number of times 5 appears is one, and the greatest number of times x appears in either product is one.

Step 3 $\text{LCD} = 2 \cdot 2 \cdot 2 \cdot 3 \cdot 5$ $\text{LCD} = 2 \cdot 2 \cdot 2 \cdot 5 \cdot x$

$$= 2^3 \cdot 3 \cdot 5 \qquad\qquad\qquad = 2^3 \cdot 5 \cdot x$$
$$= 120 \qquad\qquad\qquad\qquad = 40x$$

E X A M P L E   2   Finding the LCD

Find the LCD for $\dfrac{5}{6r^2}$ and $\dfrac{3}{4r^3}$.

Step 1   Factor each denominator.

$$6r^2 = 2 \cdot 3 \cdot r^2$$
$$4r^3 = 2 \cdot 2 \cdot r^3 = 2^2 \cdot r^3$$

Step 2   The greatest number of times 2 appears is two, the greatest number of times 3 appears is one, and the greatest number of times r appears is three; therefore,

Step 3 $$\text{LCD} = 2^2 \cdot 3 \cdot r^3 = 12r^3.$$

E X A M P L E   3   Finding the LCD

Find the LCD.

(a) $\dfrac{6}{5m}, \dfrac{4}{m^2 - 3m}$

Factor each denominator.

$$5m = 5 \cdot m$$
$$m^2 - 3m = m(m - 3)$$

Take each different factor the greatest number of times it appears as a factor.

$$\text{LCD} = 5 \cdot m \cdot (m - 3) = 5m(m - 3)$$

Since m is not a *factor* of $m - 3$, both factors, m and $m - 3$, must appear in the LCD.

(b) $\dfrac{1}{r^2 - 4r - 5}, \dfrac{1}{r^2 - r - 20}, \dfrac{1}{r^2 - 10r + 25}$

Factor each denominator.

$$r^2 - 4r - 5 = (r - 5)(r + 1)$$
$$r^2 - r - 20 = (r - 5)(r + 4)$$
$$r^2 - 10r + 25 = (r - 5)^2$$

The LCD is $(r - 5)^2(r + 1)(r + 4)$.

(c) $\dfrac{1}{q - 5}, \dfrac{3}{5 - q}$

The expression $5 - q$ can be written as $-1(q - 5)$, since

$$-1(q - 5) = -q + 5 = 5 - q.$$

Because of this, either $q - 5$ or $5 - q$ can be used as the LCD.

OBJECTIVE 2   **Rewrite rational expressions with given denominators.** Once we find the LCD, we can use the fundamental property to write equivalent rational expressions with this LCD. The next example shows how to do this with both numerical and algebraic fractions.

EXAMPLE 4 Writing a Fraction with a Given Denominator

Rewrite each expression with the indicated denominator.

(a) $\dfrac{3}{8} = \dfrac{}{40}$

(b) $\dfrac{9k}{25} = \dfrac{}{50k}$

For each example, first factor the denominator on the right. Then compare the denominator on the left with the one on the right to decide what factors are missing.

$$\frac{3}{8} = \frac{}{5 \cdot 8}$$

A factor of 5 is missing. Multiply by $\frac{5}{5}$ to get a denominator of 40.

$$\frac{3}{8} = \frac{3}{8} \cdot \frac{5}{5} = \frac{15}{40}$$
$$\downarrow$$
$$\tfrac{5}{5} = 1$$

$$\frac{9k}{25} = \frac{}{25 \cdot 2k}$$

Factors of 2 and k are missing. Get a denominator of $50k$ by multiplying by $\frac{2k}{2k}$.

$$\frac{9k}{25} = \frac{9k}{25} \cdot \frac{2k}{2k} = \frac{18k^2}{50k}$$
$$\downarrow$$
$$\tfrac{2k}{2k} = 1$$

Notice the use of the multiplicative identity property in each part of this example.

EXAMPLE 5 Writing a Fraction with a Given Denominator

Rewrite the following rational expression with the indicated denominator.

$$\frac{12p}{p^2 + 8p} = \frac{}{p^3 + 4p^2 - 32p}$$

Factor $p^2 + 8p$ as $p(p + 8)$. Compare with the denominator on the right which factors as $p(p + 8)(p - 4)$. The factor $p - 4$ is missing, so multiply $\dfrac{12p}{p(p + 8)}$ by $\dfrac{p - 4}{p - 4}$.

$$\frac{12p}{p^2 + 8p} = \frac{12p}{p(p + 8)} \cdot \frac{p - 4}{p - 4} \qquad \text{Multiplicative identity property}$$

$$= \frac{12p(p - 4)}{p(p + 8)(p - 4)} \qquad \text{Multiplication of rational expressions}$$

$$= \frac{12p^2 - 48p}{p^3 + 4p^2 - 32p} \qquad \text{Multiply the factors.}$$

 In the next section we add and subtract rational expressions, which sometimes requires the steps illustrated in Examples 4 and 5. While it is beneficial to leave the denominator in factored form, we multiplied the factors in the denominator in Example 5 to give the answer in the same form as the original problem.

6.3 EXERCISES

Choose the correct response in Exercises 1–4.

1. Suppose that the greatest common factor of a and b is 1. Then the least common denominator for $\dfrac{1}{a}$ and $\dfrac{1}{b}$ is

  (a) a.   **(b)** b.   **(c)** ab.   **(d)** 1.

2. If a is a factor of b, then the least common denominator for $\dfrac{1}{a}$ and $\dfrac{1}{b}$ is

  (a) a.   **(b)** b.   **(c)** ab.   **(d)** 1.

3. The least common denominator for $\dfrac{11}{20}$ and $\dfrac{1}{2}$ is

  (a) 40.   **(b)** 2.   **(c)** 20.   **(d)** none of these.

4. Suppose that we wish to write the fraction $\dfrac{1}{(x-4)^2(y-3)}$ with denominator $(x-4)^3(y-3)^2$. We must multiply both the numerator and the denominator by

  (a) $(x-4)(y-3)$.   **(b)** $(x-4)^2$.   **(c)** $x-4$.   **(d)** $(x-4)^2(y-3)$.

Find the LCD for the fractions in each list. See Examples 1–3.

EXERCISES

5. $\dfrac{-7}{15},\dfrac{21}{20}$ **6.** $\dfrac{9}{10},\dfrac{12}{25}$ **7.** $\dfrac{17}{100},\dfrac{23}{120},\dfrac{43}{180}$ **8.** $\dfrac{17}{250},\dfrac{-21}{300},\dfrac{127}{360}$

9. $\dfrac{9}{x^2},\dfrac{8}{x^5}$ **10.** $\dfrac{12}{m^7},\dfrac{13}{m^8}$ **11.** $\dfrac{-2}{5p},\dfrac{15}{6p}$ **12.** $\dfrac{14}{15k},\dfrac{9}{4k}$

13. $\dfrac{17}{15y^2},\dfrac{55}{36y^4}$ **14.** $\dfrac{4}{25m^3},\dfrac{-9}{10m^4}$ **15.** $\dfrac{13}{5a^2b^3},\dfrac{29}{15a^5b}$ **16.** $\dfrac{-7}{3r^4s^5},\dfrac{-22}{9r^6s^8}$

17. $\dfrac{7}{6p},\dfrac{15}{4p-8}$ **18.** $\dfrac{7}{8k},\dfrac{-23}{12k-24}$

RELATING CONCEPTS (EXERCISES 19–22)

Suppose we want to find the LCD for the two common fractions

$$\frac{1}{24} \quad \text{and} \quad \frac{1}{20}.$$

In their prime factored forms, the denominators are

$$24 = 2^3 \cdot 3$$

and

$$20 = 2^2 \cdot 5.$$

*Refer to this information as necessary and **work Exercises 19–22 in order.***

19. What is the prime factored form of the LCD of the two fractions?

20. Suppose that two algebraic fractions have denominators $(t+4)^3(t-3)$ and $(t+4)^2(t+8)$. What is the factored form of the LCD of these?

21. What is the similarity between your answers in Exercises 19 and 20?

22. Comment on the following statement: The method for finding the LCD for two algebraic fractions is the same as the method for finding the LCD for two common fractions.

Did you make the connection between finding the LCD for common fractions and for algebraic fractions?

Find the LCD for the fractions in each list. See Examples 1–3.

23. $\dfrac{37}{6r - 12}, \dfrac{25}{9r - 18}$

24. $\dfrac{-14}{5p - 30}, \dfrac{5}{6p - 36}$

25. $\dfrac{5}{12p + 60}, \dfrac{17}{p^2 + 5p}, \dfrac{16}{p^2 + 10p + 25}$

26. $\dfrac{13}{r^2 + 7r}, \dfrac{-3}{5r + 35}, \dfrac{-7}{r^2 + 14r + 49}$

27. $\dfrac{3}{8y + 16}, \dfrac{22}{y^2 + 3y + 2}$

28. $\dfrac{-2}{9m - 18}, \dfrac{-9}{m^2 - 7m + 10}$

29. $\dfrac{12}{m - 3}, \dfrac{-4}{3 - m}$

30. $\dfrac{-17}{8 - a}, \dfrac{2}{a - 8}$

31. $\dfrac{29}{p - q}, \dfrac{18}{q - p}$

32. $\dfrac{16}{z - x}, \dfrac{8}{x - z}$

33. $\dfrac{6}{a^2 + 6a}, \dfrac{-5}{a^2 + 3a - 18}$

34. $\dfrac{8}{y^2 - 5y}, \dfrac{-2}{y^2 - 2y - 15}$

35. $\dfrac{-5}{k^2 + 2k - 35}, \dfrac{-8}{k^2 + 3k - 40}, \dfrac{9}{k^2 - 2k - 15}$

36. $\dfrac{19}{z^2 + 4z - 12}, \dfrac{16}{z^2 + z - 30}, \dfrac{6}{z^2 + 2z - 24}$

37. Suppose that $(2x - 5)^2$ is the LCD for two fractions. Is $(5 - 2x)^2$ also acceptable as an LCD? Why?

38. Suppose that $(4t - 3)(5t - 6)$ is the LCD for two fractions. Is $(3 - 4t)(6 - 5t)$ also acceptable as an LCD? Why?

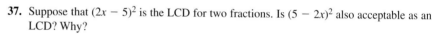

RELATING CONCEPTS (EXERCISES 39–44)

Work Exercises 39–44 in order.

39. Suppose that you want to write $\dfrac{3}{4}$ as an equivalent fraction with denominator 28. By what number must you multiply both the numerator and the denominator?

40. If you write $\dfrac{3}{4}$ as an equivalent fraction with denominator 28, by what number are you actually multiplying the fraction?

41. What property of multiplication is being used when we write a common fraction as an equivalent one with a larger denominator? (See Section 1.7.)

42. Suppose that you want to write $\dfrac{2x + 5}{x - 4}$ as an equivalent fraction with denominator $7x - 28$. By what number must you multiply both the numerator and the denominator?

43. If you write $\dfrac{2x + 5}{x - 4}$ as an equivalent fraction with denominator $7x - 28$, by what number are you actually multiplying the fraction?

44. Repeat Exercise 41, changing "a common" to "an algebraic."

Did you make the connection between writing common fractions with larger denominators and writing algebraic fractions with more denominator factors?

Write each rational expression on the left with the indicated denominator. See Examples 4 and 5.

45. $\dfrac{15m^2}{8k} = \dfrac{}{32k^4}$

46. $\dfrac{5t^2}{3y} = \dfrac{}{9y^2}$

47. $\dfrac{19z}{2z - 6} = \dfrac{}{6z - 18}$

48. $\dfrac{2r}{5r - 5} = \dfrac{}{15r - 15}$

49. $\dfrac{-2a}{9a - 18} = \dfrac{}{18a - 36}$

50. $\dfrac{-5y}{6y + 18} = \dfrac{}{24y + 72}$

51. $\dfrac{6}{k^2 - 4k} = \dfrac{}{k(k - 4)(k + 1)}$

52. $\dfrac{15}{m^2 - 9m} = \dfrac{}{m(m - 9)(m + 8)}$

53. $\dfrac{36r}{r^2 - r - 6} = \dfrac{}{(r - 3)(r + 2)(r + 1)}$

54. $\dfrac{4m}{m^2 - 8m + 15} = \dfrac{}{(m - 5)(m - 3)(m + 2)}$

55. $\dfrac{a + 2b}{2a^2 + ab - b^2} = \dfrac{}{2a^3b + a^2b^2 - ab^3}$

56. $\dfrac{m - 4}{6m^2 + 7m - 3} = \dfrac{}{12m^3 + 14m^2 - 6m}$

57. $\dfrac{4r - t}{r^2 + rt + t^2} = \dfrac{}{t^3 - r^3}$

58. $\dfrac{3x - 1}{x^2 + 2x + 4} = \dfrac{}{x^3 - 8}$

59. $\dfrac{2(z - y)}{y^2 + yz + z^2} = \dfrac{}{y^4 - z^3y}$

60. $\dfrac{2p + 3q}{p^2 + 2pq + q^2} = \dfrac{}{(p + q)(p^3 + q^3)}$

61. Write an explanation of how to find the least common denominator for a group of denominators. Give an example.

62. Write an explanation of how to write a rational expression as an equivalent rational expression with a given denominator. Give an example.

6.4 Addition and Subtraction of Rational Expressions

OBJECTIVES

1 Add rational expressions having the same denominator.

2 Add rational expressions having different denominators.

3 Subtract rational expressions.

To add and subtract rational expressions, we use our previous work on finding least common denominators and writing fractions with the LCD.

OBJECTIVE 1   Add rational expressions having the same denominator.   We find the sum of two rational expressions with a procedure similar to the one used for adding two fractions.

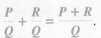

Adding Rational Expressions

If $\dfrac{P}{Q}$ and $\dfrac{R}{Q}$ are rational expressions, then

$$\frac{P}{Q} + \frac{R}{Q} = \frac{P + R}{Q}.$$

Again, the first example shows how addition of rational expressions compares with that of rational numbers.

┌ **E X A M P L E 1** Adding Rational Expressions with the Same Denominator

Add.

(a) $\dfrac{4}{7} + \dfrac{2}{7}$ **(b)** $\dfrac{3x}{x+1} + \dfrac{2x}{x+1}$

The denominators are the same, so the sum is found by adding the two numerators and keeping the same (common) denominator.

$$\frac{4}{7} + \frac{2}{7} = \frac{4+2}{7}$$

$$= \frac{6}{7}$$

$$\frac{3x}{x+1} + \frac{2x}{x+1} = \frac{3x+2x}{x+1}$$

$$= \frac{5x}{x+1}$$

OBJECTIVE 2 Add rational expressions having different denominators. Use the steps given below to add two rational expressions with different denominators. These are the same steps used to add fractions with different denominators.

Adding with Different Denominators

Step 1 **Find the LCD.** Find the least common denominator (LCD).

Step 2 **Rewrite fractions.** Rewrite each rational expression as an equivalent fraction with the LCD as the denominator.

Step 3 **Add.** Add the numerators to get the numerator of the sum. The LCD is the denominator of the sum.

Step 4 **Write in lowest terms.** Use the fundamental property to write the answer in lowest terms.

┌ **E X A M P L E 2** Adding Rational Expressions with Different Denominators

Add.

(a) $\dfrac{1}{12} + \dfrac{7}{15}$ **(b)** $\dfrac{2}{3y} + \dfrac{1}{4y}$

Step 1 First find the LCD using the methods of the previous section.

$$\text{LCD} = 2^2 \cdot 3 \cdot 5 = 60 \qquad \mid \qquad \text{LCD} = 2^2 \cdot 3 \cdot y = 12y$$

Step 2 Now rewrite each rational expression as a fraction with the LCD (either 60 or 12y) as the denominator.

$$\frac{1}{12} + \frac{7}{15} = \frac{1(5)}{12(5)} + \frac{7(4)}{15(4)}$$

$$= \frac{5}{60} + \frac{28}{60}$$

$$\frac{2}{3y} + \frac{1}{4y} = \frac{2(4)}{3y(4)} + \frac{1(3)}{4y(3)}$$

$$= \frac{8}{12y} + \frac{3}{12y}$$

Step 3   Since the fractions now have common denominators, add the numerators.

Step 4   Write in lowest terms if necessary.

$$\frac{5}{60} + \frac{28}{60} = \frac{5 + 28}{60}$$

$$= \frac{33}{60} = \frac{11}{20}$$

$$\frac{8}{12y} + \frac{3}{12y} = \frac{8 + 3}{12y}$$

$$= \frac{11}{12y}$$

E X A M P L E   3   Adding Rational Expressions

Add and write the sum in lowest terms.

$$\frac{2x}{x^2 - 1} + \frac{-1}{x + 1}$$

Step 1   Since the denominators are different, find the LCD.

$$x^2 - 1 = (x + 1)(x - 1)$$

$$x + 1 \text{ is prime.}$$

The LCD is $(x + 1)(x - 1)$.

Step 2   Rewrite each rational expression as a fraction with common denominator $(x + 1)(x - 1)$.

$$\frac{2x}{x^2 - 1} + \frac{-1}{x + 1} = \frac{2x}{(x + 1)(x - 1)} + \frac{-1(x - 1)}{(x + 1)(x - 1)} \qquad \text{Multiply second fraction by } \frac{x - 1}{x - 1}.$$

$$= \frac{2x}{(x + 1)(x - 1)} + \frac{-x + 1}{(x + 1)(x - 1)} \qquad \text{Distributive property}$$

Step 3
$$= \frac{2x - x + 1}{(x + 1)(x - 1)} \qquad \text{Add numerators; keep the same denominator.}$$

$$= \frac{x + 1}{(x + 1)(x - 1)} \qquad \text{Combine like terms in the numerator.}$$

Step 4
$$= \frac{1(x + 1)}{(x + 1)(x - 1)} \qquad \text{Identity property for multiplication}$$

$$= \frac{1}{x - 1} \qquad \text{Fundamental property of rational expressions}$$

E X A M P L E   4   Adding Rational Expressions

Add and write the sum in lowest terms.

$$\frac{2x}{x^2 + 5x + 6} + \frac{x + 1}{x^2 + 2x - 3}$$

Begin by factoring the denominators completely.

$$\frac{2x}{(x + 2)(x + 3)} + \frac{x + 1}{(x + 3)(x - 1)}$$

The LCD is $(x + 2)(x + 3)(x - 1)$. Use the fundamental property of rational expressions to rewrite each fraction with the LCD.

$$\frac{2x}{(x + 2)(x + 3)} + \frac{x + 1}{(x + 3)(x - 1)}$$

$$= \frac{2x(x - 1)}{(x + 2)(x + 3)(x - 1)} + \frac{(x + 1)(x + 2)}{(x + 3)(x - 1)(x + 2)}$$

$$= \frac{2x(x - 1) + (x + 1)(x + 2)}{(x + 2)(x + 3)(x - 1)} \qquad \text{Add numerators; keep the same denominator.}$$

$$= \frac{2x^2 - 2x + x^2 + 3x + 2}{(x + 2)(x + 3)(x - 1)} \qquad \text{Distributive property}$$

$$= \frac{3x^2 + x + 2}{(x + 2)(x + 3)(x - 1)} \qquad \text{Combine terms.}$$

Since $3x^2 + x + 2$ cannot be factored, the rational expression cannot be simplified further. It is usually best to leave the denominator in factored form since it is then easier to identify common factors in the numerator and denominator.

Rational expressions to be added or subtracted may have denominators that are opposites of each other. The next example illustrates this.

EXAMPLE 5   Adding Rational Expressions with Denominators That Are Opposites
Add and write the sum in lowest terms.

$$\frac{y}{y - 2} + \frac{8}{2 - y}$$

To get a common denominator of $y - 2$, multiply the second expression by -1 in both the numerator and the denominator.

$$\frac{y}{y - 2} + \frac{8}{2 - y} = \frac{y}{y - 2} + \frac{8(-1)}{(2 - y)(-1)} \qquad \text{Fundamental property}$$

$$= \frac{y}{y - 2} + \frac{-8}{y - 2} \qquad \text{Distributive property}$$

$$= \frac{y - 8}{y - 2} \qquad \text{Add numerators; keep the same denominator.}$$

If we had chosen to use $2 - y$ as the common denominator, the final answer would be in the form $\frac{8 - y}{2 - y}$, which is equivalent to $\frac{y - 8}{y - 2}$.

OBJECTIVE 3 Subtract rational expressions. To subtract rational expressions, use the following rule.

Subtracting Rational Expressions

If $\frac{P}{Q}$ and $\frac{R}{Q}$ are rational expressions, then

$$\frac{P}{Q} - \frac{R}{Q} = \frac{P - R}{Q}.$$

┌ EXAMPLE 6 **Subtracting Rational Expressions**

Subtract and write the difference in lowest terms.

$$\frac{2m}{m-1} - \frac{2}{m-1}$$

By the definition of subtraction,

$$\frac{2m}{m-1} - \frac{2}{m-1} = \frac{2m-2}{m-1} \qquad \text{Subtract numerators; keep the same denominator.}$$

$$= \frac{2(m-1)}{m-1} \qquad \text{Factor the numerator.}$$

$$= 2. \qquad \text{Write in lowest terms.}$$

┌ EXAMPLE 7 **Subtracting Rational Expressions with Different Denominators**

Subtract and write the difference in lowest terms.

$$\frac{9}{x-2} - \frac{3}{x} = \frac{9x}{x(x-2)} - \frac{3(x-2)}{x(x-2)} \qquad \text{The LCD is } x(x-2).$$

$$= \frac{9x - 3(x-2)}{x(x-2)} \qquad \text{Subtract numerators; keep the same denominator.}$$

$$= \frac{9x - 3x + 6}{x(x-2)} \qquad \text{Distributive property}$$

$$= \frac{6x + 6}{x(x-2)} \qquad \text{Combine like terms in the numerator.}$$

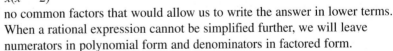

We could factor the final numerator in Example 7 to get an answer in the form $\frac{6(x+1)}{x(x-2)}$; however, the fundamental property would not apply, since there are no common factors that would allow us to write the answer in lower terms. When a rational expression cannot be simplified further, we will leave numerators in polynomial form and denominators in factored form.

┌ EXAMPLE 8 **Subtracting Rational Expressions with Denominators That Are Opposites**

Subtract and write the difference in lowest terms.

$$\frac{3x}{x-5} - \frac{2x-25}{5-x}$$

The denominators are opposites, so either may be used as the common denominator. We choose $x - 5$.

$$\frac{3x}{x - 5} - \frac{2x - 25}{5 - x} = \frac{3x}{x - 5} - \frac{(2x - 25)}{5 - x} \cdot \frac{-1}{-1} \qquad \text{Fundamental property}$$

$$= \frac{3x}{x - 5} - \frac{(-2x + 25)}{x - 5} \qquad \text{Multiply.}$$

$$= \frac{3x - (-2x + 25)}{x - 5} \qquad \text{Subtract numerators.}$$

$$= \frac{3x + 2x - 25}{x - 5} \qquad \text{Distributive property}$$

$$= \frac{5x - 25}{x - 5} \qquad \text{Combine terms.}$$

$$= \frac{5(x - 5)}{x - 5} \qquad \text{Factor.}$$

$$= 5 \qquad \text{Lowest terms}$$

 CAUTION Sign errors often occur in subtraction problems like the ones in Examples 7 and 8. Use parentheses after the subtraction sign to avoid this common error. Remember that the numerator of the fraction being subtracted must be treated as a single quantity.

EXAMPLE 9   Subtracting Rational Expressions

Subtract and write the difference in lowest terms.

$$\frac{6x}{x^2 - 2x + 1} - \frac{1}{x^2 - 1}$$

Begin by factoring.

$$\frac{6x}{x^2 - 2x + 1} - \frac{1}{x^2 - 1} = \frac{6x}{(x - 1)(x - 1)} - \frac{1}{(x - 1)(x + 1)}$$

From the factored denominators, we can now identify the common denominator, $(x - 1)(x - 1)(x + 1)$. Use the factor $x - 1$ twice, since it appears twice in the first denominator.

$$\frac{6x}{(x - 1)(x - 1)} - \frac{1}{(x - 1)(x + 1)}$$

$$= \frac{6x(x + 1)}{(x - 1)(x - 1)(x + 1)} - \frac{1(x - 1)}{(x - 1)(x - 1)(x + 1)} \qquad \text{Fundamental property}$$

$$= \frac{6x(x + 1) - 1(x - 1)}{(x - 1)(x - 1)(x + 1)} \qquad \text{Subtract.}$$

$$= \frac{6x^2 + 6x - x + 1}{(x - 1)(x - 1)(x + 1)} \qquad \text{Distributive property}$$

$$= \frac{6x^2 + 5x + 1}{(x - 1)(x - 1)(x + 1)} \qquad \text{Combine like terms.}$$

The result may be written as $\dfrac{6x^2 + 5x + 1}{(x - 1)^2(x + 1)}$.

When adding and subtracting rational expressions, several different equivalent forms of the answer often exist. If your answer does not look exactly like the one given in the back of the book, check to see if you have written an equivalent form.

6.4 EXERCISES

RELATING CONCEPTS (EXERCISES 1-8)

Work Exercises 1–8 in order. This will help you later in determining whether your answer is equivalent to the one given in the answer section if it doesn't match the form shown.

Jill worked a problem involving the sum of two rational expressions correctly. Her answer was given as $\dfrac{5}{x-3}$. Jack worked the same problem and gave his answer as $\dfrac{-5}{3-x}$. We want to decide whether Jack's answer was also correct.

1. Evaluate the fraction $\dfrac{5}{7-3}$ using the rule for order of operations.

2. Multiply the fraction given in Exercise 1 by -1 in both the numerator and the denominator, using the distributive property as necessary. Leave it in unsimplified form.

3. In Exercise 2, what number did you actually multiply the *fraction* by?

4. Simplify the result you found in Exercise 2 and compare it to the one found in Exercise 1. How do they compare?

5. Now look at the answers given by Jill and Jack above. Based on what you learned in Exercises 1–4, determine whether Jack's answer was also correct. Why or why not?

6. Jason Jordan, a perceptive algebra student, made the following comment and was praised by his teacher for his insight: "I can see that if I change the sign of each term in a fraction, the result I get is equivalent to the fraction that I started with." Explain why Jason's observation is correct.

7. Jennifer Crum, another perceptive algebra student, followed Jason's comment with one of her own: "I can see that if I put a negative sign in front of a fraction and change the signs of all terms in either the numerator or the denominator, but not both, then the result I get is equivalent to the fraction that I started with." The teacher knew that some real education was happening in the classroom. Explain why Jennifer's observation is also correct.

8. Use the concepts of Exercises 1–7 to determine whether each rational expression is equivalent or not equivalent to $\dfrac{y-4}{3-y}$.

   (a) $\dfrac{y-4}{y-3}$ **(b)** $\dfrac{-y+4}{-3+y}$ **(c)** $-\dfrac{y-4}{y-3}$

   (d) $-\dfrac{y-4}{3-y}$ **(e)** $-\dfrac{4-y}{y-3}$ **(f)** $\dfrac{y+4}{3+y}$

Did you make the connection that rational expressions can be written in equivalent forms?

Add or subtract. Write the answer in lowest terms. See Examples 1 and 6.

9. $\dfrac{4}{m} + \dfrac{7}{m}$

10. $\dfrac{5}{p} + \dfrac{11}{p}$

11. $\dfrac{a+b}{2} - \dfrac{a-b}{2}$

12. $\dfrac{x-y}{2} - \dfrac{x+y}{2}$

13. $\dfrac{x^2}{x+5} + \dfrac{5x}{x+5}$

14. $\dfrac{t^2}{t-3} + \dfrac{-3t}{t-3}$

15. $\dfrac{y^2-3y}{y+3} + \dfrac{-18}{y+3}$

16. $\dfrac{r^2-8r}{r-5} + \dfrac{15}{r-5}$

17. Explain with an example how to add or subtract rational expressions with the same denominators.

18. Explain with an example how to add or subtract rational expressions with different denominators.

Add or subtract. Write the answer in lowest terms. See Examples 2, 3, 4, and 7.

19. $\dfrac{z}{5} + \dfrac{1}{3}$

20. $\dfrac{p}{8} + \dfrac{3}{5}$

21. $\dfrac{5}{7} - \dfrac{r}{2}$

22. $\dfrac{10}{9} - \dfrac{z}{3}$

23. $-\dfrac{3}{4} - \dfrac{1}{2x}$

24. $-\dfrac{5}{8} - \dfrac{3}{2a}$

25. $\dfrac{x+1}{6} + \dfrac{3x+3}{9}$

26. $\dfrac{2x-6}{4} + \dfrac{x+5}{6}$

27. $\dfrac{x+3}{3x} + \dfrac{2x+2}{4x}$

28. $\dfrac{x+2}{5x} + \dfrac{6x+3}{3x}$

29. $\dfrac{7}{3p^2} - \dfrac{2}{p}$

30. $\dfrac{12}{5m^2} - \dfrac{5}{m}$

31. $\dfrac{x}{x-2} + \dfrac{4}{x+2} - \dfrac{8}{x^2-4}$

32. $\dfrac{2x}{x-1} + \dfrac{3}{x+1} - \dfrac{4}{x^2-1}$

33. $\dfrac{t}{t+2} + \dfrac{5-t}{t} - \dfrac{4}{t^2+2t}$

34. $\dfrac{2p}{p-3} + \dfrac{2+p}{p} - \dfrac{-6}{p^2-3p}$

35. What are the two possible LCDs that could be used for the sum

$$\dfrac{10}{m-2} + \dfrac{5}{2-m}?$$

36. If one form of the correct answer to a sum or difference of rational expressions is $\dfrac{4}{k-3}$, what would an alternate form of the answer be if the denominator is $3-k$?

Add or subtract. Write the answer in lowest terms. See Examples 5 and 8.

37. $\dfrac{4}{x-5} + \dfrac{6}{5-x}$

38. $\dfrac{10}{m-2} + \dfrac{5}{2-m}$

39. $\dfrac{-1}{1-y} + \dfrac{3-4y}{y-1}$

40. $\dfrac{-4}{p-3} - \dfrac{p+1}{3-p}$

41. $\dfrac{2}{x-y^2} + \dfrac{7}{y^2-x}$

42. $\dfrac{-8}{p-q^2} + \dfrac{3}{q^2-p}$

43. $\dfrac{x}{5x-3y} - \dfrac{y}{3y-5x}$

44. $\dfrac{t}{8t-9s} - \dfrac{s}{9s-8t}$

45. $\dfrac{3}{4p-5} + \dfrac{9}{5-4p}$

46. $\dfrac{8}{3-7y} - \dfrac{2}{7y-3}$

In these subtraction problems, the rational expression that follows the subtraction sign has a numerator with more than one term. Be very careful with signs and find each difference.

47. $\dfrac{2m}{m-n} - \dfrac{5m+n}{2m-2n}$

48. $\dfrac{5p}{p-q} - \dfrac{3p+1}{4p-4q}$

49. $\dfrac{5}{x^2-9} - \dfrac{x+2}{x^2+4x+3}$

50. $\dfrac{1}{a^2-1} - \dfrac{a-1}{a^2+3a-4}$

51. $\dfrac{2q + 1}{3q^2 + 10q - 8} - \dfrac{3q + 5}{2q^2 + 5q - 12}$

52. $\dfrac{4y - 1}{2y^2 + 5y - 3} - \dfrac{y + 3}{6y^2 + y - 2}$

Perform the indicated operations. See Examples 1–9.

53. $\dfrac{4}{r^2 - r} + \dfrac{6}{r^2 + 2r} - \dfrac{1}{r^2 + r - 2}$

54. $\dfrac{6}{k^2 + 3k} - \dfrac{1}{k^2 - k} + \dfrac{2}{k^2 + 2k - 3}$

55. $\dfrac{x + 3y}{x^2 + 2xy + y^2} + \dfrac{x - y}{x^2 + 4xy + 3y^2}$

56. $\dfrac{m}{m^2 - 1} + \dfrac{m - 1}{m^2 + 2m + 1}$

57. $\dfrac{r + y}{18r^2 + 12ry - 3ry - 2y^2} + \dfrac{3r - y}{36r^2 - y^2}$

58. $\dfrac{2x - z}{2x^2 - 4xz + 5xz - 10z^2} - \dfrac{x + z}{x^2 - 4z^2}$

Perform the indicated operations. Remember the order of operations.

59. $\left(\dfrac{-k}{2k^2 - 5k - 3} + \dfrac{3k - 2}{2k^2 - k - 1}\right)\dfrac{2k + 1}{k - 1}$

60. $\left(\dfrac{3p + 1}{2p^2 + p - 6} - \dfrac{5p}{3p^2 - p}\right)\dfrac{2p - 3}{p + 2}$

61. $\dfrac{k^2 + 4k + 16}{k + 4}\left(\dfrac{-5}{16 - k^2} + \dfrac{2k + 3}{k^3 - 64}\right)$

62. $\dfrac{m - 5}{2m + 5}\left(\dfrac{-3m}{m^2 - 25} - \dfrac{m + 4}{125 - m^3}\right)$

63. Refer to the rectangle in the figure.
  (a) Find an expression that represents its perimeter. Give the simplified form.
  (b) Find an expression that represents its area. Give the simplified form.

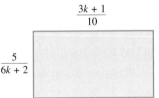

$\dfrac{3k + 1}{10}$

$\dfrac{5}{6k + 2}$

64. Refer to the triangle in the figure. Find an expression that represents its perimeter.

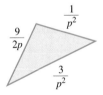

$\dfrac{1}{p^2}$

$\dfrac{9}{2p}$

$\dfrac{3}{p^2}$

6.5 Complex Fractions

OBJECTIVES

1 Simplify a complex fraction by writing it as a division problem (Method 1).

2 Simplify a complex fraction by multiplying by the least common denominator (Method 2).

The quotient of two mixed numbers in arithmetic, such as $2\frac{1}{2} \div 3\frac{1}{4}$, can be written as a fraction:

$$2\frac{1}{2} \div 3\frac{1}{4} = \frac{2\frac{1}{2}}{3\frac{1}{4}} = \frac{2 + \frac{1}{2}}{3 + \frac{1}{4}}.$$

The last expression is the quotient of expressions that involve fractions. In algebra, some rational expressions also have fractions in the numerator, or denominator, or both.

Complex Fraction

A rational expression with fractions in the numerator, denominator, or both, is called a **complex fraction.**

Examples of complex fractions include

$$\frac{2 + \dfrac{1}{2}}{3 + \dfrac{1}{4}}, \qquad \frac{\dfrac{3x^2 - 5x}{6x^2}}{2x - \dfrac{1}{x}}, \qquad \text{and} \qquad \frac{3 + x}{5 - \dfrac{2}{x}}.$$

The parts of a complex fraction are named as follows.

$$\frac{\dfrac{2}{p} - \dfrac{1}{q}}{\dfrac{3}{p} + \dfrac{5}{q}}$$

← Numerator of complex fraction
← Main fraction bar
← Denominator of complex fraction

OBJECTIVE **1** Simplify a complex fraction by writing it as a division problem (Method 1). Since the main fraction bar represents division in a complex fraction, one method of simplifying a complex fraction involves division.

Method 1

To simplify a complex fraction:

Step 1 Write both the numerator and denominator as single fractions.

Step 2 Change the complex fraction to a division problem.

Step 3 Perform the indicated division.

Once again, in this section the first example shows complex fractions from both arithmetic and algebra.

EXAMPLE 1 Simplifying Complex Fractions by Method 1

Simplify each complex fraction.

(a) $\dfrac{\dfrac{2}{3} + \dfrac{5}{9}}{\dfrac{1}{4} + \dfrac{1}{12}}$

(b) $\dfrac{6 + \dfrac{3}{x}}{\dfrac{x}{4} + \dfrac{1}{8}}$

Step 1 First, write each numerator as a single fraction.

$$\frac{2}{3} + \frac{5}{9} = \frac{2(3)}{3(3)} + \frac{5}{9} \qquad\qquad 6 + \frac{3}{x} = \frac{6}{1} + \frac{3}{x}$$

$$= \frac{6}{9} + \frac{5}{9} = \frac{11}{9} \qquad\qquad = \frac{6x}{x} + \frac{3}{x} = \frac{6x + 3}{x}$$

Do the same thing with each denominator.

$$\frac{1}{4} + \frac{1}{12} = \frac{1(3)}{4(3)} + \frac{1}{12} \qquad\qquad \frac{x}{4} + \frac{1}{8} = \frac{x(2)}{4(2)} + \frac{1}{8}$$

$$= \frac{3}{12} + \frac{1}{12} = \frac{4}{12} \qquad\qquad = \frac{2x}{8} + \frac{1}{8} = \frac{2x + 1}{8}$$

Step 2   The original complex fraction can now be written as follows.

$$\frac{\dfrac{11}{9}}{\dfrac{4}{12}} \qquad\qquad\qquad \frac{\dfrac{6x+3}{x}}{\dfrac{2x+1}{8}}$$

Step 3   Now use the rule for division and the fundamental property.

$$\frac{11}{9} \div \frac{4}{12} = \frac{11}{9} \cdot \frac{12}{4} \qquad\qquad \frac{6x+3}{x} \div \frac{2x+1}{8} = \frac{6x+3}{x} \cdot \frac{8}{2x+1}$$

$$= \frac{11 \cdot 3 \cdot 4}{3 \cdot 3 \cdot 4} \qquad\qquad\qquad = \frac{3(2x+1)}{x} \cdot \frac{8}{2x+1}$$

$$= \frac{11}{3} \qquad\qquad\qquad\qquad = \frac{24}{x}$$

┌ **E X A M P L E 2**   Simplifying a Complex Fraction by Method 1

Simplify the complex fraction.

$$\frac{\dfrac{xp}{q^3}}{\dfrac{p^2}{qx^2}}$$

Here the numerator and denominator are already single fractions, so use the division rule and then the fundamental property.

$$\frac{xp}{q^3} \div \frac{p^2}{qx^2} = \frac{xp}{q^3} \cdot \frac{qx^2}{p^2} = \frac{x^3}{q^2 p}$$

┌ **E X A M P L E 3**   Simplifying a Complex Fraction by Method 1

Simplify the complex fraction.

$$\frac{\dfrac{3}{x+2} - 4}{\dfrac{2}{x+2} + 1} = \frac{\dfrac{3}{x+2} - \dfrac{4(x+2)}{x+2}}{\dfrac{2}{x+2} + \dfrac{1(x+2)}{x+2}} \qquad \text{Write both second terms with a denominator of } x+2.$$

$$= \frac{\dfrac{3 - 4(x+2)}{x+2}}{\dfrac{2 + 1(x+2)}{x+2}} \qquad \text{Subtract in the numerator. Add in the denominator.}$$

$$= \frac{\dfrac{3 - 4x - 8}{x+2}}{\dfrac{2 + x + 2}{x+2}} \qquad \text{Distributive property}$$

$$\frac{\dfrac{3-4x-8}{x+2}}{\dfrac{2+x+2}{x+2}} = \frac{\dfrac{-5-4x}{x+2}}{\dfrac{4+x}{x+2}} \qquad \text{Combine terms.}$$

$$= \frac{-5-4x}{x+2} \cdot \frac{x+2}{4+x} \qquad \text{Multiply by the reciprocal.}$$

$$= \frac{-5-4x}{4+x} \qquad \text{Lowest terms}$$

OBJECTIVE 2 Simplify a complex fraction by multiplying by the least common denominator (Method 2). As an alternative method, a complex fraction may be simplified by a method that uses the fundamental property of rational expressions. Since any expression can be multiplied by a form of 1 to get an equivalent expression, we may multiply both the numerator and the denominator of a complex fraction by the same nonzero expression to get an equivalent complex fraction. If we choose the expression to be the LCD of all the fractions within the complex fraction, the complex fraction will be simplified. This is Method 2.

Method 2

To simplify a complex fraction:

Step 1 Find the LCD of all fractions within the complex fraction.

Step 2 Multiply both the numerator and the denominator of the complex fraction by this LCD using the distributive property as necessary. Write in lowest terms.

In the next example, Method 2 is used to simplify the complex fractions from Example 1.

EXAMPLE 4 Simplifying Complex Fractions by Method 2

Simplify each complex fraction.

(a) $\dfrac{\dfrac{2}{3}+\dfrac{5}{9}}{\dfrac{1}{4}+\dfrac{1}{12}}$

(b) $\dfrac{6+\dfrac{3}{x}}{\dfrac{x}{4}+\dfrac{1}{8}}$

Step 1 Find the LCD for all denominators in the complex fraction.

The LCD for 3, 9, 4, and 12 is 36. | The LCD for x, 4, and 8 is $8x$.

Step 2 Multiply numerator and denominator of the complex fraction by the LCD.

$$\frac{\dfrac{2}{3}+\dfrac{5}{9}}{\dfrac{1}{4}+\dfrac{1}{12}} = \frac{36\left(\dfrac{2}{3}+\dfrac{5}{9}\right)}{36\left(\dfrac{1}{4}+\dfrac{1}{12}\right)} \qquad \left| \qquad \frac{6+\dfrac{3}{x}}{\dfrac{x}{4}+\dfrac{1}{8}} = \frac{8x\left(6+\dfrac{3}{x}\right)}{8x\left(\dfrac{x}{4}+\dfrac{1}{8}\right)}$$

$$= \frac{36\left(\frac{2}{3}\right) + 36\left(\frac{5}{9}\right)}{36\left(\frac{1}{4}\right) + 36\left(\frac{1}{12}\right)}$$

$$= \frac{8x(6) + 8x\left(\frac{3}{x}\right)}{8x\left(\frac{x}{4}\right) + 8x\left(\frac{1}{8}\right)} \qquad \text{Distributive property}$$

$$= \frac{24 + 20}{9 + 3}$$

$$= \frac{48x + 24}{2x^2 + x}$$

$$= \frac{44}{12} = \frac{4 \cdot 11}{4 \cdot 3}$$

$$= \frac{24(2x + 1)}{x(2x + 1)} \qquad \text{Factor.}$$

$$= \frac{11}{3}$$

$$= \frac{24}{x} \qquad \text{Lowest terms}$$

Simplifying a Complex Fraction by Method 2

Simplify the complex fraction.

$$\frac{\dfrac{3}{5m} - \dfrac{2}{m^2}}{\dfrac{9}{2m} + \dfrac{3}{4m^2}}$$

The LCD for $5m$, m^2, $2m$, and $4m^2$ is $20m^2$. Multiply numerator and denominator by $20m^2$.

$$\frac{\dfrac{3}{5m} - \dfrac{2}{m^2}}{\dfrac{9}{2m} + \dfrac{3}{4m^2}} = \frac{20m^2\left(\dfrac{3}{5m} - \dfrac{2}{m^2}\right)}{20m^2\left(\dfrac{9}{2m} + \dfrac{3}{4m^2}\right)}$$

$$= \frac{20m^2\left(\dfrac{3}{5m}\right) - 20m^2\left(\dfrac{2}{m^2}\right)}{20m^2\left(\dfrac{9}{2m}\right) + 20m^2\left(\dfrac{3}{4m^2}\right)} \qquad \text{Distributive property}$$

$$= \frac{12m - 40}{90m + 15}$$

Either of the two methods shown in this section can be used to simplify a complex fraction. You may want to choose one method and stick with it to eliminate confusion. However, some students prefer to use Method 1 for problems like Example 2, which is the quotient of two fractions. They prefer Method 2 for problems like Examples 1, 3, 4, and 5, which have sums or differences in the numerators or denominators or both.

CONNECTIONS

Some numbers can be expressed as *continued fractions,* which are infinite complex fractions. For example, the irrational number $\sqrt{2}$ can be expressed as follows.

$$\sqrt{2} = 1 + \cfrac{1}{2 + \cfrac{1}{2 + \cfrac{1}{2 + \cfrac{1}{2 + \cdots}}}}$$

Better and better approximations of $\sqrt{2}$ can be found by using more and more terms of the fraction. We give the first three approximations here.

$$1 + \frac{1}{2} = 1.5$$

$$1 + \cfrac{1}{2 + \cfrac{1}{2}} = 1 + \cfrac{1}{\frac{5}{2}} = 1 + \frac{2}{5} = \frac{7}{5} = 1.4$$

$$1 + \cfrac{1}{2 + \cfrac{1}{2 + \cfrac{1}{2}}} = 1 + \cfrac{1}{2 + \cfrac{1}{\frac{5}{2}}} = 1 + \cfrac{1}{2 + \frac{2}{5}} = 1 + \cfrac{1}{\frac{12}{5}}$$

$$= 1 + \frac{5}{12} = \frac{17}{12} = 1.41\overline{6}$$

A calculator gives $\sqrt{2} \approx 1.414213562$ to nine decimal places.

FOR DISCUSSION OR WRITING

Give the next approximation of $\sqrt{2}$ and compare it to the calculator value shown above. How many places after the decimal agree?

6.5 EXERCISES

Note: In many problems involving complex fractions, several different equivalent forms of the answer exist. If your answer does not look exactly like the one given in the back of the book, check to see if your answer is an equivalent form.

1. Consider the complex fraction $\dfrac{\dfrac{1}{2} - \dfrac{1}{3}}{\dfrac{5}{6} - \dfrac{1}{12}}$. Answer each of the following, outlining Method 1 for simplifying this complex fraction.

(a) To combine the terms in the numerator, we must find the LCD of $\dfrac{1}{2}$ and $\dfrac{1}{3}$. What is this LCD? Determine the simplified form of the numerator of the complex fraction.

(b) To combine the terms in the denominator, we must find the LCD of $\dfrac{5}{6}$ and $\dfrac{1}{12}$. What is this LCD? Determine the simplified form of the denominator of the complex fraction.

CONCEPTS	EXAMPLES
Step 3 Write the equation. The sum of the fractional parts should equal 1 (whole job).	The equation is $$\frac{1}{6}x + \frac{1}{8}x = 1.$$
Step 4 Solve the equation.	The solution of the equation is $\frac{24}{7}$.
Steps 5 and 6 Answer the question; check the solution.	It would take them $\frac{24}{7}$ or $3\frac{3}{7}$ hours to cover the route together.

Solving Inverse Variation Problems
1. Write the variation equation $y = \frac{k}{x}$.

If a varies inversely as b, and $a = 4$ when $b = 4$, find a when $b = 6$. The equation for inverse variation is $a = \frac{k}{b}$.

2. Find k by substituting the given values of x and y into the equation.

Substitute $a = 4$ and $b = 4$.
$$4 = \frac{k}{4}$$
$$k = 16$$

3. Write the equation with the value of k from Step 2 and the given value of x or y. Solve for the remaining variable.

Let $k = 16$ and $b = 6$ in the variation equation.
$$a = \frac{16}{6} = \frac{8}{3}$$

CHAPTER 6 REVIEW EXERCISES

[6.1] *Find any values of the variable for which the rational expression is undefined.*

1. $\dfrac{4}{x-3}$ **2.** $\dfrac{y+3}{2y}$ **3.** $\dfrac{2k+1}{3k^2+17k+10}$

4. How would you determine the values of the variable for which a rational expression is undefined?

Find the numerical value of each rational expression when (a) $x = -2$ and (b) $x = 4$.

5. $\dfrac{4x-3}{5x+2}$ **6.** $\dfrac{3x}{x^2-4}$

Write each rational expression in lowest terms.

7. $\dfrac{5a^3b^3}{15a^4b^2}$ **8.** $\dfrac{m-4}{4-m}$

9. $\dfrac{4x^2-9}{6-4x}$ **10.** $\dfrac{4p^2+8pq-5q^2}{10p^2-3pq-q^2}$

Write four equivalent expressions for each of the following.

11. $-\dfrac{4x-9}{2x+3}$ **12.** $\dfrac{8-3x}{3+6x}$

[6.2] *Find each product or quotient and write the answer in lowest terms.*

13. $\dfrac{18p^3}{6} \cdot \dfrac{24}{p^4}$ **14.** $\dfrac{8x^2}{12x^5} \cdot \dfrac{6x^4}{2x}$

15. $\dfrac{x-3}{4} \cdot \dfrac{5}{2x-6}$ **16.** $\dfrac{2r+3}{r-4} \cdot \dfrac{r^2-16}{6r+9}$

17. $\dfrac{6a^2+7a-3}{2a^2-a-6} \div \dfrac{a+5}{a-2}$ **18.** $\dfrac{y^2-6y+8}{y^2+3y-18} \div \dfrac{y-4}{y+6}$

19. $\dfrac{2p^2 + 13p + 20}{p^2 + p - 12} \cdot \dfrac{p^2 + 2p - 15}{2p^2 + 7p + 5}$

20. $\dfrac{3z^2 + 5z - 2}{9z^2 - 1} \cdot \dfrac{9z^2 + 6z + 1}{z^2 + 5z + 6}$

[6.3]   *Find the least common denominator for each list of fractions.*

21. $\dfrac{4}{9y}, \dfrac{7}{12y^2}, \dfrac{5}{27y^4}$

22. $\dfrac{3}{x^2 + 4x + 3}, \dfrac{5}{x^2 + 5x + 4}$

Rewrite each rational expression with the given denominator.

23. $\dfrac{3}{2a^3} = \dfrac{}{10a^4}$

24. $\dfrac{9}{x - 3} = \dfrac{}{18 - 6x}$

25. $\dfrac{-3y}{2y - 10} = \dfrac{}{50 - 10y}$

26. $\dfrac{4b}{b^2 + 2b - 3} = \dfrac{}{(b + 3)(b - 1)(b + 2)}$

[6.4]   *Add or subtract and write each answer in lowest terms.*

27. $\dfrac{10}{x} + \dfrac{5}{x}$

28. $\dfrac{6}{3p} - \dfrac{12}{3p}$

29. $\dfrac{9}{k} - \dfrac{5}{k - 5}$

30. $\dfrac{4}{y} + \dfrac{7}{7 + y}$

31. $\dfrac{m}{3} - \dfrac{2 + 5m}{6}$

32. $\dfrac{12}{x^2} - \dfrac{3}{4x}$

33. $\dfrac{5}{a - 2b} + \dfrac{2}{a + 2b}$

34. $\dfrac{4}{k^2 - 9} - \dfrac{k + 3}{3k - 9}$

35. $\dfrac{8}{z^2 + 6z} - \dfrac{3}{z^2 + 4z - 12}$

36. $\dfrac{11}{2p - p^2} - \dfrac{2}{p^2 - 5p + 6}$

[6.5]

37. Simplify the complex fraction $\dfrac{\dfrac{a^4}{b^2}}{\dfrac{a^3}{b}}$ by

    (a) Method 1 as described in Section 6.5.

    (b) Method 2 as described in Section 6.5.

    (c) Explain which method you prefer, and why.

Simplify each complex fraction.

38. $\dfrac{\dfrac{2}{3} - \dfrac{1}{6}}{\dfrac{1}{4} + \dfrac{2}{5}}$

39. $\dfrac{\dfrac{y - 3}{y}}{\dfrac{y + 3}{4y}}$

40. $\dfrac{\dfrac{1}{p} - \dfrac{1}{q}}{\dfrac{1}{q - p}}$

41. $\dfrac{x + \dfrac{1}{w}}{x - \dfrac{1}{w}}$

42. $\dfrac{\dfrac{1}{r + t} - 1}{\dfrac{1}{r + t} + 1}$

[6.6]

43. Before even beginning the solution process, how do you know that 2 cannot be a solution to the equation found in Exercise 46 below?

Solve each equation and check your solutions.

44. $\dfrac{4 - z}{z} + \dfrac{3}{2} = \dfrac{-4}{z}$

45. $\dfrac{3y - 1}{y - 2} = \dfrac{5}{y - 2} + 1$

46. $\dfrac{3}{m - 2} + \dfrac{1}{m - 1} = \dfrac{7}{m^2 - 3m + 2}$

Solve each formula for the specified variable.

47. $m = \dfrac{Ry}{t}$ for t **48.** $x = \dfrac{3y - 5}{4}$ for y **49.** $p^2 = \dfrac{4}{3m - q}$ for m

[6.7] *Solve each problem. Use the six-step method.*

50. In a certain fraction, the denominator is 4 less than the numerator. If 3 is added to both the numerator and the denominator, the resulting fraction is equal to $\dfrac{3}{2}$. Find the original fraction.

51. The denominator of a certain fraction is 3 times the numerator. If 2 is added to the numerator and subtracted from the denominator, the resulting fraction is equal to 1. Find the original fraction.

52. On August 18, 1996, Scott Sharp won the True Value 200-mile Indy race driving a Ford with an average speed of 130.934 miles per hour. What was his time? (*Source: Sports Illustrated 1998 Sports Almanac.*)

53. A man can plant his garden in 5 hours, working alone. His daughter can do the same job in 8 hours. How long would it take them if they worked together?

54. The head gardener can mow the lawns in the city park twice as fast as his assistant. Working together, they can complete the job in $1\dfrac{1}{3}$ hours. How long would it take the head gardener working alone?

55. The longer the term of your subscription to *Monitoring Times,* the less you will have to pay per year. Is this an example of direct or inverse variation?

56. In 1994, 38,505 American deaths were due to firearms. Of this total, the approximate ratio of male deaths to female deaths was 63 to 10. How many males and how many females died as a result of using firearms? (*Source:* National Center for Health Statistics, Monthly Vital Statistics Reports.)

57. If a parallelogram has a fixed area, the height varies inversely as the base. A parallelogram has a height of 8 centimeters and a base of 12 centimeters. Find the height if the base is changed to 24 centimeters.

58. At a given hour, two steamboats leave a city in the same direction on a straight canal. One travels at 18 miles per hour, and the other travels at 25 miles per hour. In how many hours will the boats be 35 miles apart?

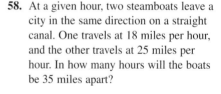

RELATING CONCEPTS (EXERCISES 59–68)

In these exercises, we summarize the various concepts involving rational expressions we have covered.

Work Exercises 59–68 in order.

Let *P*, *Q*, and *R* be rational expressions defined as follows.

$$P = \frac{6}{x + 3} \qquad Q = \frac{5}{x + 1} \qquad R = \frac{4x}{x^2 + 4x + 3}$$

59. Find the value or values for which the expression is undefined.
  (a) *P*   **(b)** *Q*   **(c)** *R*

> **RELATING CONCEPTS (EXERCISES 59-68) (CONTINUED)**
>
> **60.** Find and express in lowest terms: $(P \cdot Q) \div R$.
>
> **61.** Why is $(P \cdot Q) \div R$ not defined if $x = 0$?
>
> **62.** Find the LCD for P, Q, and R.
>
> **63.** Perform the operations and express in lowest terms: $P + Q - R$.
>
> **64.** Simplify the complex fraction $\dfrac{P + Q}{R}$.
>
> **65.** Solve the equation $P + Q = R$.
>
> **66.** How does your answer to Exercise 59 help you work Exercise 65?
>
> **67.** Suppose that a car travels 6 miles in $x + 3$ minutes. Explain why P represents the rate of the car (in miles per minute).
>
> **68.** For what value or values of x is $R = \dfrac{40}{77}$?
>
> Did you make the connections between the various operations with rational expressions?

MIXED REVIEW EXERCISES

Perform the indicated operations.

69. $\dfrac{\dfrac{5}{x - y} + 2}{3 - \dfrac{2}{x + y}}$

70. $\dfrac{4}{m - 1} - \dfrac{3}{m + 1}$

71. $\dfrac{8p^5}{5} \div \dfrac{2p^3}{10}$

72. $\dfrac{r - 3}{8} \div \dfrac{3r - 9}{4}$

73. $\dfrac{\dfrac{5}{x} - 1}{\dfrac{5 - x}{3x}}$

74. $\dfrac{4}{z^2 - 2z + 1} - \dfrac{3}{z^2 - 1}$

Solve.

75. $\dfrac{1}{k} + \dfrac{3}{r} = \dfrac{5}{z}$ for r

76. $\dfrac{5 + m}{m} + \dfrac{3}{4} = \dfrac{-2}{m}$

77. In 1996, the total number of people waiting for heart and liver transplants was 11,165. The ratio of those waiting for heart transplants to those waiting for liver transplants was approximately 2 to 1. How many of each type of candidate were there? (*Sources:* U.S. Department of Health and Human Services, Public Health Service, Division of Organ Transplantation, and United Network for Organ Sharing.)

78. Emily Falzon flew her plane 400 kilometers with the wind in the same time it took her to go 200 kilometers against the wind. The speed of the wind is 50 kilometers per hour. Find the speed of the plane in still air.

79. If x varies directly as y, and $x = 12$ when $y = 5$, find x when $y = 3$.

80. When Mario and Luigi work together on a job, they can do it in $3\dfrac{3}{7}$ days. Mario can do the job working alone in 8 days. How long would it take Luigi working alone?

47. Use the first and last data pairs in the table to write an equation relating the data. (*Hint:* See Section 7.1, Example 5.)

48. Use your equation from Exercise 47 to approximate the number of active-duty female military personnel in 1993 and 1994.

49. Use the second and third data pairs to write an equation relating the data.

50. Use your equation from Exercise 49 to approximate the number of active-duty female military personnel in 1992 and 1995. Which of the equations in Exercises 47 and 49 gives better approximations?

TECHNOLOGY INSIGHTS (EXERCISES 51-57)

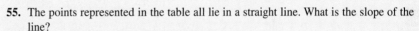

The table was generated by a graphing calculator. The expression Y_1 *represents* $f(x)$.

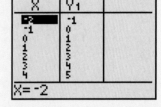

X	Y1
-2	-1
-1	0
0	1
1	2
2	3
3	4
4	5

X= -2

51. When $x = 3$, Y = _____ ?

52. What is $f(3)$?

53. When Y = 2, $x =$ _____ ?

54. If $f(x) = 2$, what is the value of x?

55. The points represented in the table all lie in a straight line. What is the slope of the line?

56. What is the y-intercept of the line?

57. Write the function in the form $y = mx + b$, for the appropriate values of m and b.

Not every function can be readily expressed as an equation. Here is an example of such a function from everyday life.

58. A chain-saw rental firm charges $7 per day or fraction of a day to rent a saw, plus a fixed fee of $4 for resharpening the blade. Let y represent the cost of renting a saw for x days. A portion of the graph is shown here.

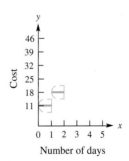

Number of days

(a) Give the domain and range of this relation.
(b) Explain how the graph can be continued.
(c) Explain how the vertical line test applies here.
(d) What do the numbers in the domain represent here? What do the numbers in the range represent?

59. Define *relation* and *function*. Compare the two definitions. How are they alike? How are they different?

60. In your own words, explain the meaning of domain and range of a function.

CHAPTER 7 GROUP ACTIVITY

▦ Taxes as Functions

Objective: Use data to determine if a relation is a function, determine domain and range, and graph the function.

Different forms of taxation fund governments. There are various types of taxes, including income, property, estate, and sales taxes. Taxes are determined in different ways. This activity looks at some of these taxes and asks you to determine if they can be modeled by linear functions. Take turns in your group graphing and answering the questions below.

A. The following table is for sales tax in Sierra County, NM.

x = Cost of Item	y = Sales Tax
$ 4.00	$ .23
$ 8.00	$ .46
$24.00	$1.38

1. Plot the points on a graph. (Determine the scale for the domain and range of your graph.)

2. Using the criteria for functions, determine if this is a function.

3. Is this a linear function?

4. What is the domain? What is the range?

5. Write the equation of the line that fits the data. Use $f(x)$ notation if applicable. (*Hint:* Start by finding the slope of the line.)

B. The table below is for income tax for a family of four in the U.S. in 1994.

x = Income	y = Tax
$10,000	−$2527
$20,000	−$358
$30,000	$2078
$40,000	$3578
$50,000	$5078

Source: 1995 Information
Please Almanac.

Complete parts (1)–(4) above for this tax.

(continued)

C. In the 1996 presidential election, one candidate proposed a flat income tax. For this activity, consider a flat tax rate of 10% of a given taxpayer's income.

   1. Write the function that would represent this tax.

   2. Using the incomes given in the table in B, determine the flat tax for each income. Make your own table.

   3. Plot these points.

   4. How does the flat tax compare to the 1994 taxes?

   5. Who might oppose/support a flat tax?

CHAPTER 7 SUMMARY

KEY TERMS

7.2 linear inequality in two variables boundary line	**7.3** components relation domain range function

NEW SYMBOLS

$f(x)$ function f of x

TEST YOUR WORD POWER

See how well you have learned the vocabulary in this chapter. Answers, with examples, are given at the bottom of the page.

1. A **relation** is
(a) any set of ordered pairs
(b) a set of ordered pairs in which each first component corresponds to exactly one second component
(c) two sets of ordered pairs that are related
(d) a graph of ordered pairs.

2. The **domain** of a relation is
(a) the set of all x- and y-values in the ordered pairs of the relation
(b) the difference between the components in an ordered pair of the relation

(c) the set of all first components in the ordered pairs of the relation
(d) the set of all second components in the ordered pairs of the relation.

3. The **range** of a relation is
(a) the set of all x- and y-values in the ordered pairs of the relation
(b) the difference between the components in an ordered pair of the relation
(c) the set of all first components in the ordered pairs of the relation
(d) the set of all second components in the ordered pairs of the relation.

4. A **function** is
(a) any set of ordered pairs
(b) a set of ordered pairs in which each first component corresponds to exactly one second component
(c) two sets of ordered pairs that are related
(d) a graph of ordered pairs.

Answers to Test Your Word Power
1. (a) *Example:* {(0, 2), (2, 4), (3, 6), (−1, 3)}   **2.** (c) *Example:* The domain in the relation given above is the set of x-values, that is, {0, 2, 3, −1}.   **3.** (d) *Example:* The range of the relation given above is the set of y-values, that is, {2, 4, 6, 3}.   **4.** (b) *Example:* The relation given above is a function since each x-value corresponds to exactly one y-value.

QUICK REVIEW

| CONCEPTS | EXAMPLES |

7.1 EQUATIONS OF A LINE

Slope-Intercept Form

$y = mx + b$

m is the slope.

$(0, b)$ is the y-intercept.

Find an equation of the line with slope 2 and y-intercept $(0, -5)$.

The equation is $y = 2x - 5$.

Point-Slope Form

$y - y_1 = m(x - x_1)$

m is the slope.

(x_1, y_1) is a point on the line.

Find an equation of the line with slope $-\frac{1}{2}$ through $(-4, 5)$.

$$y - 5 = -\frac{1}{2}(x - (-4))$$

$$y - 5 = -\frac{1}{2}(x + 4)$$

$$y - 5 = -\frac{1}{2}x - 2$$

$$y = -\frac{1}{2}x + 3$$

Standard Form

$Ax + By = C$

A, B, and C are integers and $A > 0$, $B \neq 0$.

This equation is written in standard form as

$$x + 2y = 6,$$

with $A = 1$, $B = 2$, and $C = 6$.

7.2 GRAPHING LINEAR INEQUALITIES IN TWO VARIABLES

1. Graph the line that is the boundary of the region. Make it solid if the inequality is $\leq$ or $\geq$; make it dashed if the inequality is $<$ or $>$.

2. Use any point not on the line as a test point. Substitute for x and y in the inequality. If the result is true, shade the side of the line containing the test point; if the result is false, shade the other side.

Graph $2x + y \leq 5$.

Graph the line $2x + y = 5$.

Make it solid because of $\leq$.

Use $(1, 0)$ as a test point.

$$2(1) + 0 \leq 5 \qquad ?$$
$$2 \leq 5 \qquad \text{True}$$

Shade the side of the line containing $(1, 0)$.

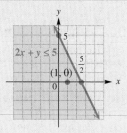

7.3 FUNCTIONS

Vertical Line Test

If a vertical line intersects a graph in more than one point, the graph is not the graph of a function.

The graph shown is not the graph of a function.

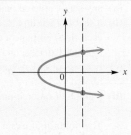

The **domain** of a function is the set of numbers that can replace x in the expression for the function. The **range** is the set of y-values that result as x is replaced by each number in the domain. To find $f(x)$ for a specific value of x, replace x by that value in the expression for the function.

Let $f(x) = (x - 1)^2$. Find

(a) the domain of f;

(b) the range of f;

(c) $f(-1)$.

(a) The domain is $(-\infty, \infty)$.

(b) The range is all y such that $y \geq 0$ or $[0, \infty)$, because y equals the square of $x - 1$, so y must be nonnegative.

(c) $f(-1) = (-1 - 1)^2 = (-2)^2 = 4$

CHAPTER 7 REVIEW EXERCISES

[7.1]   *Write an equation for each line in the form y = mx + b, if possible.*

1. $m = -1, b = \dfrac{2}{3}$

2. Through $(2, 3)$ and $(-4, 6)$

3. Through $(4, -3)$, $m = 1$

4. Through $(-1, 4)$, $m = \dfrac{2}{3}$

5. Through $(1, -1)$, $m = -\dfrac{3}{4}$

6. $m = -\dfrac{1}{4}, b = \dfrac{3}{2}$

7. Slope 0, through $(-4, 1)$

8. Through $\left(\dfrac{1}{3}, -\dfrac{5}{4}\right)$ with undefined slope

[7.2]   *Complete the graph of each linear inequality by shading the correct region.*

9. $x - y \geq 3$

10. $3x - y \leq 5$

11. $x + 2y < 6$

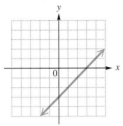
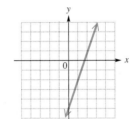
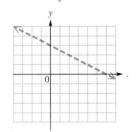

Graph each linear inequality.

12. $3x + 5y > 9$

13. $2x - 3y > -6$

14. $x - 2y \geq 0$

[7.3]   *Decide whether each relation is or is not a function. In Exercises 15 and 16, give the domain and the range.*

15. $\{(-2, 4), (0, 8), (2, 5), (2, 3)\}$

16. $\{(8, 3), (7, 4), (6, 5), (5, 6), (4, 7)\}$

17.

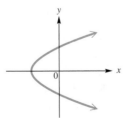

18.

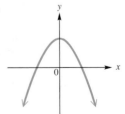

19. $2x + 3y = 12$

20. $y = x^2$

21. $x - 5y < 10$

Find the domain and range of the function.

22. $4x - 3y = 12$

23. $y = x^2 + 1$

24. $y = |x - 1|$

*Find **(a)** f(2) and **(b)** f(-1).*

25. $f(x) = 3x + 2$

26. $f(x) = 2x^2 - 1$

27. $f(x) = |x + 3|$

28. The net profits (in billions of dollars) for Coca Cola have increased in recent years as shown in the graph.

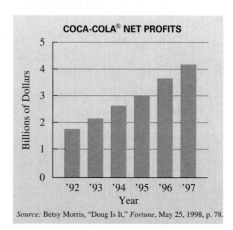

COCA-COLA® NET PROFITS

Source: Betsy Morris, "Doug Is It," *Fortune*, May 25, 1998, p. 78.

The tops of the bars appear to lie in a linear pattern. The linear equation $y = f(x) = .48x - 42.46$ gives a good approximation of the line through the centers of the tops of the bars.

(a) Use the equation to find $f(98)$ and interpret the result in terms of the application. (*Hint:* $x = 98$ represents the year 1998 and would be shown in the graph as '98.)

(b) Use the equation to find the slope of the line.

(c) Discuss the relationship between the slope of the line and the trend in net profits for Coca Cola.

MIXED REVIEW EXERCISES

Write an equation in slope-intercept form for each line described in Exercises 29–32.

29. Through $(6, -1)$ and $(5, -1)$

30. $m = -\dfrac{1}{4}; b = -\dfrac{5}{4}$

31. Through $(8, 6)$; $m = -3$

32. Through $(3, -5)$ and $(-4, -1)$

Graph each inequality.

33. $x - 2y \le 6$

34. $y < -4x$

35. $x \ge -4$

36. Find $f(2)$ if $f(x) = (3 - 2x)^2$.

37. What kind of inequality has a dashed line as its boundary? Explain why.

RELATING CONCEPTS (EXERCISES 38-44)

*Use the concepts of this chapter and Chapter 3 in **working Exercises 38–44 in order.***

38. Plot the points $(-2, 4)$ and $(3, -11)$ and draw the line joining them.

39. Find the slope of the line in Exercise 38.

40. Find the slope-intercept form of the equation of the line in Exercise 38.

41. What are the x- and y-intercepts of this line?

42. Suppose that $y = mx + b$ is the equation of the line. Graph the inequality $y > mx + b$.

43. Replace y with $f(x)$ in your answer to Exercise 40. Find $f(10)$.

44. Give the domain and the range of the function described in Exercise 43.

Did you make the connection between the equations of a line, the graphs of linear equations and inequalities, and functions?

45. The judicial branch of the U.S. government had expenditures (in billions of dollars) for the years 1994–1997 as shown in the graph. Use the graph to answer the following questions.

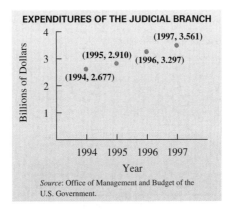

EXPENDITURES OF THE JUDICIAL BRANCH

(1997, 3.561)
(1995, 2.910)
(1996, 3.297)
(1994, 2.677)

Billions of Dollars

1994 1995 1996 1997
Year

Source: Office of Management and Budget of the U.S. Government.

(a) Let function *f* be defined by the four ordered pairs in the graph. What is $f(1995)$? If $f(x) = 3.297$, what is *x*?

(b) What are the domain and range of *f*?

(c) The points in the graph indicate that a linear equation would approximate the data quite well. Let $x = 0$ represent 1994, $x = 1$ represent 1995, and so on. Write a linear equation for *f* using the points for 1995 and 1997. Would you expect to get a slightly different equation if you used a different pair of points?

(d) Use the equation from part (c) to determine $f(1996)$. Is the result close to the actual value given in the graph?

46. The graph shows the expenditures (in billions of dollars) of the legislative branch of the U.S. government. Use the graph to respond to the following.

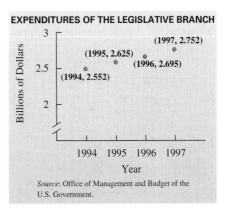

EXPENDITURES OF THE LEGISLATIVE BRANCH

(1997, 2.752)
(1995, 2.625)
(1996, 2.695)
(1994, 2.552)

Billions of Dollars

1994 1995 1996 1997
Year

Source: Office of Management and Budget of the U.S. Government.

(a) Is this the graph of a function? Could it be approximated reasonably by a linear function?

(b) Give the domain and range.

(c) Write a linear equation that approximates the given data points, using the years 1994 and 1996. Let $x = 0$ represent 1994, $x = 1$ represent 1995, and so on.

(d) Use the equation from part (c) to predict legislative branch expenditures in 1998.

CHAPTER 7 TEST

Write an equation for each line in slope-intercept form.

1. Through $(-1, 4)$ with slope 2

2.

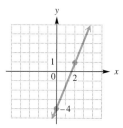

3.

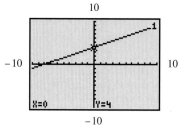   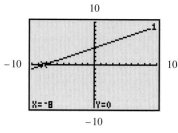

(These are two different views of the same line.)

4. Write the linear equation $y = -\dfrac{2}{3}x + 5$ in standard form.

Graph the linear inequality.

5. $x + y \leq 3$

6. $3x - y \geq 0$

7. If $f(x) = -4x + 8$, find $f(-3)$.

Decide whether each of the following represents a function. If it does, give the domain and the range.

8.

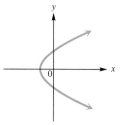

9. $y = x^2 + 2$

10. $y = 2x + 1$

11. $\{(0, 2), (0, -2), (1, 5), (2, 10)\}$

12. Why does the graph of $y = mx + b$, for real numbers m and b, satisfy the conditions for the graph of a function?

The table gives government data on consumer spending (in billions of dollars) during the years 1990 through 1996.

Personal Consumption

Year	Expenditures (in billions of dollars)
1990	3839
1991	3975
1992	4220
1993	4459
1994	4717
1995	4958
1996	5208

Source: U.S. Commerce Department, Bureau of Economic Analysis.

13. Does this set of data define a function f? If so, what is $f(1993)$?

14. If the set of data defines a function f and $f(x) = 5208$, what is x?

15. The data points are closely approximated by the line $y = 234.6x - 463,161$. What is the slope of this line?

16. What does the slope of the line tell you about the annual increase in consumer expenditures?

CUMULATIVE REVIEW EXERCISES CHAPTERS 1–7

Solve each equation.

1. $-5(8 - 2z) + 4(7 - z) = 7(8 + z) - 3$

2. $A = p + prt$ for t

3. $7x^2 + 8x + 1 = 0$

4. $\dfrac{4}{x - 2} = 4$

5. $\dfrac{2}{x - 1} = \dfrac{5}{x - 1} - \dfrac{3}{4}$

Solve each inequality and graph the solution set.

6. $-2.5x < 6.5$

7. $4(x + 3) - 5x < 12$

8. $\dfrac{2}{3}y - \dfrac{1}{6}y \le -2$

Write with only positive exponents. Assume that all variables represent positive real numbers.

9. $(x^2y^{-3})(x^{-4}y^2)$

10. $\dfrac{x^{-6}y^3z^{-1}}{x^7y^{-4}z}$

11. $(2m^{-2}n^3)^{-3}$

Perform the indicated operations.

12. $2(3x^2 - 8x + 1) - 4(x^2 - 3x - 9)$

13. $(3x + 2y)(5x - y)$

14. $(x + 2y)(x^2 - 2xy + 4y^2)$

15. $\dfrac{m^3 - 3m^2 + 5m - 3}{m - 1}$

Factor each polynomial completely.

16. $y^2 + 4yk - 12k^2$

17. $9x^4 - 25y^2$

18. $125x^4 - 400x^3y + 195x^2y^2$

19. $f^2 + 20f + 100$

20. $100x^2 + 49$

Perform each indicated operation. Express the answer in lowest terms.

21. $\dfrac{3}{2x + 6} + \dfrac{2x + 3}{2x + 6}$

22. $\dfrac{8}{x + 1} - \dfrac{2}{x + 3}$

23. $\dfrac{x^2 - 25}{3x + 6} \cdot \dfrac{4x + 8}{x^2 + 10x + 25}$

24. $\dfrac{x^2 + 2x - 3}{x^2 - 5x + 4} \cdot \dfrac{x^2 - 3x - 4}{x^2 + 3x}$

25. $\dfrac{x^2 + 5x + 6}{3x} \div \dfrac{x^2 - 4}{x^2 + x - 6}$

26. $\dfrac{6x^4y^3z^2}{8xyz^4} \div \dfrac{3x^2}{16y^2}$

Simplify each complex fraction.

27. $\dfrac{\dfrac{2}{3} - \dfrac{1}{4}}{\dfrac{1}{2} + \dfrac{1}{6}}$

28. $\dfrac{\dfrac{12}{x + 6}}{\dfrac{4}{2x + 12}}$

Find the slope of each line described.

29. Through $(-4, 5)$ and $(2, -3)$

30. Horizontal, through $(4, 5)$

Give the equation of each line described in the form $y = mx + b$.

31. Through $(4, -1)$, $m = -4$

32. Through $(0, 0)$ and $(1, 4)$

Graph each equation or inequality.

33. $-3x + 4y = 12$

34. $y \le 2x - 6$

35. $3x + 2y < 0$

Solve each problem.

36. In 1996, the U.S. government raised $484.6 billion more from individual income taxes than from corporate income taxes. The total amount raised from these two sources was $828.2 billion. How much was raised from each source? (*Source:* Office of Management and Budget.)

37. Find the measure of each angle of the triangle.

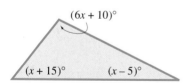

38. The length of the shorter leg of a right triangle is tripled and 4 inches is added to the result, giving the length of the hypotenuse. The longer leg is 10 inches longer than twice the shorter leg. Find the length of the shorter leg of the triangle.

39. If x varies directly as y and $x = 4$ when $y = 12$, find x when $y = 42$.

40. If a man can mow his lawn in 3 hours and his wife can do the same job in 1.5 hours, how long will it take them to do the job together?

Solve each system by any method.

34. $2x - y = -8$
$\quad\quad x + 2y = 11$

35. $4x + 5y = -8$
$\quad\quad 3x + 4y = -7$

36. $3x + 5y = 1$
$\quad\quad x = y + 3$

Use a system of equations to solve each problem.

37. Admission prices at a football game were $6 for adults and $2 for children. The total value of the tickets sold was $2528, and 454 tickets were sold. How many adults and how many children attended the game?

Kind of Ticket	Number Sold	Cost of Each (in dollars)	Total Value (in dollars)
Adult	x	6	$6x$
Child	y		
Total	454		

38. The perimeter of a triangle is 53 inches. If two sides are of equal length, and the third side measures 4 inches less than each of the equal sides, what are the lengths of the three sides?

39. The Smith family is coming to visit, and no one knows how many children they have. Janet, one of the girls, says she has as many brothers as sisters; her brother Steve says he has twice as many sisters as brothers. How many boys and how many girls are in the family?

40. Which one of the following systems of linear inequalities is graphed in the figure?

(a) $x \le 3$
$\quad\quad y \le 1$

(b) $x \le 3$
$\quad\quad y \ge 1$

(c) $x \ge 3$
$\quad\quad y \le 1$

(d) $x \ge 3$
$\quad\quad y \ge 1$

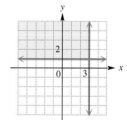

9 Roots and Radicals

The electronics industry encompasses producing, selling, and servicing the computers, TVs, VCRs, CD players, and cell phones that pervade our lives. The industry boomed for many years, but recently industry sales have leveled off. The graph, which gives the percent change in orders from year to year, shows the turbulence in the electronics industry from 1983 to 1998.*

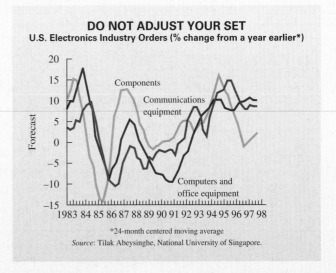

DO NOT ADJUST YOUR SET
U.S. Electronics Industry Orders (% change from a year earlier*)

*24-month centered moving average
Source: Tilak Abeysinghe, National University of Singapore.

A major reason for the downturn in electronics business is the increase in the number of companies, particularly Asian companies, that have entered the industry. This large increase in supply has caused a glut of chips, with resulting price declines. This, in turn, has contributed to the collapse of export growth throughout Eastern Asia. In the exercises for Section 9.3, we examine data on U.S. exports and imports of electronics.

*"The silicon tigers' electric shocker," *The Economist,* November 9, 1996.

9.1 Evaluating Roots

In Section 1.2 we discussed the idea of the *square* of a number. Recall that squaring a number means multiplying the number by itself.

$$\text{If } a = 7, \text{ then } a^2 = 7 \cdot 7 = 49.$$
$$\text{If } a = -5, \text{ then } a^2 = (-5) \cdot (-5) = 25.$$
$$\text{If } a = -\frac{1}{2}, \text{ then } a^2 = \left(-\frac{1}{2}\right) \cdot \left(-\frac{1}{2}\right) = \frac{1}{4}.$$

In this chapter the opposite problem is considered.

$$\text{If } a^2 = 49, \text{ then } a = \text{?}$$
$$\text{If } a^2 = 25, \text{ then } a = \text{?}$$
$$\text{If } a^2 = \frac{1}{4}, \text{ then } a = \text{?}$$

OBJECTIVE 1 Find square roots. To find a in the three statements above, we must find a number that when multiplied by itself results in the given number. The number a is called a **square root** of the number a^2.

EXAMPLE 1 Finding All Square Roots of a Number

Find all square roots of 49.

To find a square root of 49, think of a number that when multiplied by itself gives 49. One square root is 7, since $7 \cdot 7 = 49$. Another square root of 49 is -7, since $(-7)(-7) = 49$. The number 49 has two square roots, 7 and -7; one is positive and one is negative.

The positive square root of a number is written with the symbol $\sqrt{}$. For example, the positive square root of 49 is 7, written

$$\sqrt{49} = 7.$$

The symbol $-\sqrt{}$ is used for the negative square root of a number. For example, the negative square root of 49 is -7, written

$$-\sqrt{49} = -7.$$

Most calculators have a square root key, usually labeled $\boxed{\sqrt{x}}$, that allows us to find the square root of a number. For example, if we enter 49 and use the square root key, the display will show 7.

The symbol $\sqrt{}$ is called a **radical sign** and always represents the nonnegative square root. The number inside the radical sign is called the **radicand** and the entire expression, radical sign and radicand, is called a **radical.** An algebraic expression containing a radical is called a **radical expression.**

Square Roots of a

If a is a positive real number,

$$\sqrt{a} \text{ is the positive square root of } a,$$
$$-\sqrt{a} \text{ is the negative square root of } a.$$

For nonnegative a,

$$\sqrt{a} \cdot \sqrt{a} = (\sqrt{a})^2 = a \qquad \text{and} \qquad -\sqrt{a} \cdot -\sqrt{a} = (-\sqrt{a})^2 = a.$$

Also, $\sqrt{0} = 0$.

EXAMPLE 2 Finding Square Roots

Find each square root.

(a) $\sqrt{144}$

The radical $\sqrt{144}$ represents the positive square root of 144. Think of a positive number whose square is 144.

$$12^2 = 144, \qquad \text{so} \qquad \sqrt{144} = 12.$$

(b) $-\sqrt{1024}$

This symbol represents the negative square root of 1024. A calculator with a square root key can be used to find $\sqrt{1024} = 32$. Then, $-\sqrt{1024} = -32$.

(c) $\sqrt{\dfrac{4}{9}} = \dfrac{2}{3}$
                                                    (d) $-\sqrt{\dfrac{16}{49}} = -\dfrac{4}{7}$

As shown in the definition above, when the square root of a positive real number is squared, the result is that positive real number. (Also, $(\sqrt{0})^2 = 0$.)

EXAMPLE 3 Squaring Radical Expressions

Find the *square* of each radical expression.

(a) $\sqrt{13}$

$\qquad (\sqrt{13})^2 = 13$ Definition of square root

(b) $-\sqrt{29}$

$\qquad (-\sqrt{29})^2 = 29$ The square of a *negative* number is positive.

(c) $\sqrt{p^2 + 1}$

$\qquad (\sqrt{p^2 + 1})^2 = p^2 + 1$

OBJECTIVE 2 Decide whether a given root is rational, irrational, or not a real number. All numbers with square roots that are rational are called **perfect squares.** For example, 100 and $\frac{25}{4}$ are perfect squares. A number that is not a perfect square has a square root that is not a rational number. For example, $\sqrt{5}$ is not a rational number because it cannot be written as the ratio of two integers. Its decimal neither terminates nor repeats. However, $\sqrt{5}$ is a real number and corresponds to a point on the number line. As mentioned in Chapter 1, a real number that is not rational is called an *irrational number.* The number $\sqrt{5}$ is irrational. Many square roots of integers are irrational.

If a is a positive number that is not a perfect square, then $\sqrt{a}$ is irrational.

Not every number has a *real number* square root. For example, there is no real number that can be squared to get -36. (The square of a real number can never be negative.) Because of this $\sqrt{-36}$ is not a real number. A calculator may show an error message in a case like this.

If a is a negative number, then $\sqrt{a}$ is not a real number.

E X A M P L E   4   Identifying Types of Square Roots

Tell whether each square root is rational, irrational, or not a real number.

(a) $\sqrt{17}$

Since 17 is a not a perfect square, $\sqrt{17}$ is irrational.

(b) $\sqrt{64}$

The number 64 is a perfect square, 8^2, so $\sqrt{64} = 8$, a rational number.

(c) $\sqrt{85}$ is irrational.

(d) $\sqrt{81}$ is rational ($\sqrt{81} = 9$).

(e) $\sqrt{-25}$

There is no real number whose square is -25. Therefore $\sqrt{-25}$ is not a real number.

 Not all irrational numbers are square roots of integers. For example, π (approximately 3.14159) is an irrational number that is not a square root of any integer.

OBJECTIVE 3 Find decimal approximations for irrational square roots. Even if a number is irrational, a decimal that approximates the number can be found by using a calculator. For example, a calculator shows that $\sqrt{10}$ is 3.16227766, although this is only a rational number *approximation* of $\sqrt{10}$.

E X A M P L E   5   Approximating Irrational Square Roots

Find a decimal approximation for each square root. Round answers to the nearest thousandth.

(a) $\sqrt{11}$

Using the square root key of a calculator gives $\sqrt{11} \approx 3.31662479 \approx 3.317$ rounded to the nearest thousandth, where $\approx$ means "is approximately equal to."

(b) $\sqrt{39} \approx 6.245$

(c) $-\sqrt{745} \approx -27.295$

(d) $\sqrt{-180}$ is not a real number.

OBJECTIVE 4 Use the Pythagorean formula. One application of square roots uses the Pythagorean formula. Recall from Section 5.6 that by this formula, if c is the length of the hypotenuse of a right triangle, and a and b are the lengths of the two legs, then

$$c^2 = a^2 + b^2.$$

See Figure 1.

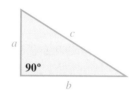

Figure 1

VIDEO

E X A M P L E 6 Using the Pythagorean Formula

Find the unknown length of the third side of each right triangle with sides a, b, and c, where c is the hypotenuse.

(a) $a = 3, b = 4$

Use the formula to find c^2 first.

$$\begin{aligned} c^2 &= a^2 + \boldsymbol{b}^2 \\ &= 3^2 + \boldsymbol{4}^2 && \text{Let } a = 3 \text{ and } b = 4. \\ &= 9 + 16 = 25 && \text{Square and add.} \end{aligned}$$

Now find the positive square root of 25 to get c.

$$c = \sqrt{25} = 5$$

(Although -5 is also a square root of 25, the length of a side of a triangle must be a positive number.)

(b) $c = 9, b = 5$

Substitute the given values in the formula $c^2 = a^2 + b^2$. Then solve for a^2.

$$\begin{aligned} 9^2 &= a^2 + \boldsymbol{5}^2 && \text{Let } c = 9 \text{ and } b = 5. \\ 81 &= a^2 + 25 && \text{Square.} \\ 56 &= a^2 && \text{Subtract 25.} \end{aligned}$$

Again, we want only the positive root $a = \sqrt{56} \approx 7.483$.

Be careful not to make the common mistake of thinking that $\sqrt{a^2 + b^2}$ equals $a + b$. As Example 6(a) shows,

$$\sqrt{9 + 16} = \sqrt{25} = 5 \neq \sqrt{9} + \sqrt{16} = 3 + 4,$$

so that, in general,

$$\sqrt{a^2 + b^2} \neq a + b.$$

CONNECTIONS

Pythagoras did not actually discover the Pythagorean formula. While he may have written the first proof, there is evidence that the Babylonians knew the concept quite well. The figure on the left illustrates the formula by using a tile pattern. In the figure, the side of the square along the hypotenuse measures 5 units, while the sides along the legs measure 3 and 4 units. If we let $a = 3$, $b = 4$, and $c = 5$, the equation of the Pythagorean formula is satisfied.

$$a^2 + b^2 = c^2$$
$$3^2 + 4^2 = 5^2 \qquad ?$$
$$25 = 25 \qquad \text{True}$$

FOR DISCUSSION OR WRITING

The diagram on the right can be used to verify the Pythagorean formula. To do so, express the area of the figure in two ways: first, as the area of the large square, and then as the sum of the areas of the smaller square and the four right triangles. Finally, set the areas equal and simplify the equation.

PROBLEM SOLVING

The Pythagorean formula can be used to solve applied problems that involve right triangles. A good way to begin the solution is to sketch the triangle and label the three sides appropriately, using a variable as needed. Then use the Pythagorean formula to write an equation. This procedure is simply Steps 1–3 of the six-step problem-solving method given in Chapter 2, and used throughout the book. In Steps 4–6, we solve the equation, answer the question(s), and check the solution(s).

EXAMPLE 7   Solving an Application

A ladder 10 feet long leans against a wall. The foot of the ladder is 6 feet from the base of the wall. How high up the wall does the top of the ladder rest?

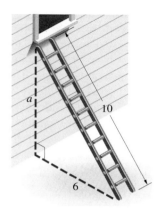

Figure 2

Steps 1 and 2   As shown in Figure 2, a right triangle is formed with the ladder as the hypotenuse. Let *a* represent the height of the top of the ladder.

Step 3   By the Pythagorean formula,

$$c^2 = a^2 + b^2.$$

$$10^2 = a^2 + 6^2 \qquad \text{Let } c = 10 \text{ and } b = 6.$$

$$100 = a^2 + 36 \qquad \text{Square.}$$

Step 4

$$64 = a^2 \qquad \text{Subtract 36.}$$

$$\sqrt{64} = a$$

$$a = 8 \qquad \sqrt{64} = 8$$

Step 5   Choose the positive square root of 64 since *a* represents a length.

Step 6   From Figure 2, we see that we must have

$$8^2 + 6^2 = 10^2 \qquad ?$$

$$64 + 36 = 100. \qquad \text{True}$$

The check shows that the top of the ladder rests 8 feet up the wall.

OBJECTIVE 5   Use the distance formula.   The Pythagorean formula is used to develop another useful formula for finding the distance between two points on the plane. Figure 3 shows the two different points (x_1, y_1) and (x_2, y_2), as well as the point (x_2, y_1). The distance between (x_2, y_2) and (x_2, y_1) is $a = y_2 - y_1$, and the distance between (x_1, y_1) and (x_2, y_1) is $b = x_2 - x_1$. From the Pythagorean formula,

$$d^2 = (x_2 - x_1)^2 + (y_2 - y_1)^2.$$

y

(x_2, y_2)

d

a

0

x

(x_1, y_1)

b

(x_2, y_1)

Figure 3

Taking the square root of each side, we get the **distance formula.**

Distance Formula

The distance between the points (x_1, y_1) and (x_2, y_2) is

$$d = \sqrt{(x_2 - x_1)^2 + (y_2 - y_1)^2}.$$

E X A M P L E   8   Using the Distance Formula

Find the distance between $(-3, 4)$ and $(2, 5)$.

Use the distance formula. Choose $(x_1, y_1) = (-3, 4)$ and $(x_2, y_2) = (2, 5)$.

$$d = \sqrt{(x_2 - x_1)^2 + (y_2 - y_1)^2}$$
$$= \sqrt{(2 - (-3))^2 + (5 - 4)^2} \qquad \text{Substitute.}$$
$$= \sqrt{5^2 + 1^2}$$
$$= \sqrt{26}$$

OBJECTIVE 6 Find higher roots. Finding the square root of a number is the inverse (reverse) of squaring a number. In a similar way, there are inverses to finding the cube of a number, or finding the fourth power of a number. These inverses are the **cube root,** written $\sqrt[3]{a}$, and the **fourth root,** written $\sqrt[4]{a}$. Similar symbols are used for higher roots. In general we have the following.

The nth root of a is written $\sqrt[n]{a}$.

In $\sqrt[n]{a}$, the number n is the **index** or **order** of the radical. It would be possible to write $\sqrt[2]{a}$ instead of $\sqrt{a}$, but the simpler symbol $\sqrt{a}$ is customary since the square root is the most commonly used root. A calculator that has a key marked $\boxed{\sqrt[x]{y}}$ or $\boxed{x^y}$ can be used to find these roots. When working with cube roots or fourth roots, it is helpful to memorize the first few *perfect cubes* ($2^3 = 8$, $3^3 = 27$, and so on) and the first few perfect fourth powers.

EXAMPLE 9 **Finding Cube Roots**

Find each cube root.

(a) $\sqrt[3]{8}$

Look for a number that can be cubed to give 8. Since $2^3 = 8$, then $\sqrt[3]{8} = 2$.

(b) $\sqrt[3]{-8}$

$\sqrt[3]{-8} = -2$ because $(-2)^3 = -8$.

As Example 9 suggests, the cube root of a positive number is positive, and the cube root of a negative number is negative. *There is only one real number cube root for each real number.*

When the index of the radical is even (square root, fourth root, and so on), the radicand must be nonnegative to get a real number root. Also, for even indexes the symbols $\sqrt{}$, $\sqrt[4]{}$, $\sqrt[6]{}$, and so on are used for the *nonnegative* roots, which are called **principal roots.** The symbols $-\sqrt{}$, $-\sqrt[4]{}$, $-\sqrt[6]{}$, and so on are used for the negative roots.

EXAMPLE 10 **Finding Higher Roots**

Find each root.

(a) $\sqrt[4]{16}$

$\sqrt[4]{16} = 2$ because 2 is positive and $2^4 = 16$.

(b) $-\sqrt[4]{16}$

From part (a), $\sqrt[4]{16} = 2$, so the negative root $-\sqrt[4]{16} = -2$.

(c) $\sqrt[4]{-16}$

To find the fourth root, the radicand must be nonnegative. There is no real number that equals $\sqrt[4]{-16}$.

(d) $\sqrt[3]{64} = 4$ since $4^3 = 64$.

(e) $-\sqrt[5]{32}$

First find $\sqrt[5]{32}$. The prime factorization of 32 as 2^5 shows that $\sqrt[5]{32} = 2$. If $\sqrt[5]{32} = 2$, then $-\sqrt[5]{32} = -2$.

9.1 EXERCISES

Decide whether each statement is true or false. If false, tell why.

1. Every nonnegative number has two square roots.

2. A negative number has negative square roots.

3. Every positive number has two real square roots.

4. Every positive number has three real cube roots.

5. The cube root of every real number has the same sign as the number itself.

6. The positive square root of a positive number is its principal square root.

Find all square roots of each number. See Example 1.

7. 16 **8.** 9 **9.** 144 **10.** 225

11. $\dfrac{25}{196}$ **12.** $\dfrac{81}{400}$ **13.** 900 **14.** 1600

Find each square root that is a real number. See Examples 2 and 4(e).

15. $\sqrt{49}$ **16.** $\sqrt{81}$ **17.** $-\sqrt{121}$ **18.** $\sqrt{196}$

19. $-\sqrt{\dfrac{144}{121}}$ **20.** $-\sqrt{\dfrac{49}{36}}$ **21.** $\sqrt{-121}$ **22.** $\sqrt{-49}$

Find the square of each radical expression. See Example 3.

23. $\sqrt{100}$ **24.** $\sqrt{36}$ **25.** $-\sqrt{19}$

26. $-\sqrt{99}$ **27.** $\sqrt{3x^2 + 4}$ **28.** $\sqrt{9y^2 + 3}$

What must be true about the variable a for each statement to be true?

29. $\sqrt{a}$ represents a positive number. **30.** $-\sqrt{a}$ represents a negative number.

31. $\sqrt{a}$ is not a real number. **32.** $-\sqrt{a}$ is not a real number.

Write rational, irrational, *or* not a real number *for each number. If the number is rational, give its exact value. If the number is irrational, give a decimal approximation to the nearest thousandth. Use a calculator as necessary. See Examples 4 and 5.*

33. $\sqrt{25}$ **34.** $\sqrt{169}$ **35.** $\sqrt{29}$ **36.** $\sqrt{33}$

37. $-\sqrt{64}$ **38.** $-\sqrt{500}$ **39.** $\sqrt{-29}$ **40.** $\sqrt{-47}$

41. Explain why the answers to Exercises 17 and 21 are different.

42. Explain why $\sqrt[3]{-8}$ and $-\sqrt[3]{8}$ represent the same number.

Use a calculator with a square root key to find each root. Round to the nearest thousandth. See Example 5.

43. $\sqrt{571}$ **44.** $\sqrt{693}$ **45.** $\sqrt{798}$

46. $\sqrt{453}$ **47.** $\sqrt{3.94}$ **48.** $\sqrt{.00895}$

Find each square root. Use a calculator and round to the nearest thousandth, if necessary. (Hint: First simplify the radicand to a single number.)

49. $\sqrt{3^2 + 4^2}$ **50.** $\sqrt{6^2 + 8^2}$ **51.** $\sqrt{8^2 + 15^2}$

52. $\sqrt{5^2 + 12^2}$ **53.** $\sqrt{2^2 + 3^2}$ **54.** $\sqrt{(-1)^2 + 5^2}$

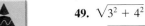

Use a calculator with a cube root key to find each root. Round to the nearest thousandth. (In Exercises 59 and 60, you may have to use the fact that if $a > 0$, $\sqrt[3]{-a} = -\sqrt[3]{a}$.)

55. $\sqrt[3]{12}$ **56.** $\sqrt[3]{74}$ **57.** $\sqrt[3]{130.6}$

58. $\sqrt[3]{251.8}$ **59.** $\sqrt[3]{-87}$ **60.** $\sqrt[3]{-95}$

Find the length of the unknown side of each right triangle with legs a and b and hypotenuse c. In Exercises 65 and 66, use a calculator and round to the nearest thousandth. See Example 6.

61. $a = 8, b = 15$ **62.** $a = 24, b = 10$ **63.** $a = 6, c = 10$

64. $b = 12, c = 13$ **65.** $a = 11, b = 4$ **66.** $a = 13, b = 9$

Use the Pythagorean formula to solve each problem. See Example 7.

67. The diagonal of a rectangle measures 25 centimeters. The width of the rectangle is 7 centimeters. Find the length of the rectangle.

25 cm

7 cm

68. The length of a rectangle is 40 meters, and the width is 9 meters. Find the measure of the diagonal of the rectangle.

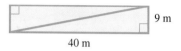

9 m

40 m

In Exercises 69–72, use these steps to solve each problem.

    Step 1 *Let x represent the unknown length.*

    Step 2 *Use the information in the problem to label the remaining sides of the triangle in the figure.*

    Step 3 *Write an equation relating the three sides of the triangle.* (Hint: *Use the Pythagorean formula.*)

    Step 4 *Solve the equation.*

    Steps 5 and 6 *Give the answer; check it.*

69. Margaret is flying a kite on 100 feet of string. How high is it above her hand (vertically) if the horizontal distance between Margaret and the kite is 60 feet?

EXERCISES

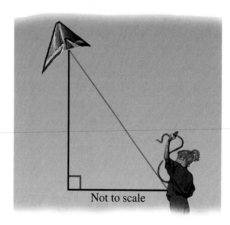

Not to scale

70. A guy wire is attached to the mast of a short-wave transmitting antenna. It is attached 96 feet above ground level. If the wire is staked to the ground 72 feet from the base of the mast, how long is the wire?

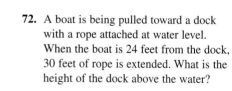

71. Two cars leave Tomball, Texas, at the same time. One travels north at 25 miles per hour and the other travels west at 60 miles per hour. How far apart are they after 3 hours? (*Hint:* Use $d = rt$ to find the distance traveled by each car.)

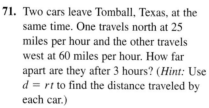

N

W

Tomball

72. A boat is being pulled toward a dock with a rope attached at water level. When the boat is 24 feet from the dock, 30 feet of rope is extended. What is the height of the dock above the water?

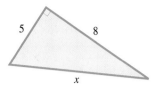 In Exercises 73 and 74, round to the nearest thousandth.

73. What is the value of x in the figure?

74. What is the value of y in the figure?

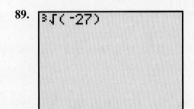

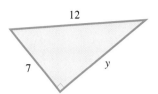

75. Use specific values for *a* and *b* different from those given in the "Caution" following Example 6 to show that $\sqrt{a^2 + b^2} \neq a + b$. Why would the values $a = 0$ and $b = 1$ *not* be satisfactory choices?

76. Explain how finding the square root of a number and squaring a number are related. Give examples.

Find the distance between each pair of points. See Example 8.

77. $(5, -3)$ and $(7, 2)$

78. $(2, 9)$ and $(-3, -3)$

79. $\left(-\dfrac{1}{4}, \dfrac{2}{3}\right)$ and $\left(\dfrac{3}{4}, -\dfrac{1}{3}\right)$

80. $\left(\dfrac{2}{5}, \dfrac{3}{2}\right)$ and $\left(-\dfrac{8}{5}, \dfrac{1}{2}\right)$

Find each root that is a real number. See Examples 9 and 10.

81. $\sqrt[3]{1000}$

82. $\sqrt[3]{8}$

83. $\sqrt[3]{-27}$

84. $\sqrt[3]{-64}$

85. $\sqrt[4]{625}$

86. $\sqrt[4]{10,000}$

87. $\sqrt[4]{-1}$

88. $\sqrt[4]{-625}$

TECHNOLOGY INSIGHTS (EXERCISES 89–92)

Exercises 89–92 illustrate the way one graphing calculator shows nth roots. Give each indicated root. Exercises 91 and 92 indicate 5th roots.

89. ³√(-27)

90. ³√(-64)

91. 5ˣ√243

92. 5ˣ√100000

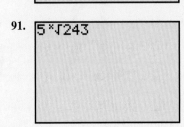

RELATING CONCEPTS (EXERCISES 93-98)

One of the many proofs of the Pythagorean formula was given in the Connections box in this section. Here is another one, attributed to the Hindu mathematician Bhāskara.

*Refer to the figures and **work Exercises 93–98 in order.***

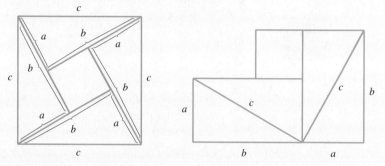

93. What is the area of the square on the left in terms of c?

94. What is the area of the small square in the middle of the figure on the left, in terms of $(b - a)$?

95. What is the sum of the areas of the two rectangles made up of triangles in the figure on the right?

96. What is the area of the small square in the figure on the right in terms of a and b?

97. The figure on the left is made up of the same square and triangles as the figure on the right. Write an equation setting the answer to Exercise 93 equal to the sum of the answers in Exercises 95 and 96.

98. Simplify the expressions you obtained in Exercise 97. What is your final result?

Did you make the connection that geometric figures can be used to prove algebraic formulas?

9.2 Multiplication and Division of Radicals

OBJECTIVES

1 Multiply radicals.

2 Simplify radicals using the product rule.

3 Simplify radical quotients using the quotient rule.

4 Use the product and quotient rules to simplify higher roots.

CONNECTIONS

The sixteenth century German radical symbol $\sqrt{}$ we use today is probably derived from the letter R. The radical symbol on the left below comes from the Latin word for root, *radix*. It was first used by Leonardo da Pisa (Fibonnaci) in 1220.

The cube root symbol shown on the right above was used by the German mathematician Christoff Rudolff in 1525. The symbol used today originated in the seventeenth century in France.

FOR DISCUSSION OR WRITING

   1. In the radical sign shown on the left, the *R* referred to above is clear. What other letter do you think is part of the symbol? What would the equivalent be in our modern notation?

   2. How is the cube root symbol on the right related to our modern radical sign?

OBJECTIVE 1 Multiply radicals. Several useful rules for finding products and quotients of radicals are developed in this section. To illustrate the rule for products, notice that

$$\sqrt{4} \cdot \sqrt{9} = 2 \cdot 3 = 6 \qquad \text{and} \qquad \sqrt{4 \cdot 9} = \sqrt{36} = 6,$$

showing that

$$\sqrt{4} \cdot \sqrt{9} = \sqrt{4 \cdot 9}.$$

This result is a particular case of the more general product rule for radicals.

Product Rule for Radicals

For nonnegative real numbers *x* and *y,*

$$\sqrt{x} \cdot \sqrt{y} = \sqrt{x \cdot y} \qquad \text{and} \qquad \sqrt{x \cdot y} = \sqrt{x} \cdot \sqrt{y}.$$

That is, the product of two radicals is the radical of the product.

In general, $\sqrt{x + y} \neq \sqrt{x} + \sqrt{y}$. To see why this is so, let $x = 16$ and $y = 9$.

$$\sqrt{16 + 9} = \sqrt{25} = 5$$

but

$$\sqrt{16} + \sqrt{9} = 4 + 3 = 7.$$

E X A M P L E   1   Using the Product Rule to Multiply Radicals

Use the product rule for radicals to find each product.

(a) $\sqrt{2} \cdot \sqrt{3} = \sqrt{2 \cdot 3} = \sqrt{6}$

(b) $\sqrt{7} \cdot \sqrt{5} = \sqrt{35}$

(c) $\sqrt{11} \cdot \sqrt{a} = \sqrt{11a}$ Assume $a \geq 0$.

OBJECTIVE 2 Simplify radicals using the product rule. A square root radical is **simplified** when no perfect square factor remains under the radical sign. This is accomplished by using the product rule as shown in Example 2.

E X A M P L E   2   Using the Product Rule to Simplify Radicals

Simplify each radical.

(a) $\sqrt{20}$

Since 20 has a perfect square factor of 4,

$$\sqrt{20} = \sqrt{4 \cdot 5} \qquad \text{4 is a perfect square.}$$
$$= \sqrt{4} \cdot \sqrt{5} \qquad \text{Product rule}$$
$$= 2\sqrt{5}. \qquad \sqrt{4} = 2$$

Thus, $\sqrt{20} = 2\sqrt{5}$. Since 5 has no perfect square factor other than 1, $2\sqrt{5}$ is called the **simplified form** of $\sqrt{20}$. Note that $2\sqrt{5}$ represents a product, where the factors are 2 and $\sqrt{5}$; also, $2\sqrt{5} \neq \sqrt{10}$.

(b) $\sqrt{72}$

Begin by looking for the largest perfect square that is a factor of 72. This number is 36, so

$$\sqrt{72} = \sqrt{36 \cdot 2} \qquad \text{36 is a perfect square.}$$
$$= \sqrt{36} \cdot \sqrt{2} \qquad \text{Product rule}$$
$$= 6\sqrt{2}. \qquad \sqrt{36} = 6$$

We could also factor 72 into its prime factors and look for pairs of like factors. Each pair of like factors produces a factor outside the radical in the simplified form.

$$\sqrt{72} = \sqrt{2 \cdot 2 \cdot 2 \cdot 3 \cdot 3} = 2 \cdot 3 \cdot \sqrt{2} = 6\sqrt{2}$$

In either case, we obtain $6\sqrt{2}$ as the simplified form of $\sqrt{72}$; our work is simpler, however, if we begin with the largest perfect square factor.

(c) $\sqrt{300} = \sqrt{100 \cdot 3} \qquad \text{100 is a perfect square.}$
$$= \sqrt{100} \cdot \sqrt{3} \qquad \text{Product rule}$$
$$= 10\sqrt{3} \qquad \sqrt{100} = 10$$

(d) $\sqrt{15}$

The number 15 has no perfect square factor (except 1), so $\sqrt{15}$ cannot be simplified further.

Sometimes the product rule can be used to simplify a product, as Example 3 shows.

E X A M P L E   3   Multiplying and Simplifying Radicals

Find each product and simplify.

(a) $\sqrt{9} \cdot \sqrt{75} = 3\sqrt{75} \qquad \sqrt{9} = 3$
$$= 3\sqrt{25 \cdot 3} \qquad \text{25 is a perfect square.}$$
$$= 3\sqrt{25} \cdot \sqrt{3} \qquad \text{Product rule}$$
$$= 3 \cdot 5\sqrt{3} \qquad \sqrt{25} = 5$$
$$= 15\sqrt{3} \qquad \text{Multiply.}$$

(b) $\sqrt{8} \cdot \sqrt{12} = \sqrt{8 \cdot 12}$ Product rule

$= \sqrt{4 \cdot 2 \cdot 4 \cdot 3}$ Factor; 4 is a perfect square.

$= \sqrt{4} \cdot \sqrt{4} \cdot \sqrt{2 \cdot 3}$ Product rule

$= 2 \cdot 2 \cdot \sqrt{6}$ $\sqrt{4} = 2$

$= 4\sqrt{6}$ Multiply.

Part (b) of Example 3 also could be simplified by using the product rule first to get

$$\sqrt{8} \cdot \sqrt{12} = \sqrt{4 \cdot 2} \cdot \sqrt{4 \cdot 3}$$
$$= 2\sqrt{2} \cdot 2\sqrt{3}$$
$$= 2 \cdot 2 \cdot \sqrt{2} \cdot \sqrt{3}$$
$$= 4\sqrt{6}.$$

Both approaches are correct.

OBJECTIVE 3 Simplify radical quotients using the quotient rule. The *quotient rule for radicals* is very similar to the product rule. It, too, can be used either way.

Quotient Rule for Radicals

If x and y are nonnegative real numbers and $y \neq 0$,

$$\frac{\sqrt{x}}{\sqrt{y}} = \sqrt{\frac{x}{y}} \quad \text{and} \quad \sqrt{\frac{x}{y}} = \frac{\sqrt{x}}{\sqrt{y}}.$$

The quotient of the radicals is the radical of the quotient.

EXAMPLE 4 Using the Quotient Rule to Simplify Radicals

Simplify each radical.

(a) $\sqrt{\frac{25}{9}} = \frac{\sqrt{25}}{\sqrt{9}} = \frac{5}{3}$ Quotient rule

(b) $\frac{\sqrt{288}}{\sqrt{2}} = \sqrt{\frac{288}{2}} = \sqrt{144} = 12$ Quotient rule

(c) $\sqrt{\frac{3}{4}} = \frac{\sqrt{3}}{\sqrt{4}} = \frac{\sqrt{3}}{2}$ Quotient rule

EXAMPLE 5 Using the Quotient Rule to Divide Radicals

Divide $27\sqrt{15}$ by $9\sqrt{3}$.

We use the quotient rule as follows.

$$\frac{27\sqrt{15}}{9\sqrt{3}} = \frac{27}{9} \cdot \frac{\sqrt{15}}{\sqrt{3}} = 3\sqrt{\frac{15}{3}} = 3\sqrt{5}$$

Some problems require both the product and quotient rules, as Example 6 shows.

E X A M P L E 6 Using Both the Product and Quotient Rules

Simplify $\sqrt{\dfrac{3}{5}} \cdot \sqrt{\dfrac{4}{5}}$.

$$\sqrt{\frac{3}{5}} \cdot \sqrt{\frac{4}{5}} = \frac{\sqrt{3}}{\sqrt{5}} \cdot \frac{\sqrt{4}}{\sqrt{5}} \qquad \text{Quotient rule}$$

$$= \frac{\sqrt{3} \cdot \sqrt{4}}{\sqrt{5} \cdot \sqrt{5}} \qquad \text{Multiply fractions.}$$

$$= \frac{\sqrt{3} \cdot 2}{\sqrt{25}} \qquad \text{Product rule; } \sqrt{4} = 2$$

$$= \frac{2\sqrt{3}}{5} \qquad \sqrt{25} = 5$$

The product and quotient rules also apply when variables appear under the radical sign, as long as all the variables represent only nonnegative numbers. For example, $\sqrt{5^2} = 5$, but $\sqrt{(-5)^2} \neq -5$.

> For a real number a, $\sqrt{a^2} = a$ only if a is nonnegative.

E X A M P L E 7 Simplifying Radicals Involving Variables

Simplify each radical. Assume all variables represent positive real numbers.

(a) $\sqrt{25m^4} = \sqrt{25} \cdot \sqrt{m^4}$ Product rule

$\qquad\quad = 5m^2$

(b) $\sqrt{64p^{10}} = 8p^5$ Product rule

(c) $\sqrt{r^9} = \sqrt{r^8 \cdot r}$

$\qquad\quad = \sqrt{r^8} \cdot \sqrt{r} = r^4\sqrt{r}$ Product rule

(d) $\sqrt{\dfrac{5}{x^2}} = \dfrac{\sqrt{5}}{\sqrt{x^2}} = \dfrac{\sqrt{5}}{x}$ Quotient rule

---

OBJECTIVE **4** Use the product and quotient rules to simplify higher roots. The product and quotient rules for radicals also work for other roots, as shown in Example 8. To simplify cube roots, look for factors that are *perfect cubes*. A **perfect cube** is a number with a rational cube root. For example, $\sqrt[3]{64} = 4$, and since 4 is a rational number, 64 is a perfect cube. Higher roots are handled in a similar manner.

Properties of Radicals

For all real numbers where the indicated roots exist,

$$\sqrt[n]{x} \cdot \sqrt[n]{y} = \sqrt[n]{xy} \qquad \text{and} \qquad \frac{\sqrt[n]{x}}{\sqrt[n]{y}} = \sqrt[n]{\frac{x}{y}}, y \neq 0.$$

┌─ **EXAMPLE 8** Simplifying Higher Roots

Simplify each radical.

(a) $\sqrt[3]{32} = \sqrt[3]{8 \cdot 4}$ 8 is a perfect cube.

$= \sqrt[3]{8} \cdot \sqrt[3]{4} = 2\sqrt[3]{4}$

(b) $\sqrt[3]{108} = \sqrt[3]{27 \cdot 4}$ 27 is a perfect cube.

$= \sqrt[3]{27} \cdot \sqrt[3]{4} = 3\sqrt[3]{4}$

(c) $\sqrt[4]{32} = \sqrt[4]{16} \cdot \sqrt[4]{2} = 2\sqrt[4]{2}$ 16 is a perfect fourth power.

(d) $\sqrt[3]{\dfrac{8}{125}} = \dfrac{\sqrt[3]{8}}{\sqrt[3]{125}} = \dfrac{2}{5}$

(e) $\sqrt[4]{\dfrac{16}{625}} = \dfrac{\sqrt[4]{16}}{\sqrt[4]{625}} = \dfrac{2}{5}$

(f) $\sqrt[3]{7} \cdot \sqrt[3]{49} = \sqrt[3]{7 \cdot 49} = \sqrt[3]{7 \cdot 7^2} = \sqrt[3]{7^3} = 7$

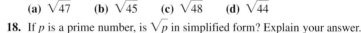

9.2 EXERCISES

Decide whether each statement is true or false for real numbers. If false, tell why.

1. $\sqrt{4} = \pm 2$ **2.** $\sqrt{(-6)^2} = -6$ **3.** $\sqrt{-6} \cdot \sqrt{6} = -6$

4. $\sqrt[3]{(-6)^3} = -6$ **5.** $\sqrt[3]{3} \cdot \sqrt[3]{2} = \sqrt[3]{6}$ **6.** $\sqrt[3]{4} \cdot \sqrt[3]{4} = 4$

7. $\sqrt{4} \cdot \sqrt{9} = \sqrt{4 \cdot 9}$ **8.** $\sqrt{4} + \sqrt{9} = \sqrt{4 + 9}$

Use the product rule for radicals to find each product. See Example 1.

9. $\sqrt{3} \cdot \sqrt{27}$ **10.** $\sqrt{2} \cdot \sqrt{8}$ **11.** $\sqrt{6} \cdot \sqrt{15}$

12. $\sqrt{10} \cdot \sqrt{15}$ **13.** $\sqrt{13} \cdot \sqrt{13}$ **14.** $\sqrt{17} \cdot \sqrt{17}$

15. $\sqrt{13} \cdot \sqrt{r}, r \geq 0$ **16.** $\sqrt{19} \cdot \sqrt{k}, k \geq 0$

17. Which one of the following radicals is simplified according to the guidelines of Objective 2?

(a) $\sqrt{47}$ **(b)** $\sqrt{45}$ **(c)** $\sqrt{48}$ **(d)** $\sqrt{44}$

18. If p is a prime number, is $\sqrt{p}$ in simplified form? Explain your answer.

Simplify each radical according to the method described in Objective 2. See Example 2.

19. $\sqrt{45}$ **20.** $\sqrt{56}$ **21.** $\sqrt{75}$ **22.** $\sqrt{18}$

23. $\sqrt{125}$ **24.** $\sqrt{80}$ **25.** $-\sqrt{700}$ **26.** $-\sqrt{600}$

27. $3\sqrt{27}$ **28.** $9\sqrt{8}$

Find each product and simplify. See Example 3.

29. $\sqrt{3} \cdot \sqrt{18}$ **30.** $\sqrt{3} \cdot \sqrt{21}$ **31.** $\sqrt{12} \cdot \sqrt{48}$

32. $\sqrt{50} \cdot \sqrt{72}$ **33.** $\sqrt{12} \cdot \sqrt{30}$ **34.** $\sqrt{30} \cdot \sqrt{24}$

35. In your own words, describe the product and quotient rules.

36. Simplify the radical $\sqrt{288}$ in two ways. First, factor 288 as $144 \cdot 2$ and then simplify completely. Second, factor 288 as $48 \cdot 6$ and then simplify completely. How do the answers compare? Make a conjecture concerning the quickest way to simplify such a radical.

Use the quotient rule and the product rule, as necessary, to simplify each radical expression. See Examples 4–6.

37. $\sqrt{\dfrac{16}{225}}$ **38.** $\sqrt{\dfrac{9}{100}}$ **39.** $\sqrt{\dfrac{7}{16}}$ **40.** $\sqrt{\dfrac{13}{25}}$

41. $\sqrt{\dfrac{5}{7}} \cdot \sqrt{35}$

42. $\sqrt{\dfrac{10}{13}} \cdot \sqrt{130}$

43. $\sqrt{\dfrac{5}{2}} \cdot \sqrt{\dfrac{125}{8}}$

44. $\sqrt{\dfrac{8}{3}} \cdot \sqrt{\dfrac{512}{27}}$

45. $\dfrac{30\sqrt{10}}{5\sqrt{2}}$

46. $\dfrac{50\sqrt{20}}{2\sqrt{10}}$

Simplify each radical. Assume that all variables represent nonnegative real numbers. See Example 7.

47. $\sqrt{m^2}$ **48.** $\sqrt{k^2}$ **49.** $\sqrt{y^4}$ **50.** $\sqrt{s^4}$

51. $\sqrt{36z^2}$ **52.** $\sqrt{49n^2}$ **53.** $\sqrt{400x^6}$ **54.** $\sqrt{900y^8}$

55. $\sqrt{z^5}$ **56.** $\sqrt{a^{13}}$ **57.** $\sqrt{x^6y^{12}}$ **58.** $\sqrt{a^8b^{10}}$

Simplify each radical. See Example 8.

59. $\sqrt[3]{40}$ **60.** $\sqrt[3]{48}$ **61.** $\sqrt[3]{54}$ **62.** $\sqrt[3]{192}$

63. $\sqrt[4]{80}$ **64.** $\sqrt[4]{243}$ **65.** $\sqrt[3]{\dfrac{8}{27}}$ **66.** $\sqrt[3]{\dfrac{64}{125}}$

67. $\sqrt[3]{-\dfrac{216}{125}}$ **68.** $\sqrt[3]{-\dfrac{1}{64}}$ **69.** $\sqrt[3]{5} \cdot \sqrt[3]{25}$ **70.** $\sqrt[3]{4} \cdot \sqrt[3]{16}$

71. $\sqrt[4]{4} \cdot \sqrt[4]{3}$ **72.** $\sqrt[4]{7} \cdot \sqrt[4]{4}$ **73.** $\sqrt[3]{4x} \cdot \sqrt[3]{8x^2}$ **74.** $\sqrt[3]{25p} \cdot \sqrt[3]{125p^3}$

75. In Example 2(a) we showed *algebraically* that $\sqrt{20}$ is equal to $2\sqrt{5}$. To give *numerical support* to this result, use a calculator to do the following:
   (a) Find a decimal approximation for $\sqrt{20}$ using your calculator. Record as many digits as the calculator shows.
   (b) Find a decimal approximation for $\sqrt{5}$ using your calculator, and then multiply the result by 2. Record as many digits as the calculator shows.
   (c) Your results in parts (a) and (b) should be the same. A mathematician would not accept this numerical exercise as *proof* that $\sqrt{20}$ is equal to $2\sqrt{5}$. Explain why.

76. On your calculator, multiply the approximations for $\sqrt{3}$ and $\sqrt{5}$. Now, predict what your calculator will show when you find an approximation for $\sqrt{15}$. What rule stated in this section justifies your answer?

The volume of a cube is found with the formula $V = s^3$, where s is the length of an edge of the cube. Use this information in Exercises 77 and 78.

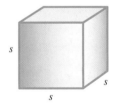

77. A container in the shape of a cube has a volume of 216 cubic centimeters. What is the depth of the container?

78. A cube-shaped box must be constructed to contain 128 cubic feet. What should the dimensions (height, width, and length) of the box be?

The volume of a sphere is found with the formula $V = \dfrac{4}{3}\pi r^3$, where r is the length of the radius of the sphere. Use this information in Exercises 79 and 80.

79. A ball in the shape of a sphere has a volume of 288π cubic inches. What is the radius of the ball?

80. Suppose that the volume of the ball described in Exercise 79 is multiplied by 8. How is the radius affected?

81. When we multiply two radicals with variables under the radical sign, such as $\sqrt{a} \cdot \sqrt{b} = \sqrt{ab}$, why is it important to know that both a and b represent nonnegative numbers?

82. Is it necessary to restrict k to a nonnegative number to say that $\sqrt[3]{k} \cdot \sqrt[3]{k} \cdot \sqrt[3]{k} = k$? Why?

RELATING CONCEPTS (EXERCISES 83-86)

An interesting way to represent the lengths corresponding to $\sqrt{2}, \sqrt{3}, \sqrt{4}, \sqrt{5}$, and so on is shown in the figure.

Work the following exercises in order.

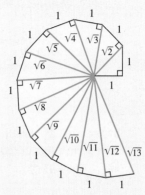

83. Use the Pythagorean formula to verify the lengths in the figure.

84. Which of the lengths indicated as radicals equal whole numbers? If the figure is continued in the same way, what would the next two whole number lengths be?

85. Find the consecutive differences between the radicands in Exercise 84. (*Hint:* The first difference is $9 - 4 = 5$.)

86. Look for a pattern that determines where these whole number lengths occur. Use this pattern to predict where the next whole number length and the one after that will occur.

Did you make the connection between the lengths of the sides of the triangles and the Pythagorean formula?

9.3 Addition and Subtraction of Radicals

OBJECTIVES

1. Add and subtract radicals.
2. Simplify radical sums and differences.
3. Simplify radical sums involving multiplication.

OBJECTIVE **1** Add and subtract radicals. We add or subtract radicals by using the distributive property. For example,

$$8\sqrt{3} + 6\sqrt{3} = (8 + 6)\sqrt{3} \qquad \text{Distributive property}$$
$$= 14\sqrt{3}.$$

Also,

$$2\sqrt{11} - 7\sqrt{11} = -5\sqrt{11}.$$

Like radicals are terms that have multiples of the *same root* of the *same number*. Only like radicals can be combined using the distributive property. In the example above, the like radicals are $2\sqrt{11}$ and $-7\sqrt{11}$. On the other hand, examples of *unlike radicals* are

$$2\sqrt{5} \quad \text{and} \quad 2\sqrt{3}, \qquad \text{Different radicands}$$

as well as

$$2\sqrt{3} \quad \text{and} \quad 2\sqrt[3]{3}. \qquad \text{Different indexes}$$

E X A M P L E   1   Adding and Subtracting Like Radicals

Add or subtract, as indicated.

(a) $3\sqrt{6} + 5\sqrt{6} = (3 + 5)\sqrt{6} = 8\sqrt{6}$ Distributive property

(b) $5\sqrt{10} - 7\sqrt{10} = (5 - 7)\sqrt{10} = -2\sqrt{10}$

(c) $\sqrt[3]{5} + \sqrt[3]{5} = 1\sqrt[3]{5} + 1\sqrt[3]{5} = (1 + 1)\sqrt[3]{5} = 2\sqrt[3]{5}$

(d) $\sqrt[4]{7} + 2\sqrt[4]{7} = 1\sqrt[4]{7} + 2\sqrt[4]{7} = 3\sqrt[4]{7}$

(e) $\sqrt{3} + \sqrt{13}$ cannot be added using the distributive property.

 In general, $\sqrt{x} + \sqrt{y} \neq \sqrt{x + y}$. In Example 1(e), it would be **incorrect** to try to simplify $\sqrt{3} + \sqrt{13}$ as $\sqrt{3} + \sqrt{13} = \sqrt{16} = 4$. Only *like radicals* can be combined.

O B J E C T I V E   2   Simplify radical sums and differences.   Sometimes we must simplify one or more radicals in a sum or difference. Doing this may result in like radicals, which we can then add or subtract.

E X A M P L E   2   Adding and Subtracting Radicals That Require Simplification

Simplify as much as possible.

(a) $3\sqrt{2} + \sqrt{8} = 3\sqrt{2} + \sqrt{4 \cdot 2}$ Factor.

$\qquad\qquad\quad = 3\sqrt{2} + \sqrt{4} \cdot \sqrt{2}$ Product rule

$\qquad\qquad\quad = 3\sqrt{2} + 2\sqrt{2}$ $\sqrt{4} = 2$

$\qquad\qquad\quad = 5\sqrt{2}$ Add like radicals.

(b) $\sqrt{18} - \sqrt{27} = \sqrt{9 \cdot 2} - \sqrt{9 \cdot 3}$ Factor.

$\qquad\qquad\quad = \sqrt{9} \cdot \sqrt{2} - \sqrt{9} \cdot \sqrt{3}$ Product rule

$\qquad\qquad\quad = 3\sqrt{2} - 3\sqrt{3}$ $\sqrt{9} = 3$

Since $\sqrt{2}$ and $\sqrt{3}$ are unlike radicals, this difference cannot be simplified further.

(c) $2\sqrt{12} + 3\sqrt{75} = 2(\sqrt{4} \cdot \sqrt{3}) + 3(\sqrt{25} \cdot \sqrt{3})$ Product rule

$\qquad\qquad\quad = 2(2\sqrt{3}) + 3(5\sqrt{3})$ $\sqrt{4} = 2$ and $\sqrt{25} = 5$

$\qquad\qquad\quad = 4\sqrt{3} + 15\sqrt{3}$ Multiply.

$\qquad\qquad\quad = 19\sqrt{3}$ Add like radicals.

(d) $3\sqrt[3]{16} + 5\sqrt[3]{2} = 3(\sqrt[3]{8} \cdot \sqrt[3]{2}) + 5\sqrt[3]{2}$ Product rule

$\qquad\qquad\quad = 3(2\sqrt[3]{2}) + 5\sqrt[3]{2}$ $\sqrt[3]{8} = 2$

$\qquad\qquad\quad = 6\sqrt[3]{2} + 5\sqrt[3]{2}$ Multiply.

$\qquad\qquad\quad = 11\sqrt[3]{2}$ Add like radicals.

O B J E C T I V E   3   Simplify radical sums involving multiplication.   Some radical expressions require both multiplication and addition (or subtraction). The order of operations presented in Chapter 1 still applies.

┌ E X A M P L E 3 Simplifying Radical Sums Involving Multiplication

Simplify each radical expression. Assume that all variables represent nonnegative real numbers.

(a) $\sqrt{5} \cdot \sqrt{15} + 4\sqrt{3} = \sqrt{5 \cdot 15} + 4\sqrt{3}$ Product rule

$\qquad\qquad\qquad\quad = \sqrt{75} + 4\sqrt{3}$ Multiply.

$\qquad\qquad\qquad\quad = \sqrt{25 \cdot 3} + 4\sqrt{3}$ 25 is a perfect square.

$\qquad\qquad\qquad\quad = \sqrt{25} \cdot \sqrt{3} + 4\sqrt{3}$ Product rule

$\qquad\qquad\qquad\quad = 5\sqrt{3} + 4\sqrt{3}$ $\sqrt{25} = 5$

$\qquad\qquad\qquad\quad = 9\sqrt{3}$ Add like radicals.

(b) $\sqrt{2} \cdot \sqrt{6k} + \sqrt{27k} = \sqrt{12k} + \sqrt{27k}$ Product rule

$\qquad\qquad\qquad\quad = \sqrt{4 \cdot 3k} + \sqrt{9 \cdot 3k}$ Factor.

$\qquad\qquad\qquad\quad = \sqrt{4} \cdot \sqrt{3k} + \sqrt{9} \cdot \sqrt{3k}$ Product rule

$\qquad\qquad\qquad\quad = 2\sqrt{3k} + 3\sqrt{3k}$ $\sqrt{4} = 2$ and $\sqrt{9} = 3$

$\qquad\qquad\qquad\quad = 5\sqrt{3k}$ Add like radicals.

(c) $\sqrt[3]{2} \cdot \sqrt[3]{16m^3} - \sqrt[3]{108m^3} = \sqrt[3]{32m^3} - \sqrt[3]{108m^3}$ Product rule

$\qquad\qquad\qquad\qquad = \sqrt[3]{(8m^3)4} - \sqrt[3]{(27m^3)4}$ Factor.

$\qquad\qquad\qquad\qquad = 2m\sqrt[3]{4} - 3m\sqrt[3]{4}$ $\sqrt[3]{8m^3} = 2m$ and $\sqrt[3]{27m^3} = 3m$

$\qquad\qquad\qquad\qquad = -m\sqrt[3]{4}$ Subtract like radicals.

 Remember that a sum or difference of radicals can be simplified only if the radicals are *like radicals*. For example, $2\sqrt{3} + 5\sqrt[3]{3}$ cannot be simplified further.

9.3 EXERCISES

Fill in each blank with the correct response.

1. $5\sqrt{2} + 6\sqrt{2} = (5 + 6)\sqrt{2} = 11\sqrt{2}$ is an example of the _____ property.

2. Like radicals have the same _____ of the same _____.

3. $\sqrt{2} + 2\sqrt{3}$ cannot be simplified because the _____ are different.

4. $2\sqrt{3} + 4\sqrt[3]{3}$ cannot be simplified because the _____ are different.

Simplify and add or subtract wherever possible. See Examples 1 and 2.

5. $14\sqrt{7} - 19\sqrt{7}$ **6.** $16\sqrt{2} - 18\sqrt{2}$ **7.** $\sqrt{17} + 4\sqrt{17}$

8. $5\sqrt{19} + \sqrt{19}$ **9.** $6\sqrt{7} - \sqrt{7}$ **10.** $11\sqrt{14} - \sqrt{14}$

11. $\sqrt{45} + 4\sqrt{20}$ **12.** $\sqrt{24} + 6\sqrt{54}$ **13.** $5\sqrt{72} - 3\sqrt{50}$

14. $6\sqrt{18} - 5\sqrt{32}$ **15.** $-5\sqrt{32} + 2\sqrt{45}$ **16.** $-4\sqrt{75} + 3\sqrt{24}$

17. $5\sqrt{7} - 3\sqrt{28} + 6\sqrt{63}$

18. $3\sqrt{11} + 5\sqrt{44} - 8\sqrt{99}$

19. $2\sqrt{8} - 5\sqrt{32} - 2\sqrt{48}$

20. $5\sqrt{72} - 3\sqrt{48} + 4\sqrt{128}$

21. $4\sqrt{50} + 3\sqrt{12} - 5\sqrt{45}$

22. $6\sqrt{18} + 2\sqrt{48} + 6\sqrt{28}$

23. $\frac{1}{4}\sqrt{288} + \frac{1}{6}\sqrt{72}$

24. $\frac{2}{3}\sqrt{27} + \frac{3}{4}\sqrt{48}$

25. The distributive property, which says $a(b + c) = ab + ac$ and $ba + ca = (b + c)a$, provides the justification for adding and subtracting like radicals. While we usually skip the step that indicates this property, we could not make the statement $2\sqrt{3} + 4\sqrt{3} = 6\sqrt{3}$ without it. Write an equation showing how the distributive property is actually used in this statement.

26. Refer to Example 1(e), and explain why $\sqrt{3} + \sqrt{7}$ cannot be further simplified. Confirm, by using calculator approximations, that $\sqrt{3} + \sqrt{7}$ is *not* equal to $\sqrt{10}$.

Perform the indicated operations. Assume that all variables represent nonnegative real numbers. See Example 3.

27. $\sqrt{6} \cdot \sqrt{2} + 9\sqrt{3}$

28. $4\sqrt{15} \cdot \sqrt{3} + 4\sqrt{5}$

29. $\sqrt{9x} + \sqrt{49x} - \sqrt{25x}$

30. $\sqrt{4a} - \sqrt{16a} + \sqrt{100a}$

31. $\sqrt{6x^2} + x\sqrt{24}$

32. $\sqrt{75x^2} + x\sqrt{108}$

33. $3\sqrt{8x^2} - 4x\sqrt{2} - x\sqrt{8}$

34. $\sqrt{2b^2} + 3b\sqrt{18} - b\sqrt{200}$

35. $-8\sqrt{32k} + 6\sqrt{8k}$

36. $4\sqrt{12x} + 2\sqrt{27x}$

37. $2\sqrt{125x^2z} + 8x\sqrt{80z}$

38. $\sqrt{48x^2y} + 5x\sqrt{27y}$

39. $4\sqrt[3]{16} - 3\sqrt[3]{54}$

40. $5\sqrt[3]{128} + 3\sqrt[3]{250}$

41. $6\sqrt[3]{8p^2} - 2\sqrt[3]{27p^2}$

42. $8k\sqrt[3]{54k} + 6\sqrt[3]{16k^4}$

43. $5\sqrt[4]{m^3} + 8\sqrt[4]{16m^3}$

44. $5\sqrt[4]{m^5} + 3\sqrt[4]{81m^5}$

45. Describe in your own words how to add and subtract radicals.

46. In the directions for Exercises 27–44, we made the assumption that all variables represent nonnegative real numbers. However, in Exercises 41 and 42, variables actually *may* represent negative numbers. Explain why this is so.

RELATING CONCEPTS (EXERCISES 47–50)

Adding and subtracting like radicals is no different than adding and subtracting other like terms.

Work Exercises 47–50 in order.

47. Combine like terms: $5x^2y + 3x^2y - 14x^2y$.

48. Combine like terms: $5(p - 2q)^2(a + b) + 3(p - 2q)^2(a + b) - 14(p - 2q)^2(a + b)$.

49. Combine like radicals: $5a^2\sqrt{xy} + 3a^2\sqrt{xy} - 14a^2\sqrt{xy}$.

50. Compare your answers in Exercises 47–49. How are they alike? How are they different?

Did you make the connection between adding and subtracting like radicals and adding and subtracting like terms?

Perform the indicated operations following the order of operations we have used throughout the book. Use a calculator and round to the nearest thousandth, if necessary.

51. $\sqrt{(-3 - 6)^2 + (2 - 4)^2}$

52. $\sqrt{(-9 - 3)^2 + (3 - 8)^2}$

53. $\sqrt{(2 - (-2))^2 + (-1 - 2)^2}$

54. $\sqrt{(3 - 1)^2 + (2 - (-1))^2}$

A common error in a problem like the one in Example 5(a) is to multiply by $\sqrt[3]{2}$ instead of $\sqrt[3]{2^2}$. Doing this would give a denominator of $\sqrt[3]{2} \cdot \sqrt[3]{2} = \sqrt[3]{4}$. Since 4 is not a perfect cube, the denominator is still not rationalized.

9.4 EXERCISES

Fill in each blank with the correct response.

1. Rationalizing the denominator means to change the denominator from a(n) _____ to a rational number.

2. To simplify $\sqrt{x^2 y}$, where $x \geq 0$, we must remove _____ from the radicand.

3. The expression $\sqrt{\dfrac{3m}{2}}$, where $m \geq 0$, is not simplified because the radical contains a(n) _____ .

4. The expression $\dfrac{2x}{\sqrt{7}}$ is not simplified because the denominator is a(n) _____ .

Rationalize each denominator. See Examples 1 and 2.

5. $\dfrac{8}{\sqrt{2}}$ 6. $\dfrac{12}{\sqrt{3}}$ 7. $\dfrac{-\sqrt{11}}{\sqrt{3}}$ 8. $\dfrac{-\sqrt{13}}{\sqrt{5}}$ 9. $\dfrac{7\sqrt{3}}{\sqrt{5}}$

10. $\dfrac{4\sqrt{6}}{\sqrt{5}}$ 11. $\dfrac{24\sqrt{10}}{16\sqrt{3}}$ 12. $\dfrac{18\sqrt{15}}{12\sqrt{2}}$ 13. $\dfrac{16}{\sqrt{27}}$ 14. $\dfrac{24}{\sqrt{18}}$

15. $\dfrac{-3}{\sqrt{50}}$ 16. $\dfrac{-5}{\sqrt{75}}$ 17. $\dfrac{63}{\sqrt{45}}$ 18. $\dfrac{27}{\sqrt{32}}$ 19. $\dfrac{\sqrt{24}}{\sqrt{8}}$

20. $\dfrac{\sqrt{36}}{\sqrt{18}}$ 21. $\sqrt{\dfrac{1}{2}}$ 22. $\sqrt{\dfrac{1}{3}}$ 23. $\sqrt{\dfrac{13}{5}}$ 24. $\sqrt{\dfrac{17}{11}}$

25. To rationalize the denominator of an expression such as $\dfrac{4}{\sqrt{3}}$, we multiply both the numerator and the denominator by $\sqrt{3}$. By what number are we actually multiplying the given expression, and what property of real numbers justifies the fact that our result is equal to the given expression?

26. In Example 1(a), we show algebraically that $\dfrac{9}{\sqrt{6}}$ is equal to $\dfrac{3\sqrt{6}}{2}$. Support this result numerically by finding the decimal approximation of $\dfrac{9}{\sqrt{6}}$ on your calculator, and then finding the decimal approximation of $\dfrac{3\sqrt{6}}{2}$. What do you notice?

Multiply and simplify each result. See Example 3.

27. $\sqrt{\dfrac{7}{13}} \cdot \sqrt{\dfrac{13}{3}}$ 28. $\sqrt{\dfrac{19}{20}} \cdot \sqrt{\dfrac{20}{3}}$ 29. $\sqrt{\dfrac{21}{7}} \cdot \sqrt{\dfrac{21}{8}}$ 30. $\sqrt{\dfrac{5}{8}} \cdot \sqrt{\dfrac{5}{6}}$

31. $\sqrt{\dfrac{1}{12}} \cdot \sqrt{\dfrac{1}{3}}$ 32. $\sqrt{\dfrac{1}{8}} \cdot \sqrt{\dfrac{1}{2}}$ 33. $\sqrt{\dfrac{2}{9}} \cdot \sqrt{\dfrac{9}{2}}$ 34. $\sqrt{\dfrac{4}{3}} \cdot \sqrt{\dfrac{3}{4}}$

Simplify each radical. Assume that all variables represent positive real numbers. See Example 4.

35. $\sqrt{\dfrac{7}{x}}$ 36. $\sqrt{\dfrac{19}{y}}$ 37. $\sqrt{\dfrac{4x^3}{y}}$ 38. $\sqrt{\dfrac{9t^3}{s}}$

39. $\sqrt{\dfrac{18x^3}{6y}}$ 40. $\sqrt{\dfrac{24t^3}{8p}}$ 41. $\sqrt{\dfrac{9a^2r^5}{7t}}$ 42. $\sqrt{\dfrac{16x^3y^2}{13z}}$

43. Which one of the following would be an appropriate choice for multiplying the numerator and the denominator of $\dfrac{\sqrt[3]{2}}{\sqrt[3]{5}}$ in order to rationalize the denominator?

(a) $\sqrt[3]{5}$ **(b)** $\sqrt[3]{25}$ **(c)** $\sqrt[3]{2}$ **(d)** $\sqrt[3]{3}$

44. In Example 5(b), we multiply the numerator and denominator of $\dfrac{\sqrt[3]{3}}{\sqrt[3]{4}}$ by $\sqrt[3]{2}$ to rationalize the denominator. Suppose we had chosen to multiply by $\sqrt[3]{16}$ instead. Would we have obtained the correct answer after all simplifications were done?

Simplify. Rationalize each denominator. Assume that variables in each denominator are nonzero. See Example 5.

45. $\sqrt[3]{\dfrac{3}{2}}$ **46.** $\sqrt[3]{\dfrac{2}{5}}$ **47.** $\dfrac{\sqrt[3]{4}}{\sqrt[3]{7}}$ **48.** $\dfrac{\sqrt[3]{5}}{\sqrt[3]{10}}$

49. $\sqrt[3]{\dfrac{3}{4y^2}}$ **50.** $\sqrt[3]{\dfrac{3}{25x^2}}$ **51.** $\dfrac{\sqrt[3]{7m}}{\sqrt[3]{36n}}$ **52.** $\dfrac{\sqrt[3]{11p}}{\sqrt[3]{49q}}$

 In Exercises 53 and 54, (a) give the answer as a simplified radical and (b) use a calculator to give the answer correct to the nearest thousandth.

53. The period p of a pendulum is the time it takes for it to swing from one extreme to the other and back again. The value of p in seconds is given by

$$p = k \cdot \sqrt{\dfrac{L}{g}},$$

where L is the length of the pendulum, g is the acceleration due to gravity, and k is a constant. Find the period when $k = 6$, $L = 9$ feet, and $g = 32$ feet per second squared.

54. The velocity v of a meteorite approaching the earth is given by

$$v = \dfrac{k}{\sqrt{d}}$$

kilometers per second, where d is its distance from the center of the earth and k is a constant. What is the velocity of a meteorite that is 6000 kilometers away from the center of the earth, if $k = 450$?

9.5 Simplifying Radical Expressions

OBJECTIVES

1 Simplify products of radical expressions.

2 Simplify quotients of radical expressions.

3 Write radical expressions with quotients in lowest terms.

The conditions for which a radical is in simplest form were listed in the previous section. Below is a set of guidelines to follow when you are simplifying radical expressions. Although the guidelines are illustrated with square roots, they apply to higher roots as well.

Simplifying Radical Expressions

1. If a radical represents a rational number, use that rational number in place of the radical.

Examples: $\sqrt{49}$ is simplified by writing 7; $\sqrt{\dfrac{169}{9}}$ by writing $\dfrac{13}{3}$.

> **Simplifying Radical Expressions (continued)**
>
> **2.** If a radical expression contains products of radicals, use the product rule for radicals, $\sqrt{x} \cdot \sqrt{y} = \sqrt{xy}$, to get a single radical.
>
> > *Examples:* $\sqrt{3} \cdot \sqrt{2}$ is simplified to $\sqrt{6}$; $\sqrt{5} \cdot \sqrt{x}$ to $\sqrt{5x}$.
>
> **3.** If a radicand has a factor that is a perfect square, express the radical as the product of the positive square root of the perfect square and the remaining radical factor.
>
> > *Examples:* $\sqrt{20}$ is simplified to $\sqrt{20} = \sqrt{4 \cdot 5} = \sqrt{4} \cdot \sqrt{5} = 2\sqrt{5}$;
> > $$\sqrt[3]{16} = \sqrt[3]{8 \cdot 2} = \sqrt[3]{8} \cdot \sqrt[3]{2} = 2\sqrt[3]{2}.$$
>
> **4.** If a radical expression contains sums or differences of radicals, use the distributive property to combine like radicals.
>
> > *Examples:* $3\sqrt{2} + 4\sqrt{2} = 7\sqrt{2}$, but $3\sqrt{2} + 4\sqrt{3}$ cannot be further simplified.
>
> **5.** Rationalize any denominator containing a radical.
>
> > *Examples:* $\dfrac{5}{\sqrt{3}}$ is rationalized as $\dfrac{5}{\sqrt{3}} = \dfrac{5\sqrt{3}}{\sqrt{3} \cdot \sqrt{3}} = \dfrac{5\sqrt{3}}{3}$;
> > $$\sqrt{\frac{3}{2}} = \frac{\sqrt{3}}{\sqrt{2}} = \frac{\sqrt{3} \cdot \sqrt{2}}{\sqrt{2} \cdot \sqrt{2}} = \frac{\sqrt{6}}{2}.$$

OBJECTIVE 1 Simplify products of radical expressions. Use the above guidelines to do this.

EXAMPLE 1 Multiplying Radical Expressions

Find each product and simplify the answers.

(a) $\sqrt{5}(\sqrt{8} - \sqrt{32})$

Simplify inside the parentheses.

$$
\begin{aligned}
\sqrt{5}(\sqrt{8} - \sqrt{32}) &= \sqrt{5}(2\sqrt{2} - 4\sqrt{2}) \\
&= \sqrt{5}(-2\sqrt{2}) && \text{Subtract like radicals.} \\
&= -2\sqrt{5 \cdot 2} && \text{Product rule; commutative property} \\
&= -2\sqrt{10} && \text{Multiply.}
\end{aligned}
$$

(b) $(\sqrt{3} + 2\sqrt{5})(\sqrt{3} - 4\sqrt{5})$

The product of these sums of radicals can be found in the same way that we found the product of binomials in Chapter 4 using the FOIL method.

$$(\sqrt{3} + 2\sqrt{5})(\sqrt{3} - 4\sqrt{5})$$

$$\qquad\quad \text{F} \qquad\quad \text{O} \qquad\quad\quad \text{I} \qquad\qquad \text{L}$$

$$
\begin{aligned}
&= \sqrt{3} \cdot \sqrt{3} + \sqrt{3}(-4\sqrt{5}) + 2\sqrt{5} \cdot \sqrt{3} + 2\sqrt{5}(-4\sqrt{5}) \\
&= 3 - 4\sqrt{15} + 2\sqrt{15} - 8 \cdot 5 && \text{Product rule} \\
&= 3 - 2\sqrt{15} - 40 && \text{Add like radicals.} \\
&= -37 - 2\sqrt{15} && \text{Combine terms.}
\end{aligned}
$$

(c) $(\sqrt{3} + \sqrt{21})(\sqrt{3} - \sqrt{7})$

$$(\sqrt{3} + \sqrt{21})(\sqrt{3} - \sqrt{7})$$
$$= \sqrt{3}(\sqrt{3}) + \sqrt{3}(-\sqrt{7}) + \sqrt{21}(\sqrt{3}) + \sqrt{21}(-\sqrt{7}) \qquad \text{FOIL}$$
$$= 3 - \sqrt{21} + \sqrt{63} - \sqrt{147} \qquad \text{Product rule}$$
$$= 3 - \sqrt{21} + \sqrt{9} \cdot \sqrt{7} - \sqrt{49} \cdot \sqrt{3} \qquad \text{Simplify radicals.}$$
$$= 3 - \sqrt{21} + 3\sqrt{7} - 7\sqrt{3} \qquad \sqrt{9} = 3 \text{ and } \sqrt{49} = 7$$

Since there are no like radicals, no terms may be combined.

Since radicals represent real numbers, the special products of binomials discussed in Chapter 4 can be used to find products of radicals. Example 2 uses the rule for the product that gives the difference of two squares,

$$(a + b)(a - b) = a^2 - b^2.$$

EXAMPLE 2   Using a Special Product with Radicals

Find each product.

(a) $(4 + \sqrt{3})(4 - \sqrt{3})$

Follow the pattern given above. Let $a = 4$ and $b = \sqrt{3}$.

$$(4 + \sqrt{3})(4 - \sqrt{3}) = 4^2 - (\sqrt{3})^2$$
$$= 16 - 3 = 13 \qquad 4^2 = 16 \text{ and } (\sqrt{3})^2 = 3$$

(b) $(\sqrt{12} - \sqrt{6})(\sqrt{12} + \sqrt{6}) = (\sqrt{12})^2 - (\sqrt{6})^2$
$$= 12 - 6 \qquad (\sqrt{12})^2 = 12 \text{ and } (\sqrt{6})^2 = 6$$
$$= 6$$

Both products in Example 2 resulted in rational numbers. The pairs of expressions in those products, $4 + \sqrt{3}$ and $4 - \sqrt{3}$, and $\sqrt{12} - \sqrt{6}$ and $\sqrt{12} + \sqrt{6}$, are called **conjugates** of each other.

OBJECTIVE 2   Simplify quotients of radical expressions. Products of radicals similar to those in Example 2 can be used to rationalize the denominators in quotients with binomial denominators, such as

$$\frac{2}{4 - \sqrt{3}}.$$

By Example 2(a), if this denominator, $4 - \sqrt{3}$, is multiplied by $4 + \sqrt{3}$, then the product $(4 - \sqrt{3})(4 + \sqrt{3})$ is the rational number 13. Multiplying numerator and denominator by $4 + \sqrt{3}$ gives

$$\frac{2}{4 - \sqrt{3}} = \frac{2(4 + \sqrt{3})}{(4 - \sqrt{3})(4 + \sqrt{3})}$$
$$= \frac{2(4 + \sqrt{3})}{13}.$$

The denominator now has been rationalized; it contains no radical signs.

Using Conjugates to Simplify a Radical Expression

To simplify a radical expression with two terms in the denominator, where at least one of those terms is a square root radical, multiply both the numerator and the denominator by the conjugate of the denominator.

E X A M P L E 3 Using Conjugates to Rationalize a Denominator

Simplify by rationalizing the denominator.

(a) $\dfrac{7}{3 + \sqrt{5}}$

We can eliminate the radical in the denominator by multiplying both numerator and denominator by $3 - \sqrt{5}$, the conjugate of $3 + \sqrt{5}$.

$$\frac{7}{3 + \sqrt{5}} = \frac{7(3 - \sqrt{5})}{(3 + \sqrt{5})(3 - \sqrt{5})} \qquad \text{Multiply by the conjugate.}$$

$$= \frac{7(3 - \sqrt{5})}{3^2 - (\sqrt{5})^2} \qquad (a + b)(a - b) = a^2 - b^2$$

$$= \frac{7(3 - \sqrt{5})}{9 - 5} \qquad 3^2 = 9 \text{ and } (\sqrt{5})^2 = 5$$

$$= \frac{7(3 - \sqrt{5})}{4} \qquad \text{Subtract.}$$

(b) $\dfrac{6 + \sqrt{2}}{\sqrt{2} - 5}$

Multiply numerator and denominator by $\sqrt{2} + 5$.

$$\frac{6 + \sqrt{2}}{\sqrt{2} - 5} = \frac{(6 + \sqrt{2})(\sqrt{2} + 5)}{(\sqrt{2} - 5)(\sqrt{2} + 5)}$$

$$= \frac{6\sqrt{2} + 30 + 2 + 5\sqrt{2}}{2 - 25} \qquad \text{FOIL}$$

$$= \frac{11\sqrt{2} + 32}{-23} \qquad \text{Combine terms.}$$

$$= -\frac{11\sqrt{2} + 32}{23} \qquad \frac{a}{-b} = -\frac{a}{b}$$

Rationalizing the denominator in the two expressions

$$\frac{7}{\sqrt{x} + 5} \qquad \text{and} \qquad \frac{7}{\sqrt{x} + \sqrt{5}}$$

involves different procedures. In the first denominator, which has one term, multiply both the numerator and the denominator by $\sqrt{x} + 5$. In the second denominator, which has two terms, use $\sqrt{x} - \sqrt{5}$.

OBJECTIVE 3 Write radical expressions with quotients in lowest terms. The final example shows this.

E X A M P L E 4 **Writing a Radical Quotient in Lowest Terms**

Write $\dfrac{3\sqrt{3} + 15}{12}$ in lowest terms.

Factor the numerator and denominator, and then divide numerator and denominator by any common factors.

$$\frac{3\sqrt{3} + 15}{12} = \frac{3(\sqrt{3} + 5)}{3 \cdot 4} = \frac{\sqrt{3} + 5}{4}$$

This technique is used in Chapter 10.

A common error is to reduce an expression like the one in Example 4 incorrectly to lowest terms before factoring. For example,

$$\frac{4 + 8\sqrt{5}}{4} \ne 1 + 8\sqrt{5}.$$

The correct simplification is $1 + 2\sqrt{5}$. Why?

9.5 EXERCISES

Based on the work so far, many simple operations involving radicals should now be performed mentally. In Exercises 1–8, perform the operations mentally, and write the answer without doing intermediate steps.

1. $\sqrt{49} + \sqrt{36}$ 　　　　2. $\sqrt{100} - \sqrt{81}$ 　　　　3. $\sqrt{2} \cdot \sqrt{8}$
4. $\sqrt{8} \cdot \sqrt{8}$ 　　　　5. $\sqrt{2}(\sqrt{32} - \sqrt{8})$ 　　　　6. $\sqrt{3}(\sqrt{27} - \sqrt{3})$
7. $\sqrt[3]{8} + \sqrt[3]{27}$ 　　　　8. $\sqrt[3]{4} - \sqrt[3]{64} + \sqrt[4]{16}$

Simplify each expression. Use the five guidelines given in the text. See Examples 1 and 2.

9. $3\sqrt{5} + 2\sqrt{45}$ 　　　　　　10. $2\sqrt{2} + 4\sqrt{18}$
11. $8\sqrt{50} - 4\sqrt{72}$ 　　　　　　12. $4\sqrt{80} - 5\sqrt{45}$
13. $\sqrt{5}(\sqrt{3} - \sqrt{7})$ 　　　　　　14. $\sqrt{7}(\sqrt{10} + \sqrt{3})$
15. $2\sqrt{5}(\sqrt{2} + 3\sqrt{5})$ 　　　　　　16. $3\sqrt{7}(2\sqrt{7} + 4\sqrt{5})$
17. $3\sqrt{14} \cdot \sqrt{2} - \sqrt{28}$ 　　　　　　18. $7\sqrt{6} \cdot \sqrt{3} - 2\sqrt{18}$
19. $(2\sqrt{6} + 3)(3\sqrt{6} + 7)$ 　　　　　　20. $(4\sqrt{5} - 2)(2\sqrt{5} - 4)$
21. $(5\sqrt{7} - 2\sqrt{3})(3\sqrt{7} + 4\sqrt{3})$ 　　　　　　22. $(2\sqrt{10} + 5\sqrt{2})(3\sqrt{10} - 3\sqrt{2})$
23. $(2\sqrt{7} + 3)^2$ 　　　　　　24. $(4\sqrt{5} + 5)^2$
25. $(5 - \sqrt{2})(5 + \sqrt{2})$ 　　　　　　26. $(3 - \sqrt{5})(3 + \sqrt{5})$
27. $(\sqrt{8} - \sqrt{7})(\sqrt{8} + \sqrt{7})$ 　　　　　　28. $(\sqrt{12} - \sqrt{11})(\sqrt{12} + \sqrt{11})$
29. $(\sqrt{2} + \sqrt{3})(\sqrt{6} - \sqrt{2})$ 　　　　　　30. $(\sqrt{3} + \sqrt{5})(\sqrt{15} - \sqrt{5})$
31. $(\sqrt{10} - \sqrt{5})(\sqrt{5} + \sqrt{20})$ 　　　　　　32. $(\sqrt{6} - \sqrt{3})(\sqrt{3} + \sqrt{18})$
33. $(5\sqrt{7} - 2\sqrt{3})(3\sqrt{7} + 3\sqrt{3})$ 　　　　　　34. $(2\sqrt{10} + 5\sqrt{2})(3\sqrt{10} - 4\sqrt{2})$

35. In Example 1(b), the original expression simplifies to $-37 - 2\sqrt{15}$. Students often try to simplify expressions like this by combining the -37 and the -2 to get $-39\sqrt{15}$, which is incorrect. Explain why.

36. If you try to rationalize the denominator of $\dfrac{2}{4 + \sqrt{3}}$ by multiplying the numerator and denominator by $4 + \sqrt{3}$, what problem arises? What should you multiply by?

Rationalize each denominator. See Example 3.

37. $\dfrac{1}{3 + \sqrt{2}}$

38. $\dfrac{1}{4 - \sqrt{3}}$

39. $\dfrac{14}{2 - \sqrt{11}}$

40. $\dfrac{19}{5 - \sqrt{6}}$

41. $\dfrac{\sqrt{2}}{2 - \sqrt{2}}$

42. $\dfrac{\sqrt{7}}{7 - \sqrt{7}}$

43. $\dfrac{\sqrt{5}}{\sqrt{2} + \sqrt{3}}$

44. $\dfrac{\sqrt{3}}{\sqrt{2} + \sqrt{3}}$

45. $\dfrac{\sqrt{12}}{\sqrt{3} + 1}$

46. $\dfrac{\sqrt{18}}{\sqrt{2} - 1}$

47. $\dfrac{\sqrt{5} + 2}{2 - \sqrt{3}}$

48. $\dfrac{\sqrt{7} + 3}{4 - \sqrt{5}}$

Write each quotient in lowest terms. See Example 4.

49. $\dfrac{6\sqrt{11} - 12}{6}$

50. $\dfrac{12\sqrt{5} - 24}{12}$

51. $\dfrac{2\sqrt{3} + 10}{16}$

52. $\dfrac{4\sqrt{6} + 24}{20}$

53. $\dfrac{12 - \sqrt{40}}{4}$

54. $\dfrac{9 - \sqrt{72}}{12}$

Simplify each radical expression. Assume all variables represent nonnegative real numbers.

55. $(\sqrt{5x} + \sqrt{30})(\sqrt{6x} + \sqrt{3})$

56. $(\sqrt{10y} - \sqrt{20})(\sqrt{2y} - \sqrt{5})$

57. $(3\sqrt{t} + \sqrt{7})(2\sqrt{t} - \sqrt{14})$

58. $(2\sqrt{z} - \sqrt{3})(\sqrt{z} - \sqrt{5})$

59. $(\sqrt{3m} + \sqrt{2n})(\sqrt{5m} - \sqrt{5n})$

60. $(\sqrt{4p} - \sqrt{3k})(\sqrt{2p} + \sqrt{9k})$

61. $\sqrt[3]{4}(\sqrt[3]{2} - 3)$

62. $\sqrt[3]{5}(4\sqrt[3]{5} - \sqrt[3]{25})$

63. $2\sqrt[4]{2}(3\sqrt[4]{8} + 5\sqrt[4]{4})$

64. $6\sqrt[4]{9}(2\sqrt[4]{9} - \sqrt[4]{27})$

65. $(\sqrt[3]{2} - 1)(\sqrt[3]{4} + 3)$

66. $(\sqrt[3]{9} + 5)(\sqrt[3]{3} - 4)$

67. $(\sqrt[3]{5} - \sqrt[3]{4})(\sqrt[3]{25} + \sqrt[3]{20} + \sqrt[3]{16})$

68. $(\sqrt[3]{4} + \sqrt[3]{2})(\sqrt[3]{16} - \sqrt[3]{8} + \sqrt[3]{4})$

▶ **RELATING CONCEPTS (EXERCISES 69–74)**

Work Exercises 69–74 in order. *They are designed to help you see why a common student error is indeed an error.*

69. Use the distributive property to write $6(5 + 3x)$ as a sum.

70. Your answer in Exercise 69 should be $30 + 18x$. Why can we not combine these two terms to get $48x$?

71. Repeat Exercise 22 from earlier in this exercise set.

72. Your answer in Exercise 71 should be $30 + 18\sqrt{5}$. Many students will, in error, try to combine these terms to get $48\sqrt{5}$. Why is this wrong?

73. Write the expression similar to $30 + 18x$ that simplifies to $48x$. Then write the expression similar to $30 + 18\sqrt{5}$ that simplifies to $48\sqrt{5}$.

74. Write a short explanation of the similarities between combining like terms and combining like radicals.

Did you make the connection that the procedure used in combining radical terms is the same as that used in combining variable terms?

Solve each problem.

75. The radius of the circular top or bottom of a tin can with a surface area S and a height h is given by

$$r = \frac{-h + \sqrt{h^2 + .64S}}{2}.$$

What radius should be used to make a can with a height of 12 inches and a surface area of 400 square inches?

76. If an investment of P dollars grows to A dollars in two years, the annual rate of return on the investment is given by

$$r = \frac{\sqrt{A} - \sqrt{P}}{\sqrt{P}}.$$

First rationalize the denominator and then find the annual rate of return (as a percent) if $50,000 increases to $58,320.

9.6 Solving Equations with Radicals

OBJECTIVES

1. Solve equations with radicals.
2. Identify equations with no solutions.
3. Solve equations that require squaring a binomial.

CONNECTIONS

The most common formula for the area of a triangle is $A = \frac{1}{2}bh$, where b is the length of the base and h is the height. What if the height is not known? What if we know only the lengths of the sides? Another formula, known as *Heron's formula*, allows us to calculate the area of a triangle if we know the lengths of the sides a, b, and c. First let s equal the *semiperimeter*, which is one-half the perimeter.

$$s = \frac{1}{2}(a + b + c)$$

The area A is given by the formula

$$A = \sqrt{s(s - a)(s - b)(s - c)}.$$

For example, the familiar 3–4–5 right triangle has area

$$A = \frac{1}{2}(3)(4) = 6 \text{ square units,}$$

using the familiar formula. Using Heron's formula,

$$s = \frac{1}{2}(3 + 4 + 5) = 6.$$

Therefore,

$$A = \sqrt{6(6 - 3)(6 - 4)(6 - 5)}$$
$$= \sqrt{36}$$
$$= 6$$

So $A = 6$ square units, as expected.

CONNECTIONS (CONTINUED)

FOR DISCUSSION OR WRITING

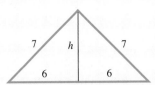

1. Use Heron's formula to find the area of a triangle with sides 7, 7, and 12.

2. The area of this triangle can be found with the formula $A = \frac{1}{2}bh$ as follows. Divide the triangle into two equal triangles as shown in the figure. Use the Pythagorean formula to find h using one of the small triangles. Note that h is the altitude of the original triangle. Now find the area using the formula $A = \frac{1}{2}bh$. Which way do you prefer?

OBJECTIVE 1 **Solve equations with radicals.** The addition and multiplication properties of equality are not enough to solve an equation with radicals such as

$$\sqrt{x + 1} = 3.$$

Solving equations that have square roots requires a new property, the **squaring property.**

Squaring Property of Equality

If both sides of a given equation are squared, all solutions of the original equation are *among* the solutions of the squared equation.

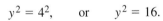

Be very careful with the squaring property. Using this property can give a new equation with *more* solutions than the original equation. For example, starting with the equation $y = 4$ and squaring each side gives

$$y^2 = 4^2, \qquad \text{or} \qquad y^2 = 16.$$

 This last equation, $y^2 = 16$, has *two* solutions, 4 or -4, while the original equation, $y = 4$, has only *one* solution, 4. Because of this possibility, checking is more than just a guard against algebraic errors when solving an equation with radicals. It is an essential part of the solution process. *All potential solutions from the squared equation must be checked in the original equation.*

EXAMPLE 1   Using the Squaring Property of Equality

Solve the equation $\sqrt{p + 1} = 3$.

Use the squaring property of equality to square both sides of the equation and then solve this new equation.

$$(\sqrt{p + 1})^2 = 3^2$$
$$p + 1 = 9 \qquad (\sqrt{p + 1})^2 = p + 1$$
$$p = 8 \qquad \text{Subtract 1.}$$

Now check this answer in the original equation.

$$\sqrt{p + 1} = 3$$
$$\sqrt{8 + 1} = 3 \quad ? \qquad \text{Let } p = 8.$$
$$\sqrt{9} = 3 \quad ?$$
$$3 = 3 \qquad \text{True}$$

Since this statement is true, the solution set of $\sqrt{p + 1} = 3$ is $\{8\}$. In this case the squared equation had just one solution, which also satisfied the original equation.

EXAMPLE 2   Using the Squaring Property with Radicals on Each Side

Solve $3\sqrt{x} = \sqrt{x + 8}$.

Squaring both sides gives

$$(3\sqrt{x})^2 = (\sqrt{x + 8})^2$$
$$3^2(\sqrt{x})^2 = (\sqrt{x + 8})^2 \qquad (ab)^2 = a^2b^2$$
$$9x = x + 8 \qquad (\sqrt{x})^2 = x; \ (\sqrt{x + 8})^2 = x + 8$$
$$8x = 8 \qquad \text{Subtract } x.$$
$$x = 1. \qquad \text{Divide by 8.}$$

Check this potential solution.

$$3\sqrt{x} = \sqrt{x + 8}$$
$$3\sqrt{1} = \sqrt{1 + 8} \quad ? \qquad \text{Let } x = 1.$$
$$3(1) = \sqrt{9} \quad ? \qquad \sqrt{1} = 1$$
$$3 = 3 \qquad \text{True}$$

The check shows that the solution set of the given equation is $\{1\}$.

OBJECTIVE 2   Identify equations with no solutions. Not all equations with radicals have a solution, as shown in Examples 3 and 4.

EXAMPLE 3   Using the Squaring Property When One Side Is Negative

Solve the equation $\sqrt{y} = -3$.

Square both sides.

$$(\sqrt{y})^2 = (-3)^2$$
$$y = 9$$

Check this proposed answer in the original equation.

$$\sqrt{y} = -3$$
$$\sqrt{9} = -3 \quad ? \qquad \text{Let } y = 9.$$
$$3 = -3 \qquad \text{False}$$

Since the statement $3 = -3$ is false, the number 9 is not a solution of the given equation and is said to be *extraneous*. In fact, $\sqrt{y} = -3$ has no solution. Since $\sqrt{y}$ represents the *nonnegative* square root of y, we might have seen immediately that there is no solution. The solution set is $\emptyset$.

RELATING CONCEPTS (EXERCISES 70–76)

In Chapter 5 we presented methods for factoring trinomials. Some trinomials cannot be factored using integer coefficients, however. There is a way to determine beforehand whether a trinomial of the form $ax^2 + bx + c$ can be factored using the discriminant.

Work Exercises 70–76 in order.

70. For the trinomial $ax^2 + bx + c$, the expression $b^2 - 4ac$ is called the discriminant. Where have you seen the discriminant before?

71. Each of the following trinomials is factorable. Find the discriminant for each one.
   (a) $18x^2 - 9x - 2$ **(b)** $5x^2 + 7x - 6$
   (c) $48x^2 + 14x + 1$ **(d)** $x^2 - 5x - 24$

72. What do you notice about the discriminants you found in Exercise 71.

73. Factor each of the trinomials in Exercise 71.

74. Each of the following trinomials is not factorable using the methods of Chapter 5. Find the discriminant for each one.
   (a) $2x^2 + x - 5$ **(b)** $2x^2 + x + 5$ **(c)** $x^2 + 6x + 6$ **(d)** $3x^2 + 2x - 9$

75. Are any of the discriminants you found in Exercise 74 perfect squares?

76. Make a conjecture (an educated guess) concerning when a trinomial of the form $ax^2 + bx + c$ is factorable. Then use your conjecture to determine whether each trinomial is factorable. (Do not actually factor.)
   (a) $42x^2 + 117x + 66$ **(b)** $99x^2 + 186x - 24$ **(c)** $58x^2 + 184x + 27$

Did you make the connection between the nature of the discriminant and whether a trinomial with that discriminant can be factored?

SUMMARY Exercises on Quadratic Equations

Four algebraic methods have now been introduced for solving quadratic equations written in the form $ax^2 + bx + c = 0$. The chart below shows some advantages and some disadvantages of each method.

Method	Advantages	Disadvantages
1. Factoring	It is usually the fastest method.	Not all equations can be solved by factoring. Some factorable polynomials are difficult to factor.
2. Square root property	It is the simplest method for solving equations of the form $(ax + b)^2 = $ a number.	Few equations are given in this form.
3. Completing the square	It can always be used. (Also, the procedure is useful in other areas of mathematics.)	It requires more steps than other methods.
4. Quadratic formula	It can always be used.	It is more difficult than factoring because of the $\sqrt{b^2 - 4ac}$ expression.

SUMMARY EXERCISES

Solve each quadratic equation by the method of your choice.

1. $s^2 = 36$

2. $x^2 + 3x = -1$

3. $y^2 - \dfrac{100}{81} = 0$

4. $81t^2 = 49$

5. $z^2 - 4z + 3 = 0$

6. $w^2 + 3w + 2 = 0$

7. $z(z - 9) = -20$

8. $x^2 + 3x - 2 = 0$

9. $(3k - 2)^2 = 9$

10. $(2s - 1)^2 = 10$

11. $(x + 6)^2 = 121$

12. $(5k + 1)^2 = 36$

13. $(3r - 7)^2 = 24$

14. $(7p - 1)^2 = 32$

15. $(5x - 8)^2 = -6$

16. $2t^2 + 1 = t$

17. $-2x^2 = -3x - 2$

18. $-2x^2 + x = -1$

19. $8z^2 = 15 + 2z$

20. $3k^2 = 3 - 8k$

21. $0 = -x^2 + 2x + 1$

22. $3x^2 + 5x = -1$

23. $5y^2 - 22y = -8$

24. $y(y + 6) + 4 = 0$

25. $(x + 2)(x + 1) = 10$

26. $16x^2 + 40x + 25 = 0$

27. $4x^2 = -1 + 5x$

28. $2p^2 = 2p + 1$

29. $3m(3m + 4) = 7$

30. $5x - 1 + 4x^2 = 0$

31. $\dfrac{r^2}{2} + \dfrac{7r}{4} + \dfrac{11}{8} = 0$

32. $t(15t + 58) = -48$

33. $9k^2 = 16(3k + 4)$

34. $\dfrac{1}{5}x^2 + x + 1 = 0$

35. $y^2 - y + 3 = 0$

36. $4m^2 - 11m + 8 = -2$

37. $-3x^2 + 4x = -4$

38. $z^2 - \dfrac{5}{12}z = \dfrac{1}{6}$

39. $5k^2 + 19k = 2k + 12$

40. $\dfrac{1}{2}n^2 - n = \dfrac{15}{2}$

41. $k^2 - \dfrac{4}{15} = -\dfrac{4}{15}k$

42. If $D > 0$ and $\dfrac{5 + \sqrt{D}}{3}$ is a solution of $ax^2 + bx + c = 0$, what must be another solution of the equation?

43. How would you respond to this statement? "Since I know how to solve quadratic equations by the factoring method, there is no reason for me to learn any other method of solving quadratic equations."

44. How many real solutions are there for a quadratic equation that has a negative number as its radicand in the quadratic formula?

10.4 Complex Numbers

OBJECTIVES

1. Write complex numbers as multiples of i.

2. Add and subtract complex numbers.

3. Multiply complex numbers.

4. Write complex number quotients in standard form.

5. Solve quadratic equations with complex number solutions.

As shown earlier in this chapter, some quadratic equations have no real number solutions. For example, the number

$$\frac{4 \pm \sqrt{-4}}{2},$$

which occurred in the solution of Example 5 in the previous section, is not a real number, because -4 appears as the radicand. For every quadratic equation to have a solution, we need a new set of numbers that includes the real numbers. This new set of numbers is defined using a new number i having the properties given below.

The Number i

$$i = \sqrt{-1} \quad \text{and} \quad i^2 = -1$$

OBJECTIVE 1 Write complex numbers as multiples of i. We can now write numbers like $\sqrt{-4}$, $\sqrt{-5}$, and $\sqrt{-8}$ as multiples of i, using a generalization of the product rule for radicals, as in the next example.

EXAMPLE 1 Simplifying Square Roots of Negative Numbers

Write each number as a multiple of i.

(a) $\sqrt{-4} = \sqrt{-1 \cdot 4} = \sqrt{-1} \cdot \sqrt{4} = i\sqrt{4} = i \cdot 2 = 2i$

(b) $\sqrt{-5} = \sqrt{-1 \cdot 5} = \sqrt{-1} \cdot \sqrt{5} = i\sqrt{5}$

(c) $\sqrt{-8} = i\sqrt{8} = i \cdot 2 \cdot \sqrt{2} = 2i\sqrt{2}$

 It is easy to mistake $\sqrt{2}\, i$ for $\sqrt{2i}$, with the i under the radical. For this reason, it is customary to write the i factor first when it is multiplied by a radical. For example, we usually write $i\sqrt{2}$ rather than $\sqrt{2}\, i$.

Numbers that are nonzero multiples of i are *imaginary numbers*. The *complex numbers* include all real numbers and all imaginary numbers.

Complex Number

A **complex number** is a number of the form $a + bi$, where a and b are real numbers. If $b \neq 0$, $a + bi$ is also an **imaginary number.**

For example, the real number 2 is a complex number since it can be written as $2 + 0i$. Also, the imaginary number $3i = 0 + 3i$ is a complex number. Other complex numbers are

$$3 - 2i, \quad 1 + i\sqrt{2}, \quad \text{and} \quad -5 + 4i.$$

In the complex number $a + bi$, a is called the **real part** and b (*not* bi) is called the **imaginary part.***

---

*Some texts *do* refer to bi as the imaginary part.

A complex number written in the form $a + bi$ (or $a + ib$) is in **standard form.** Figure 2 shows the relationships among the various types of numbers discussed in this book. (Compare this figure to Figure 8 in Chapter 1.)

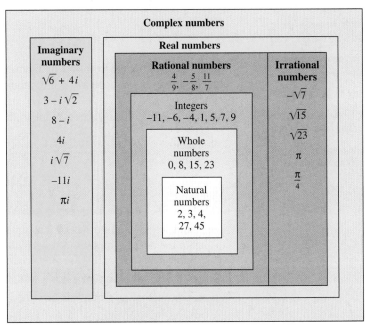

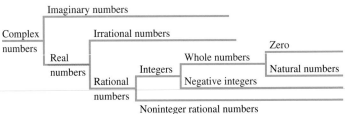

Figure 2

O B J E C T I V E   2   Add and subtract complex numbers. Adding and subtracting complex numbers is similar to adding and subtracting binomials.

Addition and Subtraction of Complex Numbers
1. To add complex numbers, add their real parts and add their imaginary parts.
2. To subtract complex numbers, change the number following the subtraction sign to its negative, and then add.

The properties of Section 1.7 (commutative, associative, etc.) also hold for operations with complex numbers.

┌ **E X A M P L E   2**   Adding and Subtracting Complex Numbers
│ Add or subtract.
│
│ **(a)** $(2 - 6i) + (7 + 4i) = (2 + 7) + (-6 + 4)i = 9 - 2i$
└ **(b)** $3i + (-2 - i) = -2 + (3 - 1)i = -2 + 2i$

(c) $(2 + 6i) - (-4 + i)$

Change $-4 + i$ to its negative, and then add.

$$(2 + 6i) - (-4 + i) = (2 + 6i) + (4 - i) \qquad \text{\small $-(-4 + i) = 4 - i$}$$
$$= (2 + 4) + (6 - 1)i \qquad \text{\small Commutative, associative, and}$$
$$\text{\small distributive properties}$$
$$= 6 + 5i$$

(d) $(-1 + 2i) - 4 = (-1 - 4) + 2i = -5 + 2i$

OBJECTIVE **3** Multiply complex numbers. We multiply complex numbers as we do polynomials. Since $i^2 = -1$ by definition, whenever i^2 appears, we replace it with -1.

EXAMPLE 3 Multiplying Complex Numbers

Find the following products.

(a) $3i(2 - 5i) = 6i - 15i^2 \qquad$ Distributive property
$$= 6i - 15(-1) \qquad \text{\small $i^2 = -1$}$$
$$= 6i + 15$$
$$= 15 + 6i \qquad \text{\small Commutative property}$$

The last step gives the result in standard form.

(b) $(4 - 3i)(2 + 5i)$

Use FOIL.

$$(4 - 3i)(2 + 5i) = 4(2) + 4(5i) + (-3i)(2) + (-3i)(5i)$$
$$= 8 + 20i - 6i - 15i^2$$
$$= 8 + 14i - 15(-1)$$
$$= 8 + 14i + 15$$
$$= 23 + 14i$$

(c) $(1 + 2i)(1 - 2i) = 1 - 2i + 2i - 4i^2$
$$= 1 - 4(-1)$$
$$= 1 + 4$$
$$= 5$$

OBJECTIVE **4** Write complex number quotients in standard form. The quotient of two complex numbers is expressed in standard form by changing the denominator into a real number. For example, to write

$$\frac{8 + i}{1 + 2i}$$

in standard form, the denominator must be a real number. As seen in Example 3(c), the product $(1 + 2i)(1 - 2i)$ is 5, a real number. This suggests multiplying the numerator and denominator of the given quotient by $1 - 2i$ as follows.

$$\frac{8 + i}{1 + 2i} = \frac{8 + i}{1 + 2i} \cdot \frac{1 - 2i}{1 - 2i}$$

$$= \frac{8 - 16i + i - 2i^2}{1 - 4i^2} \qquad \text{Multiply.}$$

$$= \frac{8 - 16i + i - 2(-1)}{1 - 4(-1)} \qquad i^2 = -1$$

$$= \frac{10 - 15i}{5} \qquad \text{Combine terms.}$$

$$= \frac{5(2 - 3i)}{5} = 2 - 3i \qquad \begin{array}{l}\text{Factor and write} \\ \text{in standard form.}\end{array}$$

Recall that this is the method used to rationalize some radical expressions in Chapter 9. The complex numbers $1 + 2i$ and $1 - 2i$ are *conjugates*. That is, the **conjugate** of the complex number $a + bi$ is $a - bi$. Multiplying the complex number $a + bi$ by its conjugate $a - bi$ gives the real number $a^2 + b^2$.

Product of Conjugates

$$(a + bi)(a - bi) = a^2 + b^2$$

The product of a complex number and its conjugate is the sum of the squares of the real and imaginary parts.

EXAMPLE 4   Dividing Complex Numbers

Write the following quotients in standard form.

(a) $\dfrac{-4 + i}{2 - i}$

Multiply numerator and denominator by $2 + i$, the conjugate of the denominator.

$$\frac{-4 + i}{2 - i} \cdot \frac{2 + i}{2 + i} = \frac{-8 - 4i + 2i + i^2}{4 - i^2}$$

$$= \frac{-8 - 2i - 1}{4 - (-1)} \qquad i^2 = -1$$

$$= \frac{-9 - 2i}{5}$$

$$= -\frac{9}{5} - \frac{2}{5}i \qquad \text{Standard form}$$

(b) $\dfrac{3 + i}{-i}$

Here, the conjugate of $0 - i$ is $0 + i$, or i.

$$\frac{3 + i}{-i} \cdot \frac{i}{i} = \frac{3i + i^2}{-i^2}$$

$$= \frac{-1 + 3i}{-(-1)} \qquad i^2 = -1; \text{ commutative property}$$

$$= -1 + 3i$$

CONNECTIONS

The complex number $a + bi$ is also written with the notation $\langle a, b \rangle$. (Note the similarity to an ordered pair.) This notation suggests a way to graph complex numbers on a plane in a manner similar to the way we graph ordered pairs. For graphing complex numbers, the x-axis is called the *real axis* and the y-axis is called the *imaginary axis*. For example, we graph the complex number $2 + 3i$ or $\langle 2, 3 \rangle$ just as we would the ordered pair $(2, 3)$, as shown in the figure. The figure also shows the graphs of the complex numbers $-1 - 4i$, $2i$, and -5.

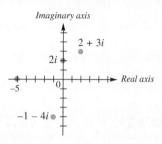

FOR DISCUSSION OR WRITING

1. Give the alternative notation for the last three complex numbers graphed above.

2. What is the real part of the complex number $-1 + 2i$? What is its imaginary part? Explain why we call the axes the real axis and the imaginary axis.

OBJECTIVE **5** Solve quadratic equations with complex number solutions. Quadratic equations that have no real solutions do have complex solutions, as shown in the next examples.

EXAMPLE 5 Solving a Quadratic Equation with Complex Solutions (Square Root Method)

Solve $(x + 3)^2 = -25$ for complex solutions.

Use the square root property.

$$(x + 3)^2 = -25$$
$$x + 3 = \sqrt{-25} \quad \text{or} \quad x + 3 = -\sqrt{-25}$$

Since $\sqrt{-25} = 5i$,

$$x + 3 = 5i \quad \text{or} \quad x + 3 = -5i$$
$$x = -3 + 5i \quad \text{or} \quad x = -3 - 5i.$$

The solution set is $\{-3 \pm 5i\}$.

EXAMPLE 6 Solving a Quadratic Equation with Complex Solutions (Quadratic Formula)

Solve $2p^2 = 4p - 5$ for complex solutions.

Write the equation as $2p^2 - 4p + 5 = 0$. Then $a = 2$, $b = -4$, and $c = 5$. The solutions are

$$p = \frac{-(-4) \pm \sqrt{(-4)^2 - 4(2)(5)}}{2(2)}$$
$$= \frac{4 \pm \sqrt{16 - 40}}{4}$$
$$= \frac{4 \pm \sqrt{-24}}{4}.$$

Since $\sqrt{-24} = i\sqrt{24} = i \cdot \sqrt{4} \cdot \sqrt{6} = i \cdot 2 \cdot \sqrt{6} = 2i\sqrt{6},$

$$p = \frac{4 \pm 2i\sqrt{6}}{4}$$

$$p = \frac{2(2 \pm i\sqrt{6})}{4} \qquad \text{Factor out a 2.}$$

$$p = \frac{2 \pm i\sqrt{6}}{2} \qquad \text{Lowest terms}$$

$$p = \frac{2}{2} \pm \frac{i\sqrt{6}}{2} \qquad \begin{array}{l}\text{Separate into real and}\\ \text{imaginary parts.}\end{array}$$

$$p = 1 \pm \frac{\sqrt{6}}{2}i. \qquad \text{Standard form}$$

The solution set is $\left\{1 \pm \frac{\sqrt{6}}{2}i\right\}$.

10.4 EXERCISES

Write each number as a multiple of i. See Example 1.

1. $\sqrt{-9}$ **2.** $\sqrt{-36}$ **3.** $\sqrt{-20}$ **4.** $\sqrt{-27}$

5. $\sqrt{-18}$ **6.** $\sqrt{-50}$ **7.** $\sqrt{-125}$ **8.** $\sqrt{-98}$

Add or subtract as indicated. See Example 2.

9. $(2 + 8i) + (3 - 5i)$ **10.** $(4 + 5i) + (7 - 2i)$

11. $(8 - 3i) - (2 + 6i)$ **12.** $(1 + i) - (3 - 2i)$

13. $(3 - 4i) + (6 - i) - (3 + 2i)$ **14.** $(5 + 8i) - (4 + 2i) + (3 - i)$

Find each product. See Example 3.

15. $(3 + 2i)(4 - i)$ **16.** $(9 - 2i)(3 + i)$ **17.** $(5 - 4i)(3 - 2i)$

18. $(10 + 6i)(8 - 4i)$ **19.** $(3 + 6i)(3 - 6i)$ **20.** $(11 - 2i)(11 + 2i)$

TECHNOLOGY INSIGHTS (EXERCISES 21-28)

Modern graphing calculators are capable of performing operations with complex numbers. The top screen shows how the TI-83 calculator can be set for complex mode $(a + bi)$. The lower left screen shows how the square root of a negative number returns the product of a real number and i, and how addition and subtraction of complex numbers is accomplished. The lower right screen shows how multiplication is performed, how the calculator returns the real part of a complex number, and how the conjugate is given.

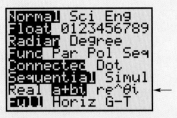

Predict the display the calculator will give for each of the following screens.

21.

    √(-169)+√(-25)

22.

    √(-36)+√(-100)

23.

    (5-i)+(-2+3i)-(7
    +2i)

24.

    3(1+2i)-4(3-i)

EXERCISES

(*Hint:* Use the order of operations.)

25.

    conj(6-4i)

26.

    imag(5+10i)

27.

    (15-5i)/(3+i)

28.

    (2+3i)²

Write each quotient in standard form. See Example 4.

EXERCISES

29. $\dfrac{17 + i}{5 + 2i}$ **30.** $\dfrac{21 + i}{4 + i}$ **31.** $\dfrac{40}{2 + 6i}$

32. $\dfrac{13}{3 + 2i}$ **33.** $\dfrac{i}{4 - 3i}$ **34.** $\dfrac{-i}{1 + 2i}$

When you first divided whole numbers, you probably learned to check your work by multiplying your answer (the quotient) by the number doing the dividing (the divisor). For example,

$$\frac{2744}{28} = 98 \text{ is true, because } 98 \times 28 = 2744.$$

This same procedure works for real numbers other than whole numbers. Does it work for complex numbers?

(continued)

▬ **RELATING CONCEPTS (EXERCISES 35-40) (CONTINUED)**

Work Exercises 35–40 in order to see whether it does or not.

35. Find the standard form of the quotient $\dfrac{-29 - 3i}{2 + 9i}$.

36. Multiply your answer from Exercise 35 by the divisor, $2 + 9i$. What is your answer? Is it equal to the original dividend (the numerator), $-29 - 3i$?

37. Find the standard form of the quotient $\dfrac{4 - 3i}{i}$.

38. Multiply your answer from Exercise 37 by the divisor, i. What is your answer? Is it equal to the original dividend?

39. Use the pattern established in Exercises 35–38 to determine whether the following is true: $\dfrac{14 - 5i}{4 + i} = 3 - 2i$.

40. State a rule, based on your observations, that tells whether the answer to a division problem involving complex numbers is correct.

Did you make the connection that checking complex number division is similar to checking real number division?

Solve each quadratic equation for complex solutions by the square root property. Write solutions in standard form. See Example 5.

41. $(a + 1)^2 = -4$ **42.** $(p - 5)^2 = -36$ **43.** $(k - 3)^2 = -5$

44. $(y + 6)^2 = -12$ **45.** $(3x + 2)^2 = -18$ **46.** $(4z - 1)^2 = -20$

Solve each quadratic equation for complex solutions by the quadratic formula. Write solutions in standard form. See Example 6.

47. $m^2 - 2m + 2 = 0$ **48.** $b^2 + b + 3 = 0$ **49.** $2r^2 + 3r + 5 = 0$

50. $3q^2 = 2q - 3$ **51.** $p^2 - 3p + 4 = 0$ **52.** $2a^2 = -a - 3$

53. $5x^2 + 3 = 2x$ **54.** $6y^2 + 2y + 1 = 0$ **55.** $2m^2 + 7 = -2m$

56. $4z^2 + 2z + 3 = 0$ **57.** $r^2 + 3 = r$ **58.** $4q^2 - 2q + 3 = 0$

Exercises 59–60 deal with quadratic equations having real number coefficients.

59. Suppose you are solving a quadratic equation by the quadratic formula. How can you tell, before completing the solution, whether the equation will have solutions that are not real numbers?

60. Refer to the solutions in Examples 5 and 6, and complete the following statement: If a quadratic equation has imaginary solutions, they are _____ of each other.

Answer true *or* false *to each of the following. If false, say why.*

61. Every real number is a complex number.

62. Every imaginary number is a complex number.

63. Every complex number is a real number.

64. Some complex numbers are imaginary.

65. Write a paragraph explaining how to add, subtract, multiply, and divide complex numbers. Give examples.

10.5 More on Graphing Quadratic Equations; Quadratic Functions

OBJECTIVES

OBJECTIVES

1 Graph quadratic equations of the form $y = ax^2 + bx + c$ ($a \neq 0$).

2 Use the vertex formula and then graph a parabola.

3 Use a graph to determine the number of real solutions of a quadratic equation.

4 Solve applications using quadratic functions.

In Section 4.1 we graphed the quadratic equation $y = x^2$. By plotting points, we obtained the graph of a parabola, shown again here in Figure 3.

x	y
3	9
2	4
1	1
0	0
-1	1
-2	4
-3	9

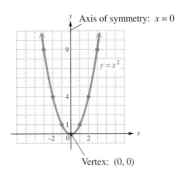

Figure 3

Recall that the lowest point on this graph is called the *vertex* of the parabola. (If the parabola opens downward, the vertex is the highest point.) The vertical line through the vertex is called the *axis,* or *axis of symmetry.* The two halves of the parabola are mirror images of each other across this axis.

We have also seen that other quadratic equations such as $y = -x^2 + 3$ and $y = (x + 2)^2$ also have parabolas as their graphs. See Figures 4 and 5.

x	y
2	-1
1	2
0	3
-1	2
-2	-1

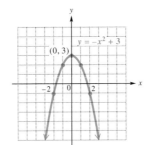

Figure 4

x	y
0	4
-1	1
-2	0
-3	1
-4	4

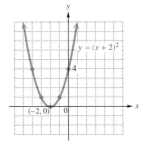

Figure 5

OBJECTIVE 1 Graph quadratic equations of the form $y = ax^2 + bx + c$ ($a \neq 0$). Every equation of the form

$$y = ax^2 + bx + c,$$

with $a \neq 0$, has a graph that is a parabola. Because of its many useful properties, the parabola occurs frequently in real-life applications. For example, if an object is thrown into the air, the path that the object follows is a parabola (ignoring wind resistance). Cross sections of radar, spotlight, and telescope reflectors also form parabolas.

When graphing parabolas, we are interested in finding the vertex, the x-intercept(s), if any, and the y-intercept. The first example shows how this is done.

E X A M P L E   1   Graphing a Parabola by Finding the Vertex and Intercepts

Graph $y = x^2 - 2x - 3$.

We want to find the vertex of the graph. Note in Figures 4 and 5 that the x-value of each vertex is exactly halfway between the x-intercepts. If a parabola has two x-intercepts this is always the case because of the symmetry of the figure. Therefore, begin by finding the x-intercepts. Let $y = 0$ in the equation and solve for x.

$$0 = x^2 - 2x - 3$$
$$0 = (x + 1)(x - 3) \qquad \text{Factor.}$$
$$x + 1 = 0 \qquad \text{or} \qquad x - 3 = 0 \qquad \text{Set each factor equal to 0.}$$
$$x = -1 \qquad \text{or} \qquad x = 3$$

There are two x-intercepts, $(-1, 0)$ and $(3, 0)$. Now find any y-intercepts. Substitute $x = 0$ in the equation.

$$y = 0^2 - 2(0) - 3 = -3$$

There is one y-intercept, $(0, -3)$.

As mentioned above, the x-value of the vertex is halfway between the x-values of the two x-intercepts. Thus, it is half their sum.

$$x = \frac{1}{2}(-1 + 3) = 1$$

Find the corresponding y-value by substituting 1 for x in the equation.

$$y = 1^2 - 2(1) - 3 = -4$$

The vertex is $(1, -4)$. The axis is the line $x = 1$. Plot the three intercepts and the vertex. Find additional ordered pairs as needed. For example, if $x = 2$,

$$y = 2^2 - 2(2) - 3 = -3,$$

leading to the ordered pair $(2, -3)$. A table of values with the ordered pairs we have found is shown with the graph in Figure 6.

	x	y
	-2	5
x-intercept	-1	0
y-intercept	0	-3
vertex	1	-4
	2	-3
x-intercept	3	0
	4	5

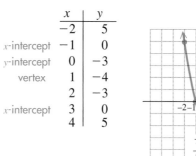

Figure 6

OBJECTIVE **2** Use the vertex formula and then graph a parabola. We can generalize from Example 1. The x-values of the x-intercepts for the equation $y = ax^2 + bx + c$, by the quadratic formula, are

$$x = \frac{-b + \sqrt{b^2 - 4ac}}{2a} \qquad \text{and} \qquad x = \frac{-b - \sqrt{b^2 - 4ac}}{2a}.$$

Thus, the x-value of the vertex is

$$x = \frac{1}{2}\left(\frac{-b + \sqrt{b^2 - 4ac}}{2a} + \frac{-b - \sqrt{b^2 - 4ac}}{2a}\right)$$

$$x = \frac{1}{2}\left(\frac{-b + \sqrt{b^2 - 4ac} - b - \sqrt{b^2 - 4ac}}{2a}\right)$$

$$x = \frac{1}{2}\left(\frac{-2b}{2a}\right) = -\frac{b}{2a}.$$

For the equation in Example 1, $y = x^2 - 2x - 3$, $a = 1$ and $b = -2$. Thus, the x-value of the vertex is

$$x = -\frac{b}{2a} = -\frac{-2}{2(1)} = 1,$$

which is the same x-value for the vertex we found in Example 1. (The x-value of the vertex is $x = -\frac{b}{2a}$ even if the graph has no x-intercepts.) A procedure for graphing quadratic equations follows.

Graphing the Parabola $y = ax^2 + bx + c$
Step 1 Find the y-intercept by letting $x = 0$.
Step 2 Find the x-intercepts, if any, by letting $y = 0$ and solving the equation.
Step 3 Find the vertex. Let $x = -\frac{b}{2a}$ and find the corresponding y-value by substituting for x in the equation.
Step 4 Plot the intercepts and the vertex.
Step 5 Find and plot additional ordered pairs near the vertex and intercepts as needed, using symmetry about the axis of the parabola.

EXAMPLE 2 Using the Steps to Graph a Parabola

Graph $y = x^2 - 4x + 1$.

Step 1 To find the y-intercept, let $x = 0$ to get $y = 0^2 - 4(0) + 1 = 1$. The y-intercept is $(0, 1)$.

Step 2 To find the x-intercept(s), if any, let $y = 0$ and solve the equation

$$x^2 - 4x + 1 = 0.$$

Using the quadratic formula with $a = 1$, $b = -4$, and $c = 1$, we get

$$x = 2 \pm \sqrt{3}.$$

A calculator shows that the x-intercepts are $(3.7, 0)$ and $(.3, 0)$ to the nearest tenth.

Step 3 The x-value of the vertex is

$$x = -\frac{b}{2a} = -\frac{-4}{2(1)} = 2.$$

The y-value of the vertex is

$$y = 2^2 - 4(2) + 1 = -3,$$

so the vertex is $(2, -3)$. The axis is the line $x = 2$.

Steps 4 and 5 A table of values of the points found so far, along with some others, is shown with the graph. Join these points with a smooth curve, as shown in Figure 7.

x	y
-1	6
0	1
$2 - \sqrt{3} \approx .3$	0
1	-2
2	-3
3	-2
$2 + \sqrt{3} \approx 3.7$	0
4	1
5	6

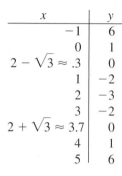

Figure 7

CONNECTIONS

As shown in the figure, the trajectory of a shell fired from a cannon is a parabola. To reach the maximum range with a cannon, it is shown in calculus that the muzzle must be set at 45°. If the muzzle is elevated above 45°, the shell goes too high and falls too soon. If the muzzle is set below 45°, the shell is rapidly pulled to Earth by gravity.

FOR DISCUSSION OR WRITING
If a shell is fired with an initial speed of 32 feet per second, and the muzzle is set at 45°, the equation of the parabolic trajectory is

$$y = x - \frac{1}{32}x^2,$$

where x is the horizontal distance in feet and y is the height in feet. At what distance is the maximum height attained and what is the maximum height? What is the maximum distance to where the shell strikes the ground?

OBJECTIVE **3** Use a graph to determine the number of real solutions of a quadratic equation. It can be veri-fied by the vertical line test (Section 7.3) that the graph of an equation of the form $y = ax^2 + bx + c$ is the graph of a function. A function defined by an equation of the form $f(x) = ax^2 + bx + c$ ($a \neq 0$) is called a **quadratic function.** The domain (pos-sible x-values) of a quadratic function is all real numbers, or $(-\infty, \infty)$; the range (the resulting y-values) can be determined after the function is graphed. In Example 2, the domain is $(-\infty, \infty)$ and the range is $[-3, \infty)$.

Look again at Figure 7, the graph of $y = f(x) = x^2 - 4x + 1$. Recall that setting y equal to 0 gives the x-intercepts, where the x-values are

$$2 + \sqrt{3} \approx 3.7 \quad \text{and} \quad 2 - \sqrt{3} \approx .3.$$

The solutions of $0 = x^2 - 4x + 1$ are the x-values of the x-intercepts of the graph of the corresponding quadratic function.

Intercepts of the Graph of a Quadratic Function

The real number solutions of a quadratic equation $ax^2 + bx + c = 0$ are the x-values of the x-intercepts of the graph of the corresponding quadratic function $f(x) = ax^2 + bx + c$.

Since the graph of a quadratic function can intersect the x-axis in two, one, or no points, this result shows why some quadratic equations have two, some have one, and some have no real solutions.

EXAMPLE 3 Determining the Number of Real Solutions from a Graph

(a) Figure 8 shows the graph of $f(x) = x^2 - 3$. The equation $0 = x^2 - 3$ has two real solutions, $\sqrt{3}$ and $-\sqrt{3}$, which correspond to the x-intercepts. The solution set is $\{\pm\sqrt{3}\}$.

(b) Figure 9 shows the graph of $f(x) = x^2 - 4x + 4$. The equation $0 = x^2 - 4x + 4$ has one real solution, 2, which corresponds to the x-intercept. The solution set is $\{2\}$.

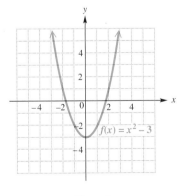

Figure 8

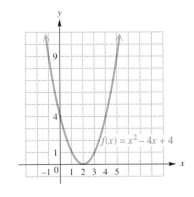

Figure 9

(c) Figure 10 shows the graph of $f(x) = x^2 + 2$. The equation $0 = x^2 + 2$ has no real solutions, since there are no x-intercepts. The solution set over the domain of real numbers is $\emptyset$. (The equation *does* have two imaginary solutions, $i\sqrt{2}$ and $-i\sqrt{2}$.)

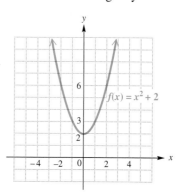

Figure 10

OBJECTIVE **4** Solve applications using quadratic functions. Because we can determine the coordinates of the vertex of the graph of a quadratic function, we are able to find the x-value that leads to the maximum or minimum y-value. This fact allows us to solve applications that lead to quadratic functions.

EXAMPLE 4 Solving a Problem Involving a Rectangular Region

A farmer wishes to enclose a rectangular region. He has 240 feet of fencing, and plans to use one side of his barn as part of the enclosure. See Figure 11. What dimensions should he use so that the enclosed region has maximum area? What will this maximum area be?

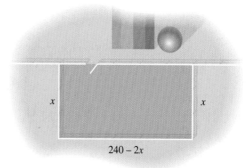

Figure 11

Let x represent the length of each of the two parallel sides of the enclosure. Then $240 - 2x$ represents the length of the third side formed from the fencing. Because area = length $\times$ width, the area of the region can be represented by the quadratic function

$$A(x) = x(240 - 2x)$$
$$A(x) = -2x^2 + 240x.$$

The graph of this function is a parabola that opens downward (because of the negative coefficient on x^2). The vertex is the highest point on the graph. Here, $a = -2$ and $b = 240$, so the x-coordinate of the vertex is

$$x = -\frac{b}{2a} = -\frac{240}{2(-2)} = 60.$$

So each of the two parallel sides of fencing should measure 60 feet. The remaining side will be $240 - 2(60) = 240 - 120 = 120$ feet long. The area of the region will be $60 \times 120 = 7200$ square feet. (This area can also be found by evaluating $A(60)$.)

Parabolic shapes are found all around us. Satellite dishes that deliver television signals are becoming more popular each year. Radio telescopes use parabolic reflectors to track incoming signals. The final example discusses how to describe a cross section of a parabolic dish using an equation.

E X A M P L E   5   Finding the Equation of a Parabolic Satellite Dish

The Parkes radio telescope has a parabolic dish shape with a diameter of 210 feet and a depth of 32 feet. (*Source:* J. Mar and H. Liebowitz, *Structure Technology for Large Radio and Radar Telescope Systems,* The MIT Press, Cambridge, MA, 1969.) Figure 12(a) shows a diagram of such a dish, and Figure 12(b) shows how a cross section of the dish can be modeled by a graph, with the vertex of the parabola at the origin of a coordinate system. Find the equation of this graph.

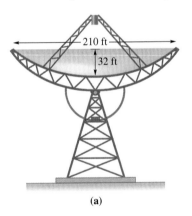

(a)

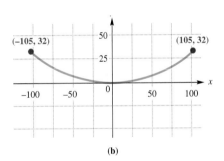

(b)

Figure 12

Because the vertex is at the origin, the equation will be of the form $y = ax^2$. As shown in Figure 12(b), one point on the graph has coordinates (105, 32). Letting $x = 105$ and $y = 32$, we can solve for a.

$$y = ax^2 \qquad \text{General equation}$$
$$32 = a(105)^2 \qquad \text{Substitute for } x \text{ and } y.$$
$$32 = 11{,}025a \qquad 105^2 = 11{,}025$$
$$a = \frac{32}{11{,}025} \qquad \text{Divide by } 11{,}025.$$

Thus the equation is $y = \frac{32}{11{,}025}x^2$.

10.5 EXERCISES

Fill in each blank with the correct response.

1. The highest or lowest point on the graph of a quadratic function is called the _____ of the graph.

2. The vertical line through the vertex of the graph of a quadratic function is called its _____.

3. The vertex of the graph of $f(x) = 2x^2 - 8x + 3$ is _____.

4. If the vertex of the graph of $f(x) = ax^2 + bx + c$ is above the x-axis and the graph opens downward, then the equation $f(x) = 0$ has _____ real solution(s).

Sketch the graph of each equation and give the coordinates of the vertex. See Examples 1 and 2.

5. $y = x^2 + 2x + 3$

6. $y = x^2 - 4x + 3$

7. $y = x^2 + 6x + 9$

8. $y = x^2 - 8x + 16$

9. $y = -x^2 + 6x - 5$

10. $y = -x^2 - 4x - 3$

11. $y = -x^2 + 4x - 4$

12. $y = -x^2 - 2x - 1$

Decide from each graph how many real solutions $f(x) = 0$ has. Then give the solution set (of real solutions). See Example 3.

13.
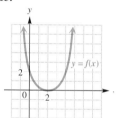

14.
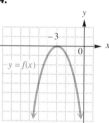

15.
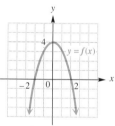

16.
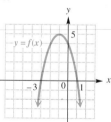

17.
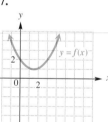

18.
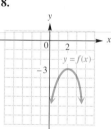

TECHNOLOGY INSIGHTS (EXERCISES 19-26)

A quadratic equation can be solved graphically using a graphing calculator. Assuming that the equation is in the form $ax^2 + bx + c = 0$, we enter $ax^2 + bx + c$ as Y_1, and then direct the calculator to find the x-intercepts of the graph. (These are also referred to as *zeros* of the function.) For example, to solve $x^2 - 5x - 6 = 0$ graphically, refer to the three screens on the next page. Notice that the displays at the bottoms of the lower two screens show the two solutions, -1 and 6.

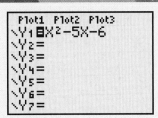

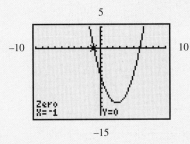
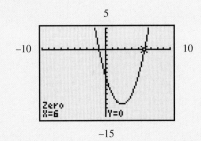

Determine the solution set of each quadratic equation by observing the corresponding screens. Then verify your answers by solving the quadratic equation using the method of your choice.

19. $x^2 - x - 6 = 0$

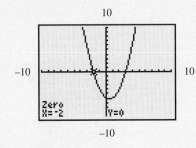
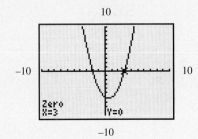

EXERCISES

20. $x^2 + 6x + 5 = 0$

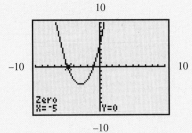
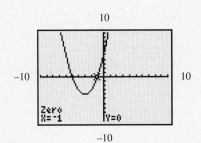

TECHNOLOGY INSIGHTS (EXERCISES 19–26) (CONTINUED)

21. $2x^2 - x - 3 = 0$

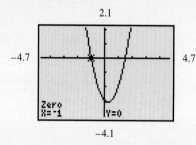

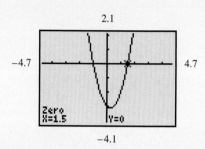

22. $4x^2 - 11x - 3 = 0$

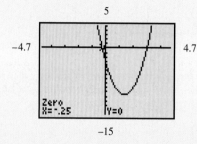

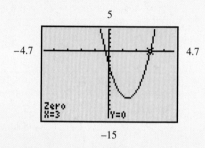

Another method of solving a quadratic equation graphically can be applied when the equation is not in standard form. For example, to solve $2x^2 = 9x + 5$, we graph the left side, $2x^2$, as Y_1 and the right side, $9x + 5$, as Y_2. Then we direct the calculator to find the coordinates of the points of intersection of the two graphs. As shown in the lower two screens below, the x-values of these points are $-.5$ and 5, giving the solution set $\{-.5, 5\}$.

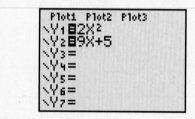

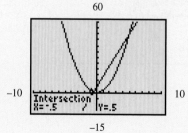

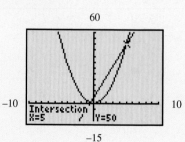

TECHNOLOGY INSIGHTS (EXERCISES 19-26) (CONTINUED)

Determine the solution set of each quadratic equation by observing the corresponding screens. Then verify your answer by solving the quadratic equation using the method of your choice.

23. $x^2 = -2x + 8$

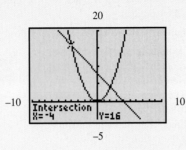 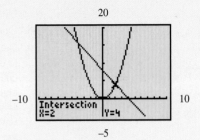

24. $x^2 = x + 6$

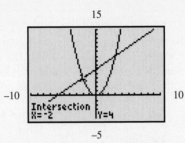 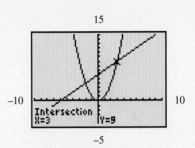

25. This table was generated by an equation of the form $Y_1 = ax^2 + bx + c$. Answer these questions by referring to the table.

    (a) What is the y-intercept of the graph?

    (b) What are the x-intercepts of the graph?

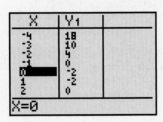

26. The graph shown here is that of the equation from Exercise 25. The calculator has the capability of finding the vertex. If the parabola opens upward, as it does in this case, the vertex is a *minimum*. The equation is $y = x^2 - x - 2$. Show *algebraically* that the point $(.5, -2.25)$ lies on the graph, by letting $x = .5$ and solving for y.

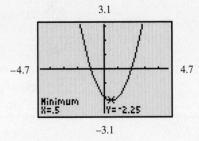

Find the domain and the range of each function graphed in the indicated exercise.

27. Exercise 13   **28.** Exercise 14   **29.** Exercise 15

30. Exercise 16   **31.** Exercise 17   **32.** Exercise 18

Given $f(x) = 2x^2 - 5x + 3$, *find each of the following.*

33. $f(0)$ **34.** $f(1)$ **35.** $f(-2)$ **36.** $f(-1)$

Use a quadratic function to solve each problem. See Examples 4 and 5.

37. Find two numbers whose sum is 80 and whose product is a maximum. (*Hint:* Let x represent one of the numbers. Then $80 - x$ represents the other. A quadratic function represents their product.)

38. Find two numbers whose sum is 300 and whose product is a maximum.

39. Delgado Community College has plans to construct a rectangular parking lot on land bordered on one side by a highway. There are 1280 feet of fencing available to fence the three other sides. Find the dimensions that will maximize the area of the region. What is this area?

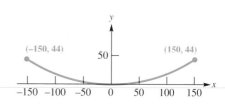

40. Repeat Exercise 39, but assume that there are 3000 feet of fencing available.

41. The U.S. Naval Research Laboratory designed a giant radio telescope that had a diameter of 300 feet and maximum depth of 44 feet. The graph depicts a cross section of this telescope. Find the equation of this parabola. (*Source:* J. Mar and H. Liebowitz, *Structure Technology for Large Radio and Radar Telescope Systems,* The MIT Press, Cambridge, MA, 1969.)

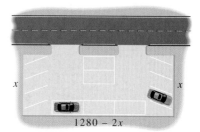

42. Suppose the telescope in Exercise 41 had a diameter of 400 feet and maximum depth of 50 feet. Find the equation of this parabola.

Exercises 43–48 use a graphing calculator to illustrate how the graph of $y = x^2$ can be transformed by using arithmetic operations.

43. In the standard viewing window of your calculator, graph the following one at a time, leaving the previous graphs on the screen as you move along.

$$Y_1 = x^2 \qquad Y_2 = 2x^2 \qquad Y_3 = 3x^2 \qquad Y_4 = 4x^2$$

Describe the effect the successive coefficients have on the parabola.

44. Repeat Exercise 43 for the following.

$$Y_1 = x^2 \qquad Y_2 = \frac{1}{2}x^2 \qquad Y_3 = \frac{1}{4}x^2 \qquad Y_4 = \frac{1}{8}x^2$$

45. Graph the pair of parabolas $Y_1 = x^2$ and $Y_2 = -x^2$ on the same screen. In your own words, describe how the graph of Y_2 can be obtained from the graph of Y_1.

46. Graph $Y_1 = -x^2$, $Y_2 = -2x^2$, $Y_3 = -3x^2$ and $Y_4 = -4x^2$ on the same screen. Make a conjecture about what happens when the coefficient of x^2 is negative.

47. In the standard viewing window of your calculator, graph the following one at a time, leaving the previous graphs on the screen as you move along.

$$Y_1 = x^2 \qquad Y_2 = x^2 + 3 \qquad Y_3 = x^2 - 6$$

Describe the effect that adding or subtracting a constant has on the parabola.

48. Repeat Exercise 47 for the following.

$$Y_1 = x^2 \qquad Y_2 = (x + 3)^2 \qquad Y_3 = (x - 6)^2$$

49. Write an explanation of how to graph a quadratic equation such as $y = x^2 - 4x - 5$.

50. The equation $x = y^2 - 2y + 3$ has a parabola with a horizontal axis of symmetry as its graph. Sketch the graph of this equation by letting y take on the values $-1, 0, 1, 2,$ and 3. Then write an explanation of why this is not the graph of a function.

CHAPTER 10 GROUP ACTIVITY

How Is a Radio Telescope Like a Wok?

Objective: Determine appropriate domains and ranges when graphing parabolas on a graphing calculator.

Example 5 of Section 10.5 explored the parabolic shape of a radio telescope. This activity will use that example and another to show the importance of setting appropriate domains and ranges on a graphing calculator. One student should write the answers to the questions while the other one does the graphing on a calculator. When you start part B, switch tasks.

A. In Example 5 of Section 10.5, an equation was found to represent the shape of the Parkes radio telescope. Using this equation, graph the parabola on a graphing calculator.

1. Use a standard viewing window, that is, Xmin $= -10$, Xmax $= 10$, Ymin $= -10$, Ymax $= 10$. Describe your graph. Does its shape look similar to the picture of the telescope in Section 10.5?

2. Change the domain and range settings to Xmin $= -50$, Xmax $= 50$, Ymin $= -5$, Ymax $= 50$. How does the graph look now? You may want to change Xscl and Yscl to 5.

3. Continue to adjust the domain and range settings until the calculator graph is close in shape to the picture of the telescope in Section 10.5. What are your settings?

B. A wok used in oriental cooking is also a parabola. Just as the shape of a parabola focuses radio waves from space, a wok focuses heat and oil for cooking.

(continued)

1. Find an equation that would model the cross section of a wok with diameter 14 inches and depth 4 inches.

2. Graph this equation using a standard viewing window on a calculator. Describe the graph of the equation. Describe the differences between this graph and the graph in part A, Exercise 1.

3. Adjust the settings for domain and range until the graph looks similar to the cross section of a wok.

4. How are a wok and a radio telescope alike?

5. How is the range (or domain) of a function different from the range (or domain) setting on a graphing calculator?

CHAPTER 10 SUMMARY

KEY TERMS

10.1 quadratic equation	real part	conjugate (of a	axis (of symmetry)
10.3 discriminant	imaginary part	complex number)	quadratic function
10.4 complex number	standard form (of a	**10.5** parabola	
imaginary number	complex number)	vertex	

NEW SYMBOLS

± positive or negative (plus or minus)	**i** $i = \sqrt{-1}$ and $i^2 = -1$

TEST YOUR WORD POWER

See how well you have learned the vocabulary in this chapter. Answers, with examples, are given at the bottom of the page.

1. A **quadratic equation** is an equation that can be written in the form
(a) $Ax + By = C$
(b) $ax^2 + bx + c = 0$
(c) $Ax + B = 0$
(d) $y = mx + b$.

2. A **complex number** is defined as
(a) a real number that includes a complex fraction
(b) a nonzero multiple of i
(c) a number of the form $a + bi$, where a and b are real numbers
(d) the square root of -1.

3. An **imaginary number** is
(a) a complex number $a + bi$ where $b \neq 0$
(b) a number that does not exist
(c) a complex number $a + bi$ where $b = 0$
(d) any real number.

4. A **parabola** is the graph of
(a) any equation in two variables
(b) a linear equation
(c) an equation of degree three
(d) a quadratic equation.

5. The **vertex** of a parabola is
(a) the point where the graph intersects the y-axis
(b) the point where the graph intersects the x-axis
(c) the lowest point on a parabola that opens up or the highest point on a parabola that opens down
(d) the origin.

6. The **axis** of a vertical parabola is
(a) either the x-axis or the y-axis
(b) the vertical line through the vertex
(c) the horizontal line through the vertex
(d) the x-axis.

Answers to Test Your Word Power

1. (b) *Examples:* $z^2 + 6z + 9 = 0$, $y^2 - 2y = 8$, $(x + 3)(x - 1) = 5$ **2.** (c) *Examples:* -5 (or $-5 + 0i$), $7i$ (or $0 + 7i$), $\sqrt{2}, -4i$
3. (a) *Examples:* $2i$, $-13i$, $3 + i\sqrt{6}$ **4.** (d) *Examples:* See Figures 3–10 in Section 10.5. **5.** (c) *Example:* The graph of $y = (x + 3)^2$ has
vertex $(-3, 0)$, which is the lowest point on the graph. **6.** (b) *Example:* The axis of $y = (x + 3)^2$ is the line $x = -3$.

QUICK REVIEW

| CONCEPTS | EXAMPLES |

10.1 SOLVING QUADRATIC EQUATIONS BY THE SQUARE ROOT PROPERTY

Square Root Property of Equations

If k is positive, and if $a^2 = k$, then $a = \sqrt{k}$ or $a = -\sqrt{k}$.

Solve $(2x + 1)^2 = 5$.

$$2x + 1 = \pm\sqrt{5}$$
$$2x = -1 \pm \sqrt{5}$$
$$x = \frac{-1 \pm \sqrt{5}}{2}$$

Solution set: $\left\{ \dfrac{-1 \pm \sqrt{5}}{2} \right\}$

10.2 SOLVING QUADRATIC EQUATIONS BY COMPLETING THE SQUARE

Completing the Square

1. If the coefficient of the squared term is 1, go to Step 2. If it is not 1, divide each side of the equation by this coefficient.

2. Make sure that all variable terms are on one side of the equation and all constant terms are on the other.

3. Take half the coefficient of x, square it, and add the square to each side of the equation. Factor the variable side and combine terms on the other.

4. Use the square root property to solve the equation.

Solve $2x^2 + 4x - 1 = 0$.

$$x^2 + 2x - \frac{1}{2} = 0$$

$$x^2 + 2x = \frac{1}{2}$$

$$x^2 + 2x + 1 = \frac{1}{2} + 1$$

$$(x + 1)^2 = \frac{3}{2}$$

$$x + 1 = \pm \sqrt{\frac{3}{2}} = \pm \frac{\sqrt{6}}{2}$$

$$x = -1 \pm \frac{\sqrt{6}}{2}$$

$$x = \frac{-2 \pm \sqrt{6}}{2}$$

Solution set: $\left\{ \dfrac{-2 \pm \sqrt{6}}{2} \right\}$

10.3 SOLVING QUADRATIC EQUATIONS BY THE QUADRATIC FORMULA

Quadratic Formula

The solutions of $ax^2 + bx + c = 0$, $(a \neq 0)$, are

$$x = \frac{-b \pm \sqrt{b^2 - 4ac}}{2a}.$$

The discriminant for the quadratic equation is $b^2 - 4ac$.
If $b^2 - 4ac > 0$, there are two real solutions.
If $b^2 - 4ac = 0$, there is one real solution.
If $b^2 - 4ac < 0$, there are no real solutions.

Solve $3x^2 - 4x - 2 = 0$.

$$x = \frac{-(-4) \pm \sqrt{(-4)^2 - 4(3)(-2)}}{2(3)} \qquad a = 3, b = -4, c = -2$$

$$x = \frac{4 \pm \sqrt{16 + 24}}{6}$$

$$x = \frac{4 \pm \sqrt{40}}{6} = \frac{4 \pm 2\sqrt{10}}{6} \qquad \begin{array}{l} \text{Discriminant: 40} \\ \text{There are two} \\ \text{real solutions.} \end{array}$$

$$x = \frac{2(2 \pm \sqrt{10})}{6} = \frac{2 \pm \sqrt{10}}{3}$$

Solution set: $\left\{ \dfrac{2 \pm \sqrt{10}}{3} \right\}$

10.4 COMPLEX NUMBERS

The number $i = \sqrt{-1}$ and $i^2 = -1$.
For the positive number b,

$$\sqrt{-b} = i\sqrt{b}.$$

Simplify: $\sqrt{-19}$.

$$\sqrt{-19} = \sqrt{-1 \cdot 19} = i\sqrt{19}$$

(continued)

CONCEPTS	EXAMPLES

Addition

Add complex numbers by adding the real parts and adding the imaginary parts.

Add: $(3 + 6i) + (-9 + 2i)$.

$$(3 + 6i) + (-9 + 2i) = (3 - 9) + (6 + 2)i$$
$$= -6 + 8i$$

Subtraction

To subtract complex numbers, change the number following the subtraction sign to its negative and add.

Subtract: $(5 + 4i) - (2 - 4i)$.

$$(5 + 4i) - (2 - 4i) = (5 + 4i) + (-2 + 4i)$$
$$= (5 - 2) + (4 + 4)i$$
$$= 3 + 8i$$

Multiplication

Multiply complex numbers in the same way polynomials are multiplied. Replace i^2 with -1.

Multiply: $(7 + i)(3 - 4i)$.

$$(7 + i)(3 - 4i)$$
$$= 7(3) + 7(-4i) + i(3) + i(-4i) \qquad \text{FOIL}$$
$$= 21 - 28i + 3i - 4i^2$$
$$= 21 - 25i - 4(-1) \qquad\qquad i^2 = -1$$
$$= 21 - 25i + 4$$
$$= 25 - 25i$$

Division

Divide complex numbers by multiplying the numerator and the denominator by the conjugate of the denominator.

Divide: $\dfrac{2}{6 + i}$.

$$\frac{2}{6 + i} = \frac{2}{6 + i} \cdot \frac{6 - i}{6 - i}$$

$$= \frac{2(6 - i)}{36 - i^2}$$

$$= \frac{12 - 2i}{36 + 1}$$

$$= \frac{12 - 2i}{37}$$

$$= \frac{12}{37} - \frac{2}{37}i \qquad \text{Standard form}$$

Complex Solutions

A quadratic equation may have nonreal, complex solutions. This occurs when the discriminant is negative. The quadratic formula will give complex solutions in such cases.

Solve for all complex solutions of

$$x^2 + x + 1 = 0.$$

Here, $a = 1$, $b = 1$, and $c = 1$.

$$x = \frac{-1 \pm \sqrt{1^2 - 4(1)(1)}}{2(1)}$$

$$x = \frac{-1 \pm \sqrt{1 - 4}}{2}$$

$$x = \frac{-1 \pm \sqrt{-3}}{2} \qquad \text{Discriminant: } -3$$

$$x = \frac{-1 \pm i\sqrt{3}}{2} \qquad \text{Two nonreal solutions}$$

Solution set: $\left\{ -\dfrac{1}{2} \pm \dfrac{\sqrt{3}}{2}i \right\}$

10.5 MORE ON GRAPHING QUADRATIC EQUATIONS; QUADRATIC FUNCTIONS

Graphing $y = ax^2 + bx + c$
1. Find the y-intercept.

Graph $y = 2x^2 - 5x - 3$.
$y = 2(0)^2 - 5(0) - 3 = -3$
The y-intercept is $(0, -3)$.

CONCEPTS	EXAMPLES

2. Find any x-intercepts.

$$0 = 2x^2 - 5x - 3$$
$$0 = (2x + 1)(x - 3)$$

$$2x + 1 = 0 \qquad \text{or} \qquad x - 3 = 0$$
$$2x = -1 \qquad \text{or} \qquad x = 3$$
$$x = -\frac{1}{2} \qquad \text{or} \qquad x = 3$$

The x-intercepts are $\left(-\frac{1}{2}, 0\right)$ and $(3, 0)$.

3. Find the vertex: $x = -\dfrac{b}{2a}$; find y by substituting this value for x in the equation.

For the vertex:

$$x = -\frac{b}{2a} = -\frac{-5}{2(2)} = \frac{5}{4}$$

$$y = 2\left(\frac{5}{4}\right)^2 - 5\left(\frac{5}{4}\right) - 3$$

$$y = 2\left(\frac{25}{16}\right) - \frac{25}{4} - 3$$

$$y = \frac{25}{8} - \frac{50}{8} - \frac{24}{8} = -\frac{49}{8} = -6\frac{1}{8}.$$

The vertex is $\left(\frac{5}{4}, -\frac{49}{8}\right)$.

4. Plot the intercepts and the vertex.
5. Find and plot additional ordered pairs near the vertex and intercepts as needed.

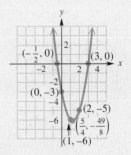

The number of real solutions of the equation

$$ax^2 + bx + c = 0$$

can be determined from the number of x-intercepts of the graph of

$$y = ax^2 + bx + c.$$

The figure shows that the equation

$$2x^2 - 5x - 3 = 0$$

has two real solutions.

CHAPTER 10 REVIEW EXERCISES

[10.1]   *Solve each equation by using the square root property. Give only real number solutions. Express all radicals in simplest form.*

1. $y^2 = 144$ **2.** $x^2 = 37$ **3.** $m^2 = 128$ **4.** $(k + 2)^2 = 25$

5. $(r - 3)^2 = 10$ **6.** $(2p + 1)^2 = 14$ **7.** $(3k + 2)^2 = -3$

8. Which one of the following equations has two real solutions?

   (a) $x^2 = 0$ **(b)** $x^2 = -4$ **(c)** $(x + 5)^2 = -16$ **(d)** $(x + 6)^2 = 25$

[10.2] *Solve each equation by completing the square. Give only real number solutions.*

9. $m^2 + 6m + 5 = 0$

10. $p^2 + 4p = 7$

11. $-x^2 + 5 = 2x$

12. $2y^2 - 3 = -8y$

13. $5k^2 - 3k - 2 = 0$

14. $(4a + 1)(a - 1) = -7$

Solve each problem.

15. If an object is thrown upward on Earth from a height of 50 feet, with an initial velocity of 32 feet per second, then its height after t seconds is given by $h = -16t^2 + 32t + 50$, where h is in feet. After how many seconds will it reach a height of 30 feet?

16. Find the lengths of the three sides of the right triangle shown.

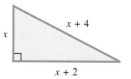

17. What must be added to $x^2 + kx$ to make it a perfect square?

[10.3]

18. Consider the equation $x^2 - 9 = 0$.
   (a) Solve the equation by factoring.
   (b) Solve the equation by the square root property.
   (c) Solve the equation by the quadratic formula.
   (d) Compare your answers. If a quadratic equation can be solved by both the factoring and the quadratic formula methods, should you always get the same results? Explain.

Solve each equation by using the quadratic formula. Give only real number solutions.

19. $x^2 - 2x - 4 = 0$

20. $3k^2 + 2k = -3$

21. $2p^2 + 8 = 4p + 11$

22. $-4x^2 + 7 = 2x$

23. $\frac{1}{4}p^2 = 2 - \frac{3}{4}p$

24. What is the value of the discriminant for $3x^2 - x - 2 = 0$? How many real solutions does this equation have?

[10.4]

25. Write an explanation of the method used to divide complex numbers.

26. Use the fact that $i^2 = -1$ to complete each of the following. Do them in order.
   (a) Since $i^3 = i^2 \cdot i$, $i^3 =$ _____ .
   (b) Since $i^4 = i^3 \cdot i$, $i^4 =$ _____ .
   (c) Since $i^{48} = (i^4)^{12}$, $i^{48} =$ _____ .

Perform the indicated operations.

27. $(3 + 5i) + (2 - 6i)$

28. $(-2 - 8i) - (4 - 3i)$

29. $(6 - 2i)(3 + i)$

30. $(2 + 3i)(2 - 3i)$

31. $\dfrac{1 + i}{1 - i}$

32. $\dfrac{5 + 6i}{2 + 3i}$

33. $\dfrac{1}{7 - i}$

34. What is the conjugate of the real number a?

35. Is it possible to multiply a complex number by its conjugate and get an imaginary product? Explain.

Find the complex solutions of each quadratic equation.

36. $(m + 2)^2 = -3$

37. $(3p - 2)^2 = -8$

38. $3k^2 = 2k - 1$

39. $h^2 + 3h = -8$

40. $4q^2 + 2 = 3q$

41. $9z^2 + 2z + 1 = 0$

[10.5]   *Sketch the graph of each equation and identify the vertex.*

42. $y = x^2 - 2x + 1$
43. $y = -x^2 + 2x + 3$
44. $y = x^2 + 4x + 2$

Decide from the graph how many real number solutions the equation $f(x) = 0$ has. Determine the solution set (of real solutions) for $f(x) = 0$ from the graph.

45.

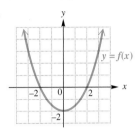

46.

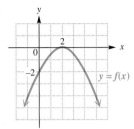

47.

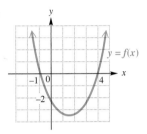

48.

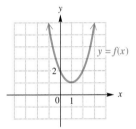

TECHNOLOGY INSIGHTS (EXERCISES 49 AND 50)

Follow the directions for Exercises 45–48.

49.

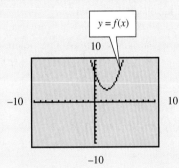

50.

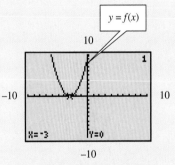

The parabola is tangent to the *x*-axis (that is, it touches at only one point).

MIXED REVIEW EXERCISES

Solve by any method. Give only real number solutions.

51. $(2t - 1)(t + 1) = 54$
52. $(2p + 1)^2 = 100$

53. $(k + 2)(k - 1) = 3$
54. $6t^2 + 7t - 3 = 0$

55. $2x^2 + 3x + 2 = x^2 - 2x$
56. $x^2 + 2x + 5 = 7$

57. $m^2 - 4m + 10 = 0$
58. $k^2 - 9k + 10 = 0$

59. $(3x + 5)^2 = 0$
60. $\frac{1}{2}r^2 = \frac{7}{2} - r$

61. $x^2 + 4x = 1$
62. $7x^2 - 8 = 5x^2 + 8$

63. Becky and Brad are the owners of Cole's Baseball Cards. They have found that the price p, in dollars, of a particular Brad Radke baseball card depends on the demand d, in hundreds, for the card, according to the formula $p = (d - 2)^2$. What demand produces a price of $5 for the card?

64. Use a calculator and the Pythagorean formula to find the lengths of the sides of the triangle to the nearest thousandth.

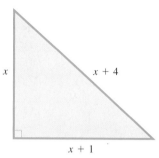

x $x + 4$

$x + 1$

RELATING CONCEPTS (EXERCISES 65-70)

In courses such as Intermediate Algebra and College Algebra, we learn that if r and s are solutions of the equation $x^2 + bx + c = 0$, then $x^2 + bx + c$ factors as $(x - r)(x - s)$. For example, since 2 and 5 are solutions of $x^2 - 7x + 10 = 0$, we have

$$x^2 - 7x + 10 = (x - 2)(x - 5).$$

In Chapter 5 we learned various methods of factoring polynomials using integer coefficients. Now, with the property stated above, we can factor any trinomial of the form $x^2 + bx + c$.

Work Exercises 65–70 in order.

65. Solve the quadratic equation $x^2 - 2x - 2 = 0$ using the quadratic formula. It has two real solutions, both of which are irrational numbers.

66. Suppose that r and s represent your solutions from Exercise 65. Write the trinomial $x^2 - 2x - 2$ in the factored form $(x - r)(x - s)$. Use parentheses carefully.

67. Regroup the terms in the factors obtained in Exercise 66 so that in each factor, the first two terms are grouped. These "binomials within binomials" should be the same in each factor.

68. Multiply your factors in Exercise 67 by using the special product $(a + b)(a - b) = a^2 - b^2$. Then simplify by using the special product $(a + b)^2 = a^2 + 2ab + b^2$. Combine terms.

69. Compare your answer in Exercise 68 to the trinomial given in Exercise 66. They should be the same.

70. Use the method of Exercises 65–69 to factor $x^2 - 4x - 1$. (This process is called *factoring over the real numbers*.)

Did you make the connection between factors of a polynomial and solutions of the corresponding equation?

CHAPTER 10 TEST

*Items marked * require knowledge of complex numbers.*

Solve by using the square root property.

1. $x^2 = 39$

2. $(y + 3)^2 = 64$

3. $(4x + 3)^2 = 24$

Solve by completing the square.

4. $x^2 - 4x = 6$

5. $2x^2 + 12x - 3 = 0$

6. **(a)** Find the discriminant for the quadratic equation $4x^2 - 20x + 25 = 0$.
   (b) Based on your answer to part (a), how many real solutions does the equation have?

Solve by the quadratic formula.

7. $2x^2 + 5x - 3 = 0$

8. $3w^2 + 2 = 6w$

9. $4x^2 + 8x + 11 = 0$

10. $t^2 - \dfrac{5}{3}t + \dfrac{1}{3} = 0$

Solve by the method of your choice.

11. $p^2 - 2p - 1 = 0$

12. $(2x + 1)^2 = 18$

13. $(x - 5)(2x - 1) = 1$

14. $t^2 + 25 = 10t$

Solve each problem.

15. If an object is propelled into the air from ground level on Earth with an initial velocity of 64 feet per second, its height s (in feet) after t seconds is given by the formula $s = -16t^2 + 64t$. After how many seconds will the object reach a height of 64 feet?

16. Use the formula from Kepler's third law of planetary motion, $P^2 = a^3$, to determine P, in years, for Pluto, if $a = 39.4$ AU.

Perform the indicated operations.

17. $(3 + i) + (-2 + 3i) - (6 - i)$

18. $(6 + 5i)(-2 + i)$

19. $(3 - 8i)(3 + 8i)$

20. $\dfrac{15 - 5i}{7 + i}$

Sketch each graph and identify the vertex.

21. $y = x^2 - 6x + 9$

22. $y = -x^2 - 2x - 4$

23. $f(x) = x^2 + 6x + 7$

24. Refer to the equation in Exercise 23, and do the following:
   (a) Determine the number of real solutions of $x^2 + 6x + 7 = 0$ by looking at the graph.
   (b) Use the quadratic formula to find the exact values of the real solutions. Give the solution set.
   (c) Use a calculator to find approximations for the solutions. Round your answers to the nearest thousandth.

25. Find two numbers whose sum is 400 and whose product is a maximum.

CUMULATIVE REVIEW EXERCISES CHAPTERS 1–10

Note: This cumulative review exercise set can be considered a final examination for the course.

Perform the indicated operations.

1. $\dfrac{-4 \cdot 3^2 + 2 \cdot 3}{2 - 4 \cdot 1}$

2. $|-3| - |1 - 6|$

3. $-9 - (-8)(2) + 6 - (6 + 2)$

4. $-4r + 14 + 3r - 7$

5. $13k - 4k + k - 14k + 2k$

6. $5(4m - 2) - (m + 7)$

Solve each equation.

7. $x - 5 = 13$

8. $3k - 9k - 8k + 6 = -64$

9. $2(m - 1) - 6(3 - m) = -4$

Solve each problem.

10. Find the measures of the marked angles.

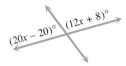

11. The perimeter of a basketball court is 288 feet. The width of the court is 44 feet less than the length. What are the dimensions of the court?

L

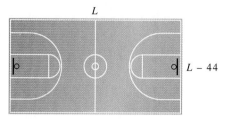

$L - 44$

12. Solve the formula $P = 2L + 2W$ for L.

Solve each inequality and graph the solution set.

13. $-8m < 16$

14. $-9p + 2(8 - p) - 6 \geq 4p - 50$

15. Graph the equation $2x + 3y = 6$.

16. Find the slope of the line passing through the points $(5, -2)$ and $(1, 2)$.

17. Subtract: $(5x^5 - 9x^4 + 8x^2) - (9x^2 + 8x^4 - 3x^5)$.

18. Complete the table and graph $y = x^2 - 3$.

x	y
-2	
-1	
0	
1	
2	

Simplify each expression. Write answers with positive exponents.

19. $(3^2 \cdot x^{-4})^{-1}$

20. $\left(\dfrac{b^{-3}c^4}{b^5c^3}\right)^{-2}$

21. $\left(\dfrac{5}{3}\right)^{-3}$

Perform each indicated operation.

22. $(2x + 3)(x - 6)$

23. $(2x - 5)(x^3 + 3x^2 - 2x - 4)$

24. $\dfrac{3x^3 + 10x^2 - 7x + 4}{x + 4}$

25. (a) The number of possible hands in contract bridge is about 6,350,000,000. Write this number in scientific notation.

(b) The body of a 150-pound person contains about 2.3×10^{-4} pounds of copper. Write this number without using exponents.

Factor.

26. $16x^3 - 48x^2y$

27. $2a^2 - 5a - 3$

28. $16x^4 - 1$

29. $25m^2 - 20m + 4$

30. Solve by factoring: $x^2 + 3x - 54 = 0$.

31. If an object is dropped in Earth's atmosphere, the distance d in feet it falls in t seconds is given by the formula $d = 16t^2$. How long will it take for an object to fall 100 feet?

82. $(x - 1)(x^2 + x + 1)(x + 1)(x^2 - x + 1)$ **83.** $(x^2 - 1)(x^4 + x^2 + 1)$ **84.** $(x - 1)(x + 1)(x^4 + x^2 + 1)$
85. The result in Exercise 82 is completely factored.   **86.** Show that $x^4 + x^2 + 1 = (x^2 + x + 1)(x^2 - x + 1)$.
87. difference of squares   **88.** $(x - 3)(x^2 + 3x + 9)(x + 3)(x^2 - 3x + 9)$

SECTION 5.5 (PAGE 323)

CONNECTIONS   **Page 322: 1.** 4 seconds   **2.** 64 feet

EXERCISES   **1.** $ax^2 + bx + c$ **3.** factoring   **5.** $\left\{-\dfrac{8}{3}, -7\right\}$ **7.** $\{0, -4\}$ **9.** To solve $2x(3x - 4) = 0$, set each

variable factor equal to 0 to get $x = 0$ or $3x - 4 = 0$. Then solve the second equation to get $x = \dfrac{4}{3}$.
11. Because $(x - 9)^2 = (x - 9)(x - 9) = 0$ leads to two solutions of 9, we call 9 a double solution.   **13.** $\{-2, -1\}$
15. $\{1, 2\}$ **17.** $\{-8, 3\}$ **19.** $\{-1, 3\}$ **21.** $\{-2, -1\}$ **23.** $\{-4\}$ **25.** $\left\{-2, \dfrac{1}{3}\right\}$ **27.** $\left\{-\dfrac{4}{3}, \dfrac{1}{2}\right\}$
29. $\left\{-\dfrac{2}{3}\right\}$ **31.** $\{-3, 3\}$ **33.** $\left\{-\dfrac{7}{4}, \dfrac{7}{4}\right\}$ **35.** $\{-11, 11\}$ **37.** Another solution is -11.   **39.** $\{0, 7\}$
41. $\left\{0, \dfrac{1}{2}\right\}$ **43.** $\{2, 5\}$ **45.** $\left\{-4, \dfrac{1}{2}\right\}$ **47.** $\left\{-12, \dfrac{11}{2}\right\}$ **49.** $\{-1, 3\}$ **51.** $\left\{-\dfrac{5}{2}, \dfrac{1}{3}, 5\right\}$ **53.** $\left\{-\dfrac{7}{2}, -3, 1\right\}$
55. $\left\{-\dfrac{7}{3}, 0, \dfrac{7}{3}\right\}$ **57.** $\{-2, 0, 4\}$ **59.** $\{-5, 0, 4\}$ **61.** $\{-3, 0, 5\}$ **63.** $\left\{-\dfrac{4}{3}, -1, \dfrac{1}{2}\right\}$ **65.** $\left\{-\dfrac{2}{3}, 4\right\}$
67. To use the zero-factor property, one side of the equation must be 0. Before solving this equation, multiply the factors on the left side, and subtract 1 from both sides so the right side is 0.   **69.** $\{-.5, .1\}$ **71.** $\{-1.1, 2.5\}$

SECTION 5.6 (PAGE 330)

CONNECTIONS   **Page 329:**   The diagonal of the floor should be $\sqrt{208} \approx 14.4$ ft, or about 14 feet, 5 inches. The carpenter is off by 3 inches and must correct the error to avoid major construction problems.

EXERCISES   **1.** a variable; the unknown   **3.** an equation   **5. (a)** $80 = (x + 8)(x - 8)$ **(b)** 12 **(c)** length: 20 units; width: 4 units   **7. (a)** $60 = \dfrac{1}{2}(3x + 6)(x + 5)$ **(b)** 3 **(c)** base: 15 units; height: 8 units   **9.** length: 7 inches; width: 4 inches   **11.** length: 11 inches; width: 8 inches   **13.** base: 12 inches; height: 5 inches   **15.** height: 13 inches; width: 10 inches   **17.** 12 centimeters   **19.** 8 feet   **21.** 6 meters   **23.** 1 second   **25.** 3 seconds   **27.** 1 second and 3 seconds   **29.** 4 seconds   **31.** 9 centimeters   **33.** 256 feet   **35.** 5.7 seconds   **37.** 8.0 seconds
39. $-3, -2$ or $4, 5$ **41.** 7, 9, 11   **43.** $-2, 0, 2$ or $6, 8, 10$ **45. (a)** 1990 **(b)** 8.7 billion passengers; This approximation is .3 billion less than the value in the table. **(c)** 25 **(d)** 7.9 billion passengers; This approximation is .1 billion less than the table value. **(e)** 1995   **47. (a)** approximately 17,000 injuries **(b)** approximately .7 or 7 to 10; This indicates that in 1998, on the average, 1.4 people were injured per accident.   **49.** c^2 **50.** b^2 **51.** a^2 **52.** $a^2 + b^2 = c^2$; This is the Pythagorean formula.

SECTION 5.7 (PAGE 339)

CONNECTIONS   **Page 339:**   The profit is found by subtracting cost from revenue, represented by $R - C$. If $R - C = x^2 - x - 12$, the solution of $R - C > 0$ is $x < -3$ or $x > 4$. Only $x > 4$ makes sense, since x must be positive in the context of the problem.

EXERCISES   **1. (c)**   **3. (a)** true **(b)** true **(c)** false **(d)** true   **5.** $(-3, 3)$

7. $(-\infty, -6] \cup [7, \infty)$ **9.** $(-\infty, -3) \cup (-2, \infty)$

11. $[-1, 5]$ **13.** $\left(-1, \dfrac{2}{5}\right)$ **15.** $\left(-\dfrac{1}{2}, \dfrac{4}{3}\right)$

17. (1, 6)   **19.** $\left(-\infty, -\dfrac{1}{2}\right) \cup \left(\dfrac{1}{3}, \infty\right)$

21. $\left(-\dfrac{2}{3}, -\dfrac{1}{4}\right)$ **23.** $(-\infty, -2) \cup (2, \infty)$

25. $(-\infty, -4) \cup (4, \infty)$ **27.** $[-1.5, .8]$ **29.** $(-\infty, 1.5] \cup [3.5, \infty)$

31. $\left[-2, \dfrac{1}{3}\right] \cup [4, \infty)$ **33.** $(-\infty, -1) \cup (2, 4)$

35. 2 seconds and 14 seconds   **37.** between 0 and 2 seconds or between 14 and 16 seconds

CHAPTER 5 REVIEW EXERCISES   (PAGE 345)

1. $7(t + 2)$ **3.** $(x - 4)(2y + 3)$ **5.** $(x + 3)(x + 2)$ **7.** $(q + 9)(q - 3)$ **9.** $(r + 8s)(r - 12s)$
11. $8p(p + 2)(p - 5)$ **13.** $p^5(p - 2q)(p + q)$ **15.** r and $6r$, $2r$ and $3r$ **17.** $(2k - 1)(k - 2)$
19. $(3r + 2)(2r - 3)$ **21.** $(v + 3)(8v - 7)$ **23.** $-3(x + 2)(2x - 5)$ **25.** (b)   **27.** $(n + 7)(n - 7)$
29. $(7y + 5w)(7y - 5w)$ **31.** prime   **33.** $(3t - 7)^2$ **35.** $(5k + 4x)(25k^2 - 20kx + 16x^2)$ **37.** $\{-3, -1\}$
39. $\{3, 5\}$ **41.** $\left\{-\dfrac{8}{9}, \dfrac{8}{9}\right\}$ **43.** $\{-1, 6\}$ **45.** $\{6\}$ **47.** length: 10 meters; width: 4 meters   **49.** length:
6 meters; height: 5 meters   **51.** 112 feet   **53.** 256 feet   **55.** 2 inches   **57.** **(a)** \$17.4 million
(b) $-\$48.6$ million   **(c)** The equation is based on the data for 1994 to 1996. To use it to predict much beyond 1996 may lead
to incorrect results, since it is possible that the conditions the equation was based on will change.
59. $(-\infty, -4] \cup \left[\dfrac{1}{2}, \infty\right)$ **61.** $(-\infty, -4] \cup \left[\dfrac{3}{2}, \infty\right)$ **63.** $(-\infty, -5) \cup (7, \infty)$ **65.** $(3k + 5)(k + 2)$
67. $(y^2 + 25)(y + 5)(y - 5)$ **69.** $8abc(3b^2c - 7ac^2 + 9ab)$ **71.** $6xyz(2xz^2 + 2y - 5x^2yz^3)$ **73.** $(2r + 3q)(6r - 5q)$
75. $(7t + 4)^2$ **77.** $\{-5, 2\}$ **79.** **(a)** 417,000 vehicles   **(b)** The estimate may be unreliable because the conditions that
prevailed in the years 1995–1997 may have changed, causing either a greater increase or a greater decrease in the numbers of
alternative-fueled vehicles.   **81.** width: 10 meters; length: 17 meters   **83.** 34 miles   **85.** 25 miles

CHAPTER 5 TEST   (PAGE 349)

[5.1–5.4] **1.** (d)   **2.** $m^2n(2mn + 3m - 5n)$ **3.** $(x + 3)(x - 8)$ **4.** $(2x + 3)(x - 1)$ **5.** $(5z - 1)(2z - 3)$
6. prime   **7.** prime   **8.** $(2 - a)(6 + b)$ **9.** $(3y + 8)(3y - 8)$ **10.** $(2x - 7y)^2$ **11.** $-2(x + 1)^2$
12. $3t^2(2t + 9)(t - 4)$ **13.** $(r - 5)(r^2 + 5r + 25)$ **14.** $8(k + 2)(k^2 - 2k + 4)$ **15.** The product
$(p + 3)(p + 3) = p^2 + 6p + 9$, which does not equal $p^2 + 9$. The binomial $p^2 + 9$ is a prime polynomial.   [5.5] **16.** $\left\{6, \dfrac{1}{2}\right\}$
17. $\left\{-\dfrac{2}{5}, \dfrac{2}{5}\right\}$ **18.** $\{10\}$ **19.** $\{-3, 0, 3\}$ [5.6] **20.** $1\dfrac{1}{2}$ seconds and $4\dfrac{1}{2}$ seconds   **21.** **(a)** $3x - 7$ **(b)** $2x - 1$
(c) 17 feet   **22.** July 1, 1995: 341,000; July 1, 1996: 368,000   [5.7] **23.** $\left(-\dfrac{5}{2}, \dfrac{1}{3}\right)$

24. $(-\infty, -4] \cup [6, \infty)$ [5.5] **25.** Another solution is $-\dfrac{2}{3}$.

CUMULATIVE REVIEW EXERCISES CHAPTERS 1–5   (PAGE 351)

[2.2] **1.** $\{0\}$ **2.** $\{.05\}$ **3.** $\left\{-\dfrac{9}{5}\right\}$ [2.4] **4.** $P = \dfrac{A}{1 + rt}$ [2.3] **5.** 110° and 70°   **6.** exports: \$741 million;
imports: \$426 million   [2.5] **7.** 35 pounds   [3.1] **8.** **(a)** negative; positive   **(b)** negative; negative
9. (2, 27,000), (5, 63,000)   [3.2] **10.** x-intercept: $\left(-\dfrac{1}{4}, 0\right)$; y-intercept: (0, 3)

11.   [3.3] **12.** 209.2; A slope of 209.2 means that the number of radio stations increased on the average by about 209 per year.

[4.2, 4.5] **13.** 4   **14.** $\dfrac{16}{9}$ **15.** 256   **16.** $\dfrac{1}{p^2}$ [4.1] **17.** $-4k^2 - 4k + 8$

[4.3] **18.** $6m^8 - 15m^6 + 3m^4$ **19.** $3y^3 + 8y^2 + 12y - 5$ [4.4] **20.** $4p^2 - 9q^2$ [4.6] **21.** $4x^3 + 6x^2 - 3x + 10$

22. $6p^2 + 7p + 1 + \dfrac{7}{2p - 2}$ [5.2–5.3] **23.** $(2a - 1)(a + 4)$ **24.** $(2m + 3)(5m + 2)$ **25.** $(5x + 3y)(3x - 2y)$

[5.4] **26.** $(3x + 1)^2$ **27.** $-2(4t + 7z)^2$ **28.** $(5r + 9t)(5r - 9t)$ [5.1] **29.** $25(4x^2 + 1)$

[5.3] **30.** $2pq(3p + 1)(p + 1)$ [5.1] **31.** $(2x + y)(a - b)$ [5.5] **32.** $\left\{\dfrac{3}{2}, -2, 6\right\}$ **33.** $\left\{-\dfrac{2}{3}, \dfrac{1}{2}\right\}$

[5.7] **34.** $(-\infty, -2] \cup \left[\dfrac{3}{2}, \infty\right)$ [5.6] **35.** $-4, -2$ or $8, 10$ **36.** 5 meters, 12 meters, 13 meters

---

CHAPTER 6 RATIONAL EXPRESSIONS

SECTION 6.1 (PAGE 360)

CONNECTIONS   **Page 360:**   **1.** $3x^2 + 11x + 8$ cannot be factored, so this quotient cannot be reduced. By long division the quotient is $3x + 5 + \dfrac{-2}{x + 2}$.   **2.** The numerator factors as $(x - 2)(x^2 + 2x + 4)$, so by reducing, the quotient is $x - 2$. Long division gives the same quotient.

EXERCISES   **1.** (a) $3; -5$ (b) $q; -1$ **5.** 0 **7.** $-\dfrac{5}{3}$ **9.** $-3, 2$ **11.** never undefined   **13.** (a) 1 (b) $\dfrac{17}{12}$

15. (a) 0 (b) $-\dfrac{10}{3}$ **17.** (a) $\dfrac{9}{5}$ (b) undefined   **19.** (a) $\dfrac{2}{7}$ (b) $\dfrac{13}{3}$ **21.** Any number divided by itself is 1,

provided the number is not 0. This expression is equal to $\dfrac{1}{x + 2}$ for all values of x except -2 and 2.   **23.** $3r^2$ **25.** $\dfrac{2}{5}$

27. $\dfrac{x - 1}{x + 1}$ **29.** $\dfrac{7}{5}$ **31.** $m - n$ **33.** $\dfrac{3(2m + 1)}{4}$ **35.** $\dfrac{3m}{5}$ **37.** $\dfrac{3r - 2s}{3}$ **39.** $\dfrac{z - 3}{z + 5}$ **41.** $k - 3$

43. $\dfrac{x + 1}{x - 1}$ **45.** -1 **47.** $-(m + 1)$ **49.** -1 **51.** already in lowest terms   **53.** $x^2 + 3$ **Answers may vary**

in Exercises 55–59.   **55.** $\dfrac{-(x + 4)}{x - 3}, \dfrac{-x - 4}{x - 3}, \dfrac{x + 4}{-(x - 3)}, \dfrac{x + 4}{-x + 3}$ **57.** $\dfrac{-(2x - 3)}{x + 3}, \dfrac{-2x + 3}{x + 3}, \dfrac{2x - 3}{-(x + 3)}, \dfrac{2x - 3}{-x - 3}$

59. $-\dfrac{3x - 1}{5x - 6}, \dfrac{-(3x - 1)}{5x - 6}, \dfrac{-3x + 1}{-5x + 6}, \dfrac{3x - 1}{-5x + 6}$ **61.** $\dfrac{m + n}{2}$ **63.** $-\dfrac{b^2 + ba + a^2}{a + b}$ **65.** $\dfrac{z + 3}{z}$ **67.** $x + 3$

69. $x + 5$ **71.** $x - 3$

SECTION 6.2 (PAGE 367)

EXERCISES   **1.** (a) B (b) D (c) C (d) A   **3.** $\dfrac{3a}{2}$ **5.** $-\dfrac{4x^4}{3}$ **7.** $\dfrac{2}{c + d}$ **9.** 5   **11.** $-\dfrac{3}{2t^4}$ **13.** $\dfrac{1}{4}$

15. -4 causes the denominator in the first fraction to equal 0.   **16.** -5 causes the denominator in the second fraction to equal 0.   **17.** -7 causes the numerator in the divisor to equal 0, meaning that we would be dividing by 0. This is undefined.

18. We *are* allowed to divide 0 by a nonzero number.   **21.** $\dfrac{10}{9}$ **23.** $-\dfrac{3}{4}$ **25.** $-\dfrac{9}{2}$ **27.** $\dfrac{p + 4}{p + 2}$

29. $\dfrac{(k-1)^2}{(k+1)(2k-1)}$ **31.** $\dfrac{4k-1}{3k-2}$ **33.** $\dfrac{m+4p}{m+p}$ **35.** $\dfrac{m+6}{m+3}$ **37.** $\dfrac{y+3}{y+4}$ **39.** $\dfrac{m}{m+5}$ **41.** $\dfrac{r+6s}{r+s}$

43. $\dfrac{(q-3)^2(q+2)^2}{q+1}$ **45.** $\dfrac{x+10}{10}$ **47.** $\dfrac{3-a-b}{2a-b}$ **49.** $-\dfrac{(x+y)^2(x^2-xy+y^2)}{3y(y-x)(x-y)}$ or $\dfrac{(x+y)^2(x^2-xy+y^2)}{3y(x-y)^2}$

51. $\dfrac{5xy^2}{4q}$

SECTION 6.3 (PAGE 373)

EXERCISES **1.** (c) **3.** (c) **5.** 60 **7.** 1800 **9.** x^5 **11.** $30p$ **13.** $180y^4$ **15.** $15a^5b^3$ **17.** $12p(p-2)$
19. $2^3 \cdot 3 \cdot 5$ **20.** $(t+4)^3(t-3)(t+8)$ **21.** The similarity is that 2 is replaced by $t+4$, 3 is replaced by $t-3$, and 5 is replaced by $t+8$. **22.** The procedure used is the same. The only difference is that for algebraic fractions, the factors may contain variables, while in a common fraction, the factors are constants. **23.** $18(r-2)$ **25.** $12p(p+5)^2$
27. $8(y+2)(y+1)$ **29.** $m-3$ or $3-m$ **31.** $p-q$ or $q-p$ **33.** $a(a+6)(a-3)$
35. $(k+3)(k-5)(k+7)(k+8)$ **37.** Yes, because $(2x-5)^2 = (5-2x)^2$. **39.** 7 **40.** 1 **41.** identity
property of multiplication **42.** 7 **43.** 1 **44.** identity property of multiplication **45.** $\dfrac{60m^2k^3}{32k^4}$ **47.** $\dfrac{57z}{6z-18}$

49. $\dfrac{-4a}{18a-36}$ **51.** $\dfrac{6(k+1)}{k(k-4)(k+1)}$ **53.** $\dfrac{36r(r+1)}{(r-3)(r+2)(r+1)}$ **55.** $\dfrac{ab(a+2b)}{2a^3b+a^2b^2-ab^3}$

57. $\dfrac{(t-r)(4r-t)}{t^3-r^3}$ **59.** $\dfrac{2y(z-y)(y-z)}{y^4-z^3y}$ or $\dfrac{-2y(y-z)^2}{y^4-z^3y}$

SECTION 6.4 (PAGE 381)

EXERCISES **1.** $\dfrac{5}{4}$ **2.** $\dfrac{-5}{-7+3}$ **3.** 1 **4.** $\dfrac{-5}{-4} = \dfrac{5}{4}$; The two answers are equal. **5.** Jack's answer was correct. His

answer can be obtained by multiplying Jill's correct answer by $\dfrac{-1}{-1} = 1$, the identity element for multiplication.

6. Changing the sign of each term in the fraction is a way of multiplying by $\dfrac{-1}{-1}$ or 1, the identity element for multiplication.

7. Putting a negative sign in front of a fraction gives the opposite of the fraction. Changing the signs in either the numerator or the denominator also gives the opposite. Thus we have multiplied by $(-1)(-1) = 1$, the identity element for multiplication.
8. **(a)** not equivalent **(b)** equivalent **(c)** equivalent **(d)** not equivalent **(e)** not equivalent **(f)** not equivalent

9. $\dfrac{11}{m}$ **11.** b **13.** x **15.** $y-6$ **19.** $\dfrac{3z+5}{15}$ **21.** $\dfrac{10-7r}{14}$ **23.** $\dfrac{-3x-2}{4x}$ **25.** $\dfrac{x+1}{2}$ **27.** $\dfrac{5x+9}{6x}$

29. $\dfrac{7-6p}{3p^2}$ **31.** $\dfrac{x+8}{x+2}$ **33.** $\dfrac{3}{t}$ **35.** $m-2$ or $2-m$ **37.** $\dfrac{-2}{x-5}$ or $\dfrac{2}{5-x}$ **39.** -4 **41.** $\dfrac{-5}{x-y^2}$ or $\dfrac{5}{y^2-x}$

43. $\dfrac{x+y}{5x-3y}$ or $\dfrac{-x-y}{3y-5x}$ **45.** $\dfrac{-6}{4p-5}$ or $\dfrac{6}{5-4p}$ **47.** $\dfrac{-(m+n)}{2(m-n)}$ **49.** $\dfrac{-x^2+6x+11}{(x+3)(x-3)(x+1)}$

51. $\dfrac{-5q^2-13q+7}{(3q-2)(q+4)(2q-3)}$ **53.** $\dfrac{9r+2}{r(r+2)(r-1)}$ **55.** $\dfrac{2x^2+6xy+8y^2}{(x+y)(x+y)(x+3y)}$ or $\dfrac{2x^2+6xy+8y^2}{(x+y)^2(x+3y)}$

57. $\dfrac{15r^2+10ry-y^2}{(3r+2y)(6r-y)(6r+y)}$ **59.** $\dfrac{2k^2-10k+6}{(k-3)(k-1)^2}$ **61.** $\dfrac{7k^2+31k+92}{(k-4)(k+4)^2}$ **63. (a)** $\dfrac{9k^2+6k+26}{5(3k+1)}$ **(b)** $\dfrac{1}{4}$

SECTION 6.5 (PAGE 388)

CONNECTIONS **Page 388:** 1.413793103; 2

EXERCISES **1. (a)** $6; \dfrac{1}{6}$ **(b)** $12; \dfrac{3}{4}$ **(c)** $\dfrac{1}{6} \div \dfrac{3}{4}$ **(d)** $\dfrac{2}{9}$ **3.** Choice (d) is correct, because every sign has been changed in

the fraction. **7.** -6 **9.** $\dfrac{1}{xy}$ **11.** $\dfrac{2a^2b}{3}$ **13.** $\dfrac{m(m+2)}{3(m-4)}$ **15.** $\dfrac{2}{x}$ **17.** $\dfrac{8}{x}$ **19.** $\dfrac{a^2-5}{a^2+1}$ **21.** $\dfrac{31}{50}$

23. $\dfrac{y^2 + x^2}{xy(y - x)}$ **25.** $\dfrac{40 - 12p}{85p}$ **27.** $\dfrac{5y - 2x}{3 + 4xy}$ **29.** $\dfrac{a - 2}{2a}$ **31.** $\dfrac{z - 5}{4}$ **33.** $\dfrac{-m}{m + 2}$ **35.** $\dfrac{3m(m - 3)}{(m - 1)(m - 8)}$

37. division   **39.** $\dfrac{\frac{3}{8} + \frac{5}{6}}{2}$ **40.** $\dfrac{29}{48}$ **41.** $\dfrac{29}{48}$ **42.** Answers will vary.   **43.** $\dfrac{5}{3}$ **45.** $\dfrac{13}{2}$ **47.** $\dfrac{19r}{15}$

SECTION 6.6 (PAGE 396)

EXERCISES   **1.** expression; $\dfrac{43}{40}x$ **3.** equation; $\left\{\dfrac{40}{43}\right\}$ **5.** expression; $-\dfrac{1}{10}y$ **7.** equation; $\{-10\}$ **9.** $-2, 0$

11. $-3, 4, -\dfrac{1}{2}$ **13.** $-9, 1, -2, 2$ **17.** $\left\{\dfrac{1}{4}\right\}$ **19.** $\left\{-\dfrac{3}{4}\right\}$ **21.** $\{-15\}$ **23.** $\{7\}$ **25.** $\{-15\}$ **27.** $\{-5\}$

29. $\{-6\}$ **31.** $\emptyset$ **33.** $\{5\}$ **35.** $\{4\}$ **37.** $\{1\}$ **39.** $\{4\}$ **41.** $\{5\}$ **43.** $\{-2, 12\}$ **45.** $\emptyset$ **47.** $\{3\}$

49. $\{3\}$ **51.** $\left\{-\dfrac{1}{5}, 3\right\}$ **53.** $\left\{-\dfrac{1}{2}, 5\right\}$ **55.** $\{3\}$ **57.** $\left\{-\dfrac{1}{3}, 3\right\}$ **59.** $\left\{-6, \dfrac{1}{2}\right\}$ **61.** $\{6\}$

63. Transform the equation so that the terms with k are on one side and the remaining term is on the other.   **65.** $F = \dfrac{ma}{k}$

67. $a = \dfrac{kF}{m}$ **69.** $R = \dfrac{E - Ir}{I}$ or $R = \dfrac{E}{I} - r$ **71.** $A = \dfrac{h(B + b)}{2}$ **73.** $a = \dfrac{2S - ndL}{nd}$ or $a = \dfrac{2S}{nd} - L$

75. $y = \dfrac{xz}{x + z}$ **77.** $z = \dfrac{3y}{5 - 9xy}$ or $z = \dfrac{-3y}{9xy - 5}$ **79.** The solution set is $\{-5\}$. The number 3 must be rejected.

80. The simplified form is $x + 5$.   **(a)** It is the same as the actual solution.   **(b)** It is the same as the rejected solution.

81. The solution set is $\{-3\}$. The number 1 must be rejected.   **82.** The simplified form is $\dfrac{x + 3}{2(x + 1)}$.   **(a)** It is the same as

the actual solution.   **(b)** It is the same as the rejected solution.   **83.** Answers will vary.   **84.** If we transform so that 0 is on one side, then perform the operation(s) and *reduce to lowest terms* before solving, no rejected values will appear.

85. $-\dfrac{3}{10}$ **87.** An error message would occur.   **89.** An error message would occur.

SUMMARY: EXERCISES ON OPERATIONS AND EQUATIONS WITH RATIONAL EXPRESSIONS (PAGE 401)

1. operation; $\dfrac{10}{p}$ **3.** operation; $\dfrac{1}{2x^2(x + 2)}$ **5.** operation; $\dfrac{y + 2}{y - 1}$ **7.** equation; $\{39\}$ **9.** operation; $\dfrac{13}{3(p + 2)}$

11. equation; $\left\{\dfrac{1}{7}, 2\right\}$ **13.** operation; $\dfrac{7}{12z}$ **15.** operation; $\dfrac{3m + 5}{(m + 3)(m + 2)(m + 1)}$ **17.** equation; $\emptyset$

19. operation; $\dfrac{t + 2}{2(2t + 1)}$

SECTION 6.7 (PAGE 409)

CONNECTIONS   **Page 405:**   54.5 miles per hour; 57.4 miles per hour; yes; More time is spent at the slower speed, and thus the average speed is less than the average of the two speeds.

EXERCISES   **1. (a)** an amount   **(b)** $5 + x$ **(c)** $\dfrac{5 + x}{6} = \dfrac{13}{3}$ **3.** $\dfrac{12}{18}$ **5.** $\dfrac{1386}{97}$ **7.** 1989: 33,579 (thousand); 1990:

34,203 (thousand)   **9.** female: 144,000; male: 576,000   **11.** 24.15 kilometers per hour   **13.** 3.429 hours

15. 7.91 meters per second   **17.** $\dfrac{500}{x - 10} = \dfrac{600}{x + 10}$ **19.** $\dfrac{D}{R} = \dfrac{d}{r}$ **21.** 8 miles per hour   **23.** $18\dfrac{1}{2}$ miles per hour

25. N'Deti: 12.02 miles per hour; McDermott: 8.78 miles per hour   **27.** $\dfrac{1}{10}$ job per hour   **29.** $\dfrac{1}{8}x + \dfrac{1}{6}x = 1$ or

$\dfrac{1}{8} + \dfrac{1}{6} = \dfrac{1}{x}$ **31.** $4\dfrac{4}{17}$ hours   **33.** $5\dfrac{5}{11}$ hours   **35.** 3 hours   **37.** $2\dfrac{7}{10}$ hours   **39.** $9\dfrac{1}{11}$ minutes   **41.** direct

43. direct   **45.** direct   **47.** inverse   **49.** 9   **51.** $\dfrac{16}{5}$ **53.** $\dfrac{4}{9}$ **55.** increases   **57.** $106\dfrac{2}{3}$ miles per hour

59. 25 kilograms per hour   **61.** 20 pounds per square foot   **63.** 15 feet   **65.** 144 feet   **67.** direct   **69.** inverse

71. approximately 2364 in 1995 and 716 in 1985 (The actual numbers were 2361 and 719.)

CHAPTER 6 REVIEW EXERCISES (PAGE 421)

1. 3   **3.** $-5, -\dfrac{2}{3}$ **5. (a)** $\dfrac{11}{8}$ **(b)** $\dfrac{13}{22}$ **7.** $\dfrac{b}{3a}$ **9.** $\dfrac{-(2x+3)}{2}$

Answers may vary in Exercise 11.

11. $\dfrac{-(4x-9)}{2x+3}, \dfrac{-4x+9}{2x+3}, \dfrac{4x-9}{-(2x+3)}, \dfrac{4x-9}{-2x-3}$ **13.** $\dfrac{72}{p}$ **15.** $\dfrac{5}{8}$ **17.** $\dfrac{3a-1}{a+5}$ **19.** $\dfrac{p+5}{p+1}$ **21.** $108y^4$

23. $\dfrac{15a}{10a^4}$ **25.** $\dfrac{15y}{50-10y}$ **27.** $\dfrac{15}{x}$ **29.** $\dfrac{4k-45}{k(k-5)}$ **31.** $\dfrac{-2-3m}{6}$ **33.** $\dfrac{7a+6b}{(a-2b)(a+2b)}$

35. $\dfrac{5z-16}{z(z+6)(z-2)}$ **37. (a)** $\dfrac{a}{b}$ **(b)** $\dfrac{a}{b}$ **(c)** Answers will vary.   **39.** $\dfrac{4(y-3)}{y+3}$ **41.** $\dfrac{xw+1}{xw-1}$ **43.** It would cause

the first and third denominators to equal 0.   **45.** $\emptyset$ **47.** $t = \dfrac{Ry}{m}$ **49.** $m = \dfrac{4+p^2q}{3p^2}$ **51.** $\dfrac{2}{6}$ **53.** $3\dfrac{1}{13}$ hours

55. inverse   **57.** 4 centimeters   **59. (a)** -3 **(b)** -1 **(c)** $-3, -1$ **60.** $\dfrac{15}{2x}$ **61.** If $x = 0$, the divisor R is

equal to 0, and division by 0 is undefined.   **62.** $(x+3)(x+1)$ **63.** $\dfrac{7}{x+1}$ **64.** $\dfrac{11x+21}{4x}$ **65.** $\emptyset$ **66.** We

know that -3 is not allowed because P and R are undefined for $x = -3$.   **67.** Rate is equal to distance divided by time.

Here, distance is 6 miles and time is $x + 3$ minutes, so rate $= \dfrac{6}{x+3}$, which is the expression for P.   **68.** $\dfrac{6}{5}, \dfrac{5}{2}$

69. $\dfrac{(5+2x-2y)(x+y)}{(3x+3y-2)(x-y)}$ **71.** $8p^2$ **73.** 3   **75.** $r = \dfrac{3kz}{5k-z}$ or $r = \dfrac{-3kz}{z-5k}$ **77.** approximately 7443 for hearts

and 3722 for livers (The actual numbers were 7467 and 3698.)   **79.** $\dfrac{36}{5}$

CHAPTER 6 TEST (PAGE 425)

[6.1] **1.** $-2, 4$ **2. (a)** $\dfrac{11}{6}$ **(b)** undefined   **3.** (Answers may vary.) $\dfrac{-(6x-5)}{2x+3}, \dfrac{-6x+5}{2x+3}, \dfrac{6x-5}{-(2x+3)}, \dfrac{6x-5}{-2x-3}$

4. $-3x^2y^3$ **5.** $\dfrac{3a+2}{a-1}$ [6.2] **6.** $\dfrac{25}{27}$ **7.** $\dfrac{3k-2}{3k+2}$ **8.** $\dfrac{a-1}{a+4}$ [6.3] **9.** $150p^5$ **10.** $(2r+3)(r+2)(r-5)$

11. $\dfrac{240p^2}{64p^3}$ **12.** $\dfrac{21}{42m-84}$ [6.4] **13.** 2   **14.** $\dfrac{-14}{5(y+2)}$ **15.** $\dfrac{-x^2+x+1}{3-x}$ or $\dfrac{x^2-x-1}{x-3}$

16. $\dfrac{-m^2+7m+2}{(2m+1)(m-5)(m-1)}$ [6.5] **17.** $\dfrac{2k}{3p}$ **18.** $\dfrac{-2-x}{4+x}$ [6.6] **19.** $-1, 4$ **20.** $\left\{-\dfrac{1}{2}\right\}$ **21.** $D = \dfrac{dF-k}{F}$ or

$D = \dfrac{k-dF}{-F}$ [6.7] **22.** 3 miles per hour   **23.** $2\dfrac{2}{9}$ hours   **24.** 27 days

CUMULATIVE REVIEW EXERCISES CHAPTERS 1-6 (PAGE 426)

[1.2, 1.5, 1.6] **1.** 2   [2.2] **2.** $\{17\}$ [2.4] **3.** $b = \dfrac{2A}{h}$ [2.5] **4.** $\left\{-\dfrac{2}{7}\right\}$ [2.7] **5.** $[-8, \infty)$ **6.** $(4, \infty)$

[3.1] **7. (a)** $(-3, 0)$ **(b)** $(0, -4)$ [3.2] **8.**   [4.1] **9.**

[4.2, 4.5]   **10.** $\dfrac{1}{2^4 x^7}$ **11.** $\dfrac{1}{m^6}$ **12.** $\dfrac{q}{4p^2}$ [4.1]   **13.** $k^2 + 2k + 1$ [4.2]   **14.** $72x^6 y^7$ [4.4]   **15.** $4a^2 - 4ab + b^2$

[4.3]   **16.** $3y^3 + 8y^2 + 12y - 5$ [4.6]   **17.** $6p^2 + 7p + 1 + \dfrac{3}{p-1}$ [4.7]   **18.** 1.4×10^5 seconds

[5.3]   **19.** $(4t + 3v)(2t + v)$ **20.** prime   [5.4]   **21.** $(4x^2 + 1)(2x + 1)(2x - 1)$ [5.5]   **22.** $\{-3, 5\}$

23. $\left\{5, -\dfrac{1}{2}, \dfrac{2}{3}\right\}$ [5.6]   **24.** -2 or -1 **25.** 6 meters   **26.** 30 (The maximum percent is 30%.)   [6.1]   **27.** (a)

28. (d)   [6.4]   **29.** $\dfrac{4}{q}$ **30.** $\dfrac{3r + 28}{7r}$ **31.** $\dfrac{7}{15(q-4)}$ **32.** $\dfrac{-k-5}{k(k+1)(k-1)}$ [6.2]   **33.** $\dfrac{7(2z+1)}{24}$

[6.5]   **34.** $\dfrac{195}{29}$ [6.6]   **35.** $4, 0$ **36.** $\left\{\dfrac{21}{2}\right\}$ **37.** $\{-2, 1\}$ [6.7]   **38.** 150 miles   **39.** $1\dfrac{1}{5}$ hours   **40.** 20 pounds

---

CHAPTER 7 EQUATIONS OF LINES, INEQUALITIES, AND FUNCTIONS

SECTION 7.1   (PAGE 435)

CONNECTIONS   **Page 429:**   **1.** $3500, $5000; $35,000, $50,000   **2.** The loss in value each year; the depreciation when the item is brand new ($D = 0$)

EXERCISES   **1.** D   **3.** B   **5.** The slope m of a vertical line is undefined, so it is not possible to write the equation of the line in the form $y = mx + b$.   **7.** $y = 3x - 3$ **9.** $y = -x + 3$ **11.** $y = 4x - 3$ **13.** $y = 3$

15.   **17.**   **19.** 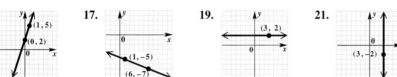   **21.**   **23.** the y-axis

25. $y = 2x - 7$ **27.** $y = \dfrac{2}{3}x + \dfrac{19}{3}$ **29.** $y = -\dfrac{4}{5}x + \dfrac{9}{5}$ **31.** $4 = -3(-2) + b$ **32.** $b = -2$

33. $y = -3x - 2$ **34.** The equations are the same.   **35.** $y = x$ (There are other forms as well.)   **37.** $y = x - 3$

39. $y = -\dfrac{5}{7}x - \dfrac{54}{7}$ **41.** $y = -\dfrac{2}{3}x - 2$ **43.** $y = \dfrac{1}{3}x + \dfrac{4}{3}$ **45. (a)** $(5, 42), (15, 61), (25, 76)$

(b) yes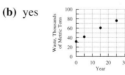

(c) $y = 1.76x + 32$ **(d)** $y = 49.6$, so there will be about 49,600 metric tons in 2005.   **47. (a)** $400   **(b)** $.25
(c) $y = .25x + 400$ **(d)** $425   **(e)** 1500   **49. (a)** $(1, 822.3), (2, 774.9)$ **(b)** $y = -47.4x + 869.7$ **(c)** The slope represents the change in sales from 1996 to 1997. The negative slope indicates that sales *decreased*.   **51.** $y = -3x + 6$
53. $Y_1 = \dfrac{3}{4}x + 1$ **55.** $(0, 32); (100, 212)$ **56.** $\dfrac{9}{5}$ **57.** $F - 32 = \dfrac{9}{5}(C - 0)$ or $F - 212 = \dfrac{9}{5}(C - 100)$
58. $F = \dfrac{9}{5}C + 32$ **59.** $C = \dfrac{5}{9}(F - 32)$ or $C = \dfrac{5}{9}F - \dfrac{160}{9}$ **60.** $86°$ **61.** $10°$ **62.** $-40°$

SECTION 7.2   (PAGE 445)

CONNECTIONS   **Page 444:**   **1.** Answers will vary.   **2. (a)** $(5, \infty)$ **(b)** $(-\infty, 5)$ **(c)** $\left[3, \dfrac{11}{3}\right)$

EXERCISES   **1.** $>$ **3.** $\geq$ **5.** $\leq$ **7.** false   **9.** true   **11.**   **13.**

15.   **19.**   **21.**   **23.**   **25.**

27.   **29.**   **31.** Every point in quadrant IV has a positive x-value and a negative y-value.

Substituting into $y > x$ would imply that a negative number is greater than a positive number, which is always false. Thus, the graph of $y > x$ cannot lie in quadrant IV.   **33.** A   **35.** C

In part (b) of Exercises 37 and 39, other answers are possible.

37. (a)   **(b)** $(500, 0), (200, 400)$ **39. (a)**   **(b)** $(0, 3153), (2, 3050.8), (10, 2642)$

41. (a) $\{-2\}$ **(b)** $(-2, \infty)$ **(c)** $(-\infty, -2)$ **43. (a)** $\{-4\}$ **(b)** $(-\infty, -4)$ **(c)** $(-4, \infty)$

SECTION 7.3   (PAGE 454)

EXERCISES   **1.** 3; 3; (1, 3)   **3.** 5; 5; (3, 5)   **5.** The graph consists of the four points $(0, 2), (1, 3), (2, 4)$, and $(3, 5)$.
7. not a function; domain: $\{-4, -2, 0\}$; range: $\{3, 1, 5, -8\}$ **9.** function; domain: $\{A, B, C, D, E\}$; range: $\{2, 3, 6, 4\}$
11. function   **13.** not a function   **15.** function   **17.** function   **19.** function   **21.** not a function
23. domain: $(-\infty, \infty)$; range: $[2, \infty)$ **25.** domain: $(-\infty, \infty)$; range: $(-\infty, \infty)$ **27.** domain: $[0, \infty)$; range: $[0, \infty)$

29. $(2, 4)$ **30.** $(-1, -4)$ **31.** $\dfrac{8}{3}$ **32.** $f(x) = \dfrac{8}{3}x - \dfrac{4}{3}$ **33. (a)** 11   **(b)** 3   **(c)** -9 **35. (a)** 4   **(b)** 2

(c) 14   **37. (a)** 2   **(b)** 0   **(c)** 3   **39. (a)** 4   **(b)** 2   **41.** $\{(1970, 9.6), (1980, 14.1), (1990, 19.8), (1997, 25.8)\}$; yes
43. $g(1980) = 14.1$; $g(1990) = 19.8$ **45.** For the year 2000, the function predicts 29.3 million foreign-born residents in the

United States.   **47.** $y = -\dfrac{14}{3}x + 9506$ **49.** $y = -4x + 8176$ **51.** 4   **53.** 1   **55.** 1   **57.** $y = x + 1$

CHAPTER 7 REVIEW EXERCISES   (PAGE 461)

1. $y = -x + \dfrac{2}{3}$ **3.** $y = x - 7$ **5.** $y = -\dfrac{3}{4}x - \dfrac{1}{4}$ **7.** $y = 1$ **9.**   **11.**

13.   **15.** not a function; domain: $\{-2, 0, 2\}$; range: $\{4, 8, 5, 3\}$ **17.** not a function   **19.** function

21. not a function   **23.** domain: $(-\infty, \infty)$; range: $[1, \infty)$ **25.** **(a)** 8 **(b)** -1 **27.** **(a)** 5 **(b)** 2   **29.** $y = -1$
31. $y = -3x + 30$ **33.**   **35.**   **37.** Either a $>$ inequality or a $<$ inequality has a

dashed boundary line to indicate that the points on the line are not included in the solution set.   **38.**

39. -3 **40.** $y = -3x - 2$ **41.** $\left(-\dfrac{2}{3}, 0\right), (0, -2)$ **42.**   **43.** -32 **44.** domain: $(-\infty, \infty)$;

range: $(-\infty, \infty)$ **45.** **(a)** \$2.910 billion; 1996   **(b)** domain: $\{1994, 1995, 1996, 1997\}$; range: $\{2.677, 2.910, 3.297, 3.561\}$
(c) $y = .3255x + 2.5845$; yes   **(d)** \$3.2355 billion; Yes, the result is reasonably close.

CHAPTER 7 TEST   (PAGE 464)

[7.1] **1.** $y = 2x + 6$ **2.** $y = \dfrac{5}{2}x - 4$ **3.** $y = \dfrac{1}{2}x + 4$ **4.** $2x + 3y = 15$ [7.2] **5.**

6.   [7.3] **7.** 20   **8.** not a function   **9.** function; domain: $(-\infty, \infty)$; range: $[2, \infty)$

10. function; domain: $(-\infty, \infty)$; range: $(-\infty, \infty)$ **11.** not a function   **12.** Every vertical line intersects the graph (a line)
in only one point.   **13.** yes; \$4459 billion or \$4,459,000,000,000   **14.** $x = 1996$ **15.** 234.6 (billion per year)
16. Consumer expenditures are increasing by \$234.6 billion each year.

CUMULATIVE REVIEW EXERCISES CHAPTERS 1-7   (PAGE 465)

[2.2] **1.** $\{-65\}$ [2.4] **2.** $t = \dfrac{A - p}{pr}$ [5.5] **3.** $\left\{-1, -\dfrac{1}{7}\right\}$ [6.6] **4.** $\{3\}$ **5.** $\{5\}$

[2.7] **6.** $(-2.6, \infty)$ **7.** $(0, \infty)$ **8.** $(-\infty, -4]$

[4.1–4.5] **9.** $\dfrac{1}{x^2 y}$ **10.** $\dfrac{y^7}{x^{13}z^2}$ **11.** $\dfrac{m^6}{8n^9}$ **12.** $2x^2 - 4x + 38$ **13.** $15x^2 + 7xy - 2y^2$ **14.** $x^3 + 8y^3$
[4.6] **15.** $m^2 - 2m + 3$ [5.1–5.4] **16.** $(y + 6k)(y - 2k)$ **17.** $(3x^2 - 5y)(3x^2 + 5y)$ **18.** $5x^2(5x - 13y)(5x - 3y)$
19. $(f + 10)^2$ **20.** prime   [6.4] **21.** 1   **22.** $\dfrac{6x + 22}{(x + 1)(x + 3)}$ or $\dfrac{2(3x + 11)}{(x + 1)(x + 3)}$ [6.2] **23.** $\dfrac{4(x - 5)}{3(x + 5)}$

24. $\dfrac{x+1}{x}$ **25.** $\dfrac{(x+3)^2}{3x}$ **26.** $\dfrac{4xy^4}{z^2}$ [6.5] **27.** $\dfrac{5}{8}$ **28.** 6   [3.3] **29.** $-\dfrac{4}{3}$ **30.** 0

[7.1] **31.** $y = -4x + 15$ **32.** $y = 4x$ [3.2] **33.**   [7.2] **34.**   **35.**

[2.3] **36.** corporate income taxes: $171.8 billion; individual income taxes: $656.4 billion   [2.6] **37.** 15°, 35°, 130°
[5.6] **38.** 7 inches   [2.5] **39.** 14   [6.7] **40.** 1 hour

CHAPTER 8 LINEAR SYSTEMS

SECTION 8.1 (PAGE 473)

EXERCISES   **1. (a)** B   **(b)** C   **(c)** D   **(d)** A   **3.** yes   **5.** no   **7.** yes   **9.** yes   **11.** no   **13. (a)**; The
ordered pair solution must be in quadrant II, and $(-4, -4)$ is in quadrant III.   **15.** $\{(4, 2)\}$

17. $\{(0, 4)\}$ **19.** $\{(4, -1)\}$

In Exercises 21–29, we do not show the graphs.
21. $\{(1, 3)\}$ **23.** $\{(0, 2)\}$ **25.** $\emptyset$ (inconsistent system)   **27.** infinite number of solutions (dependent equations)
29. $\{(4, -3)\}$ **33.** Yes, it is possible. For example, the system $\begin{aligned} x + y &= 5 \\ x - y &= -1 \\ 2x - y &= 1 \end{aligned}$ has the single solution $(2, 3)$.

35. about 350 million for each format   **37.** 200 million   **39.** between 1988 and 1990   **41. (a)** neither
(b) intersecting lines   **(c)** one solution   **43. (a)** dependent   **(b)** one line   **(c)** infinite number of solutions
45. (a) inconsistent   **(b)** parallel lines   **(c)** no solution   **47.** 40   **49.** (40, 30)   **51.** B   **53.** A
55. $\{(-1, 5)\}$ **57.** $\{2\}$ **58.** 5   **59.** $\{(2, 5)\}$ **60.** The x-coordinate, 2, is equal to the solution

of the equation.   **61.** The y-coordinate, 5, is equal to the value we obtained on both sides when checking.   **62.** 5; 3; 5

SECTION 8.2 (PAGE 481)

EXERCISES   **1.** No, it is not correct, because the solution set is $\{(3, 0)\}$. The y-value in the ordered pair must also be determined.
3. $\{(3, 9)\}$ **5.** $\{(7, 3)\}$ **7.** $\{(0, 5)\}$ **9.** $\{(-4, 8)\}$ **11.** $\{(3, -2)\}$ **13.** infinite number of solutions

15. $\left\{\left(\dfrac{1}{3}, -\dfrac{1}{2}\right)\right\}$ **17.** $\emptyset$ **19.** infinite number of solutions   **21.** The first student had less work to do, because the

coefficient of y in the first equation is -1. The second student had to divide by 2, introducing fractions into the expression for x.
23. $\{(2, -3)\}$ **25.** $\{(3, 2)\}$ **27.** $\{(-2, 1)\}$ **29.** 1993   **31.** To find the total cost, multiply the number of bicycles
(x) by the cost per bicycle ($400), and add the fixed cost ($5000). Thus, $y_1 = 400x + 5000$ gives this total cost (in dollars).

32. $y_2 = 600x$ **33.** $y_1 = 400x + 5000$
$y_2 = 600x$;   solution set: $\{(25, 15,000)\}$ **34.** 25; 15,000; 15,000

35. $\{(2, 4)\}$

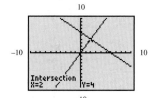

37. $\{(1, 5)\}$

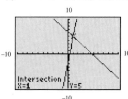

39. $\{(5, -3)\}$; The equations to input are $y_1 = \dfrac{5 - 4x}{5}$ and $y_2 = \dfrac{1 - 2x}{3}$.

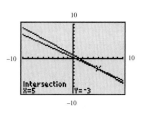

41. Adjust the viewing window so that it does appear.

SECTION 8.3 (PAGE 489)

EXERCISES   **1.** true   **3.** true   **5.** $\{(4, 6)\}$ **7.** $\{(-1, -3)\}$ **9.** $\{(-2, 3)\}$ **11.** $\left\{\left(-\dfrac{2}{3}, \dfrac{17}{2}\right)\right\}$ **13.** $\{(3, -6)\}$

15. $\{(7, 4)\}$ **17.** $\{(0, 3)\}$ **19.** $\{(3, 0)\}$ **21.** $\left\{\left(-\dfrac{32}{23}, -\dfrac{17}{23}\right)\right\}$ **25.** $\{(-3, 4)\}$ **27.** infinite number of
solutions   **29.** $\{(0, 6)\}$ **31.** $\emptyset$ **33. (a)** $\{(1, 4)\}$ **(b)** $\{(1, 4)\}$ **(c)** Answers will vary.   **35.** Yes, they should both
get the same answer, since both procedures are mathematically valid.   **37.** $\{(0, 3)\}$ **39.** $\{(24, -12)\}$ **41.** $\{(3, 2)\}$
43. $1141 = 1991a + b$ **44.** $1339 = 1996a + b$ **45.** $1991a + b = 1141$
$1996a + b = 1339$;   solution set: $\{(39.6, -77,702.6)\}$
46. $y = 39.6x - 77,702.6$ **47.** 1220.2 (million); This is slightly less than the actual figure.
48. It is not realistic to expect the data to lie in a perfectly straight line; as a result, the quantity obtained from an equation
determined in this way will probably be "off" a bit. One cannot put too much faith in models such as this one, because not all
data points are linear in nature.

SECTION 8.4 (PAGE 496)

EXERCISES   **1.** (d)   **3.** (b)   **5.** (c)   **7.** the second number; $x - y = 48$; The two numbers are 73 and 25.
9. Boyz II Men: 134; Bruce Springsteen & the E St. Band: 40   **11.** Terminal Tower: 708 feet; Society Center: 948 feet
13. 46 ones; 28 tens   **15.** 2 copies of *Godzilla*; 5 Aerosmith compact discs   **17.** \$2500 at 4%; \$5000 at 5%
19. Japan: \$17.19; Switzerland: \$13.15   **21.** 80 liters of 40% solution; 40 liters of 70% solution   **23.** 30 pounds at \$6 per
pound; 60 pounds at \$3 per pound   **25.** 60 barrels at \$40 per barrel; 40 barrels at \$60 per barrel   **27.** boat: 10 miles per
hour; current: 2 miles per hour   **29.** plane: 470 miles per hour; wind: 30 miles per hour   **31.** car leaving Cincinnati:
55 miles per hour; car leaving Toledo: 70 miles per hour   **33.** Roberto: 3 miles per hour; Juana: 2.5 miles per hour

SECTION 8.5 (PAGE 504)

EXERCISES   **1.** C   **3.** B   **5.**

7.

9.

11.

13.

15.

17.

19.

21.

23.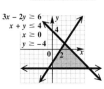

25. $(4, 0), (4, 5), (9, 5)$

27. $(-3, 3), (5, 3), (5, -5)$ **29.** D   **31.** A

CHAPTER 8 REVIEW EXERCISES (PAGE 510)

1. yes   **3.** $\{(3, 1)\}$ **5.** No, this is not correct. A false statement indicates that the solution set is $\emptyset$.   **7.** $\{(2, 1)\}$
9. infinite number of solutions   **11.** His answer was incorrect since the system has infinitely many solutions (as indicated by
the true statement $0 = 0$).   **13.** (c)   **15.** $\{(7, 1)\}$ **17.** infinite number of solutions   **19.** $\{(-4, 1)\}$ **21.** $\{(9, 2)\}$
25. *How Stella Got Her Groove Back*: 782,699; *The Deep End of the Ocean*: 840,263   **27.** Texas Commerce Tower:
75 stories; First Interstate Plaza: 71 stories   **29.** length: 27 meters; width: 18 meters   **31.** 25 pounds of $1.30 candy;
75 pounds of $.90 candy   **33.** $7000 at 3%; $11,000 at 4%   **35.** plane: 250 miles per hour; wind: 20 miles per hour

37.

39.

41. $\left\{\left(\dfrac{28}{5}, \dfrac{16}{5}\right)\right\}$ **42.** $\left\{\left(\dfrac{28}{5}, \dfrac{16}{5}\right)\right\}$ **43.** They are the same. It makes

no difference which method we use.   **44.** $y_1 = 2x - 8$ **45.** $y_2 = \dfrac{12 - x}{2}$ or $y_2 = 6 - \dfrac{1}{2}x$ **46.** $\left\{\dfrac{28}{5}\right\}$; The solution is

the x-value found in Exercises 41 and 42.   **47.** We get $\dfrac{16}{5} = \dfrac{16}{5}$. This value is the y-value found in Exercises 41 and 42.

48. 2   **49.** $-\dfrac{1}{2}$ **50.** They are perpendicular.   **51.**

52. $\left\{\left(\dfrac{28}{5}, \dfrac{16}{5}\right)\right\}$ **53.** $\{(4, 8)\}$

55. $\{(2, 0)\}$ **57.**

59. Great Smoky Mountains: 9.3 million; Rocky Mountain National Park: 2.9 million

61. slower car: 38 miles per hour; faster car: 68 miles per hour   **63.** Yes. Let x represent each of the two equal side lengths.
Then $x + 5$ is the length of the third side. The equation to solve is $x + x + (x + 5) = 29$, giving $x = 8$. The side lengths are 8,
8, and 13 inches.

CHAPTER 8 TEST (PAGE 514)

[8.1] **1.** (a) no   (b) no   (c) yes   **2.** $\{(4, 1)\}$ [8.2] **3.** $\{(1, -6)\}$ **4.** $\{(-35, 35)\}$ [8.3] **5.** $\{(5, 6)\}$
6. $\{(-1, 3)\}$ **7.** $\{(-1, 3)\}$ **8.** $\emptyset$ **9.** $\{(0, 0)\}$ **10.** $\{(-15, 6)\}$ [8.1–8.3] **11.** infinite number of solutions
12. It has no solution.   [8.4] **13.** Memphis and Atlanta: 371 miles; Minneapolis and Houston: 671 miles

14. Disneyland: 15.0 million; Magic Kingdom: 13.8 million   **15.** $33\frac{1}{3}$ liters of 25% solution; $16\frac{2}{3}$ liters of 40% solution

16. slower car: 45 miles per hour; faster car: 60 miles per hour   **17.** *The Lion King*: 30 million; *Snow White*: 28 million

[8.5] **18.**   **19.** $2x - y > 6$ $4y + 12 \geq -3x$ **20.** A

CUMULATIVE REVIEW EXERCISES CHAPTERS 1-8   (PAGE 515)

[1.7] **1.** $-1, 1, -2, 2, -4, 4, -5, 5, -8, 8, -10, 10, -20, 20, -40, 40$ [1.3] **2.** 1   [1.7] **3.** commutative property

4. distributive property   **5.** inverse property   [1.6] **6.** 46   [2.2] **7.** $\left\{-\dfrac{13}{11}\right\}$ **8.** $\left\{\dfrac{9}{11}\right\}$

[2.4] **9.** width: $8\frac{1}{4}$ inches; length: $10\frac{3}{4}$ inches   [2.7] **10.** $\left(-\dfrac{11}{2}, \infty\right)$ [3.1] **11.**   [3.3] **12.** $-\dfrac{4}{3}$

13. $-\dfrac{1}{4}$ [4.1] **14.** $14x^2 - 5x + 23$ [4.3] **15.** $6xy + 12x - 14y - 28$ [4.6] **16.** $3k^2 - 4k + 1$

[4.7] **17.** 3.65×10^{10} [4.5] **18.** x^6y [5.3] **19.** $(5m - 4p)(2m + 3p)$ [5.4] **20.** $(8t - 3)^2$

[5.5] **21.** $\left\{-\dfrac{1}{3}, \dfrac{3}{2}\right\}$ **22.** $\{-11, 11\}$ [6.1] **23.** $-1, \dfrac{5}{2}$ [6.4] **24.** $\dfrac{7}{x + 2}$ [6.2] **25.** $\dfrac{3}{4k - 3}$

[6.6] **26.** $\left\{-\dfrac{1}{4}, 3\right\}$ [5.7] **27.** $[-1, 6]$ [7.1] **28.** $y = 3x - 11$ **29.** $y = 4$ **30.** (a) $x = 9$ (b) $y = -1$

31. $y = 103.25x + 3502$; The slope represents the average yearly increase in health benefit cost during the period.

[7.2] **32.**   [7.3] **33.** 9   [8.1–8.3] **34.** $\{(-1, 6)\}$ **35.** $\{(3, -4)\}$ **36.** $\{(2, -1)\}$

[8.4] **37.** 405 adults and 49 children   **38.** 19 inches, 19 inches, 15 inches   **39.** 4 girls and 3 boys   [8.5] **40.** (b)

CHAPTER 9 ROOTS AND RADICALS

SECTION 9.1   (PAGE 526)

CONNECTIONS   **Page 523:**   The area of the large square is $(a + b)^2$ or $a^2 + 2ab + b^2$. The sum of the areas of the smaller square and the four right triangles is $c^2 + 2ab$. Set these equal to each other and subtract $2ab$ from both sides to get $a^2 + b^2 = c^2$.

EXERCISES   **1.** false; Zero has only one square root.   **3.** true   **5.** true   **7.** $-4, 4$ **9.** $-12, 12$ **11.** $-\dfrac{5}{14}, \dfrac{5}{14}$

13. $-30, 30$ **15.** 7   **17.** -11 **19.** $-\dfrac{12}{11}$ **21.** not a real number   **23.** 100   **25.** 19   **27.** $3x^2 + 4$

29. a must be positive.   **31.** a must be negative.   **33.** rational; 5   **35.** irrational; 5.385   **37.** rational; -8
39. not a real number   **41.** The answer to Exercise 17 is the negative square root of a positive number. However, in Exercise 21, the square root of a negative number is not a real number.   **43.** 23.896   **45.** 28.249   **47.** 1.985   **49.** 5
51. 17   **53.** 3.606   **55.** 2.289   **57.** 5.074   **59.** -4.431 **61.** $c = 17$ **63.** $b = 8$ **65.** $c = 11.705$
67. 24 centimeters   **69.** 80 feet   **71.** 195 miles   **73.** 9.434   **75.** Answers will vary. For example, if $a = 2$ and

$b = 7$, $\sqrt{a^2 + b^2} = \sqrt{53}$, while $a + b = 9$. $\sqrt{53} \neq 9$. If $a = 0$ and $b = 1$, then $\sqrt{a^2 + b^2}$ is equal to $a + b$ (which is not true in general).   **77.** $\sqrt{29}$ **79.** $\sqrt{2}$ **81.** 10   **83.** -3 **85.** 5   **87.** not a real number   **89.** -3 **91.** 3
93. c^2 **94.** $(b - a)^2$ **95.** $2ab$ **96.** $(b - a)^2 = a^2 - 2ab + b^2$ **97.** $c^2 = 2ab + (a^2 - 2ab + b^2)$
98. $c^2 = a^2 + b^2$

SECTION 9.2   (PAGE 535)

CONNECTIONS   **Page 530: 1.** x; $\sqrt{x}$ **2.** The last part of it ($\surd$) is used as part of the radical symbol $\sqrt{}$.

EXERCISES   **1.** false; $\sqrt{4}$ represents only the principal (positive) square root.   **3.** false; $\sqrt{-6}$ is not a real number.
5. true   **7.** true   **9.** 9   **11.** $3\sqrt{10}$ **13.** 13   **15.** $\sqrt{13r}$ **17.** (a)   **19.** $3\sqrt{5}$ **21.** $5\sqrt{3}$ **23.** $5\sqrt{5}$
25. $-10\sqrt{7}$ **27.** $9\sqrt{3}$ **29.** $3\sqrt{6}$ **31.** 24   **33.** $6\sqrt{10}$ **37.** $\dfrac{4}{15}$ **39.** $\dfrac{\sqrt{7}}{4}$ **41.** 5   **43.** $\dfrac{25}{4}$ **45.** $6\sqrt{5}$
47. m **49.** y^2 **51.** $6z$ **53.** $20x^3$ **55.** $z^2\sqrt{z}$ **57.** x^3y^6 **59.** $2\sqrt[3]{5}$ **61.** $3\sqrt[3]{2}$ **63.** $2\sqrt[4]{5}$ **65.** $\dfrac{2}{3}$
67. $-\dfrac{6}{5}$ **69.** 5   **71.** $\sqrt[4]{12}$ **73.** $2x\sqrt[3]{4}$

In Exercise 75, the number of displayed digits will vary among calculator models. Also, less sophisticated models may exhibit round-off error in the final decimal place.
75. (a) 4.472135955   **(b)** 4.472135955   **(c)** The numerical results are not a proof because both answers are approximations and they might differ if calculated to more decimal places.
77. 6 centimeters   **79.** 6 inches   **81.** The product rule for radicals requires that both a and b must be nonnegative. Otherwise $\sqrt{a}$ and $\sqrt{b}$ would not be real numbers (except when $a = b = 0$).   **83.** To verify the first length, we show, in the first triangle, that $1^2 + 1^2 = 2$, so the hypotenuse has length $\sqrt{2}$. Similarly, in the second triangle, $(\sqrt{2})^2 + 1^2 = 3$, so the hypotenuse has length $\sqrt{3}$, and so on.   **84.** $\sqrt{4}$, $\sqrt{9}$, and so on; $\sqrt{16} = 4$ and $\sqrt{25} = 5$ **85.** Look at the radicands: $9 - 4 = 5$, $16 - 9 = 7$, $25 - 16 = 9$, and so on.   **86.** The differences between consecutive whole number lengths increase by 2 each time, so we can predict that the next one will be $\sqrt{25 + 11} = \sqrt{36}$, and the one after that will be $\sqrt{36 + 13} = \sqrt{49}$.

SECTION 9.3   (PAGE 539)

EXERCISES   **1.** distributive   **3.** radicands   **5.** $-5\sqrt{7}$ **7.** $5\sqrt{17}$ **9.** $5\sqrt{7}$ **11.** $11\sqrt{5}$ **13.** $15\sqrt{2}$
15. $-20\sqrt{2} + 6\sqrt{5}$ **17.** $17\sqrt{7}$ **19.** $-16\sqrt{2} - 8\sqrt{3}$ **21.** $20\sqrt{2} + 6\sqrt{3} - 15\sqrt{5}$ **23.** $4\sqrt{2}$
25. $2\sqrt{3} + 4\sqrt{3} = (2 + 4)\sqrt{3} = 6\sqrt{3}$ **27.** $11\sqrt{3}$ **29.** $5\sqrt{x}$ **31.** $3x\sqrt{6}$ **33.** 0   **35.** $-20\sqrt{2k}$
37. $42x\sqrt{5z}$ **39.** $-\sqrt[3]{2}$ **41.** $6\sqrt[3]{p^2}$ **43.** $21\sqrt[4]{m^3}$ **47.** $-6x^2y$ **48.** $-6(p - 2q)^2(a + b)$ **49.** $-6a^2\sqrt{xy}$
50. The answers are alike because the numerical coefficient of the three answers is the same: -6. Also, the first variable factor is raised to the second power, and the second variable factor is raised to the first power. The answers are different because the variables are different: x and y, then $p - 2q$ and $a + b$, and then a and $\sqrt{xy}$.   **51.** 9.220   **53.** 5   **55.** $22\sqrt{2}$
57. (a) 1991   **(b)** 1997

SECTION 9.4   (PAGE 545)

EXERCISES   **1.** radical   **3.** fraction   **5.** $4\sqrt{2}$ **7.** $\dfrac{-\sqrt{33}}{3}$ **9.** $\dfrac{7\sqrt{15}}{5}$ **11.** $\dfrac{\sqrt{30}}{2}$ **13.** $\dfrac{16\sqrt{3}}{9}$ **15.** $\dfrac{-3\sqrt{2}}{10}$
17. $\dfrac{21\sqrt{5}}{5}$ **19.** $\sqrt{3}$ **21.** $\dfrac{\sqrt{2}}{2}$ **23.** $\dfrac{\sqrt{65}}{5}$ **25.** 1; identity property for multiplication   **27.** $\dfrac{\sqrt{21}}{3}$ **29.** $\dfrac{3\sqrt{14}}{4}$
31. $\dfrac{1}{6}$ **33.** 1   **35.** $\dfrac{\sqrt{7x}}{x}$ **37.** $\dfrac{2x\sqrt{xy}}{y}$ **39.** $\dfrac{x\sqrt{3xy}}{y}$ **41.** $\dfrac{3ar^2\sqrt{7rt}}{7t}$ **43.** (b)   **45.** $\dfrac{\sqrt[3]{12}}{2}$ **47.** $\dfrac{\sqrt[3]{196}}{7}$
49. $\dfrac{\sqrt[3]{6y}}{2y}$ **51.** $\dfrac{\sqrt[3]{42mn^2}}{6n}$ **53. (a)** $\dfrac{9\sqrt{2}}{4}$ seconds   **(b)** 3.182 seconds

SECTION 9.5 (PAGE 550)

EXERCISES **1.** 13 **3.** 4 **5.** 4 **7.** 5 **9.** $9\sqrt{5}$ **11.** $16\sqrt{2}$ **13.** $\sqrt{15} - \sqrt{35}$ **15.** $2\sqrt{10} + 30$
17. $4\sqrt{7}$ **19.** $57 + 23\sqrt{6}$ **21.** $81 + 14\sqrt{21}$ **23.** $37 + 12\sqrt{7}$ **25.** 23 **27.** 1 **29.** $2\sqrt{3} - 2 + 3\sqrt{2} - \sqrt{6}$
31. $15\sqrt{2} - 15$ **33.** $87 + 9\sqrt{21}$ **35.** Because multiplication must be performed before addition, it is incorrect to add -37 and -2. Since $-2\sqrt{15}$ cannot be simplified, the expression cannot be written in a simpler form, and the final answer is $-37 - 2\sqrt{15}$. **37.** $\dfrac{3 - \sqrt{2}}{7}$ **39.** $-4 - 2\sqrt{11}$ **41.** $1 + \sqrt{2}$ **43.** $-\sqrt{10} + \sqrt{15}$ **45.** $3 - \sqrt{3}$

47. $2\sqrt{5} + \sqrt{15} + 4 + 2\sqrt{3}$ **49.** $\sqrt{11} - 2$ **51.** $\dfrac{\sqrt{3} + 5}{8}$ **53.** $\dfrac{6 - \sqrt{10}}{2}$ **55.** $x\sqrt{30} + \sqrt{15x} + 6\sqrt{5x} + 3\sqrt{10}$
57. $6t - 3\sqrt{14t} + 2\sqrt{7t} - 7\sqrt{2}$ **59.** $m\sqrt{15} + \sqrt{10mn} - \sqrt{15mn} - n\sqrt{10}$ **61.** $2 - 3\sqrt[3]{4}$ **63.** $12 + 10\sqrt[4]{8}$
65. $-1 + 3\sqrt[3]{2} - \sqrt[3]{4}$ **67.** 1 **69.** $30 + 18x$ **70.** They are not like terms. **71.** $30 + 18\sqrt{5}$ **72.** They are not like radicals. **73.** Make the first term $30x$, so that $30x + 18x = 48x$; make the first term $30\sqrt{5}$, so that $30\sqrt{5} + 18\sqrt{5} = 48\sqrt{5}$. **74.** When combining like terms, we add (or subtract) the coefficients of the common factors of the terms: $2xy + 5xy = 7xy$. When combining like radicals, we add (or subtract) the coefficients of the common radical: $2\sqrt{ab} + 5\sqrt{ab} = 7\sqrt{ab}$.

75. 4 inches **76.** $\dfrac{\sqrt{AP} - P}{P}$; 8%

SECTION 9.6 (PAGE 557)

CONNECTIONS **Page 552:** **1.** $6\sqrt{13} \approx 21.63$ (to the nearest hundredth) **2.** $h = \sqrt{13}$; $6\sqrt{13} \approx 21.63$

EXERCISES **1.** {49} **3.** {7} **5.** {85} **7.** {−45} **9.** $\left\{ -\dfrac{3}{2} \right\}$ **11.** ∅ **13.** {121} **15.** {8} **17.** {1}
19. {6} **21.** ∅ **23.** {5} **25.** Since $\sqrt{x}$ must be greater than or equal to zero for any replacement for x, it cannot equal -8, a negative number. **29.** {12} **31.** {5} **33.** {0, 3} **35.** {−1, 3} **37.** {8} **39.** {4} **41.** {8}
43. {9} **45.** We cannot square term by term. The left side must be squared as a binomial in the first step. **47.** 158.6 feet
49. (a) 70.5 miles per hour **(b)** 59.8 miles per hour **(c)** 53.9 miles per hour **51. (a)** 1991 **(b)** 1997; Here, $f(3) \approx 97.5$. Thus, 1997 is the year when about 3 million calls were made. **(c)** about 2.1 million **53.** 4 **55.** −2 **57.** −1

SECTION 9.7 (PAGE 565)

CONNECTIONS **Page 564:** **1.** $(\sqrt{x} - 3)(\sqrt{x} + 1)$ **2.** $(x + \sqrt{10})(x - \sqrt{10})$ **3.** $(\sqrt{x} + 2)^2$
4. $x\sqrt{x} + 3\sqrt{x} + \dfrac{5\sqrt{x}}{x}$

EXERCISES **1.** (a) **3.** (c) **5.** 5 **7.** 4 **9.** 2 **11.** 2 **13.** 8 **15.** 9 **17.** 8 **19.** 4 **21.** −4
23. −4 **25.** $\dfrac{1}{343}$ **27.** $\dfrac{1}{36}$ **29.** $-\dfrac{1}{32}$ **31.** 2^3 **33.** $\dfrac{1}{6^{1/2}}$ **35.** $\dfrac{1}{15^{1/2}}$ **37.** $11^{1/7}$ **39.** 8^3 **41.** $6^{1/2}$
43. $\dfrac{5^3}{2^3}$ **45.** $\dfrac{1}{2^{8/5}}$ **47.** $6^{2/9}$ **49.** z **51.** $m^2 n^{1/6}$ **53.** $\dfrac{a^{2/3}}{b^{4/9}}$ **55.** 2 **57.** 2 **59.** $\sqrt{a}$ **61.** $\sqrt[3]{k^2}$
63. $\sqrt{2} = 2^{1/2}$ and $\sqrt[3]{2} = 2^{1/3}$ **64.** $2^{1/2} \cdot 2^{1/3}$ **65.** 6 **66.** $2^{3/6} \cdot 2^{2/6}$ **67.** $2^{5/6}$ **68.** $\sqrt[6]{2^5}$ or $\sqrt[6]{32}$ **69.** 2
71. 1.883 **73.** 3.971 **75.** 9.100 **81. (a)** $d = 1.22x^{1/2}$ **(b)** 211.31 miles **83.** Because $(7^{1/2})^2 = 7$ and $(\sqrt{7})^2 = 7$, they should be equal, so we define $7^{1/2}$ to be $\sqrt{7}$.

CHAPTER 9 REVIEW EXERCISES (PAGE 571)

1. −7, 7 **3.** −14, 14 **5.** −15, 15 **7.** 4 **9.** 10 **11.** not a real number **13.** $\dfrac{7}{6}$ **15.** a must be negative.

17. irrational; 4.796 **19.** rational; −5 **21.** $5\sqrt{3}$ **23.** $4\sqrt{10}$ **25.** 12 **27.** $16\sqrt{6}$ **29.** $-\dfrac{11}{20}$ **31.** $\dfrac{\sqrt{7}}{13}$

33. $\dfrac{2}{15}$ **35.** 8   **37.** p **39.** r^9 **41.** $a^7b^{10}\sqrt{ab}$ **43.** Yes, because both approximations are .7071067812.

45. $21\sqrt{3}$ **47.** 0   **49.** $2\sqrt{3} + 3\sqrt{10}$ **51.** $6\sqrt{30}$ **53.** 0   **55.** $11k^2\sqrt{2n}$ **57.** $\sqrt{5}$ **59.** $\dfrac{\sqrt{30}}{15}$

61. $\sqrt{10}$ **63.** $\dfrac{r\sqrt{x}}{4x}$ **65.** $\dfrac{\sqrt[3]{98}}{7}$ **67.** $-\sqrt{15} - 9$ **69.** $22 - 16\sqrt{3}$ **71.** -2 **73.** $-2 + \sqrt{5}$

75. $\dfrac{-2 + 6\sqrt{2}}{17}$ **77.** $\dfrac{-\sqrt{10} + 3\sqrt{5} + \sqrt{2} - 3}{7}$ **79.** $\dfrac{3 + 2\sqrt{6}}{3}$ **81.** $3 + 4\sqrt{3}$ **83.** $\emptyset$ **85.** $\{1\}$ **87.** $\{6\}$

89. $\{-2\}$ **91. (a)** billions of dollars in exports   **(b)** years   **(c)** 1992   **(d)** Yes, because $13 billion is between $7.5 billion and $19.6 billion, so the year should be between 1990 and 1993.   **(e)** $31.2 billion; yes   **93.** 9   **95.** 7^3 or 343

97. $x^{3/4}$ **99.** 16   **101.** $\dfrac{5 - \sqrt{2}}{23}$ **103.** $5y\sqrt{2}$ **105.** $-\sqrt{10} - 5\sqrt{15}$ **107.** $\dfrac{2 + \sqrt{13}}{2}$ **109.** $7 - 2\sqrt{10}$

111. -11 **113.** $\{7\}$ **115.** $\{8\}$ **117.** $-\dfrac{5\sqrt{7}}{5\sqrt{14}}$ **119.** $-\dfrac{\sqrt{2}}{2}$

CHAPTER 9 TEST   (PAGE 574)

[9.1] **1.** $-14, 14$ **2. (a)** irrational   **(b)** 11.916   [9.2–9.5] **3.** 6   **4.** $-3\sqrt{3}$ **5.** $\dfrac{8\sqrt{2}}{5}$ **6.** $2\sqrt[3]{4}$ **7.** $4\sqrt{6}$

8. $9\sqrt{7}$ **9.** $-5\sqrt{3x}$ **10.** $2y\sqrt[3]{4x^2}$ **11.** 31   **12.** $6\sqrt{2} + 2 - 3\sqrt{14} - \sqrt{7}$ **13.** $11 + 2\sqrt{30}$

[9.1] **14. (a)** $6\sqrt{2}$ inches   **(b)** 8.485 inches   [9.4] **15.** $\dfrac{5\sqrt{14}}{7}$ **16.** $\dfrac{\sqrt{6x}}{3x}$ **17.** $-\sqrt[3]{2}$ **18.** $\dfrac{-12 - 3\sqrt{3}}{13}$

[9.6] **19.** $\{3\}$ **20.** $\left\{\dfrac{1}{4}, 1\right\}$ [9.7] **21.** 16   **22.** -25 **23.** 5   **24.** $\dfrac{1}{3}$ [9.6] **25.** 12 is not a solution. A check shows that it does not satisfy the original equation.

CUMULATIVE REVIEW EXERCISES CHAPTERS 1–9   (PAGE 574)

[1.5–1.6] **1.** 54   **2.** 6   **3.** 3   **4.** 18   **5.** 15   **6.** 4.223   [2.2] **7.** $\{3\}$ [2.7] **8.** $[-16, \infty)$ **9.** $(5, \infty)$

[2.4] **10.** 207 cubic inches   [3.2] **11.**

12.

[3.3] **13.** $-\dfrac{5}{6}$ [4.2] **14.** $12x^{10}y^2$

[4.5] **15.** $\dfrac{y^{15}}{5832}$ [4.1] **16.** $3x^3 + 11x^2 - 13$ [4.6] **17.** $4t^2 - 8t + 5$ [5.2–5.4] **18.** $(m + 8)(m + 4)$

19. $(5t^2 + 6)(5t^2 - 6)$ **20.** $(6a + 5b)(2a - b)$ **21.** $(9z + 4)^2$ [5.5] **22.** $\{3, 4\}$ **23.** $\{-2, -1\}$ [6.1] **24.** $2, -7$

[6.2] **25.** $\dfrac{x + 1}{x}$ **26.** $(t + 5)(t + 3)$ [6.4] **27.** $\dfrac{y^2}{(y + 1)(y - 1)}$ **28.** $\dfrac{-2x - 14}{(x + 3)(x - 1)}$ [6.5] **29.** -21

[7.2] **30.**

[8.1–8.3] **31.** $\{(3, -7)\}$ **32.** infinite number of solutions   [8.4] **33.** from Chicago: 57 mph;

from Des Moines: 50 mph   **34.** CNN: 67.8 million; ESPN: 67.9 million   [9.3–9.5] **35.** $29\sqrt{3}$ **36.** $-\sqrt{3} + \sqrt{5}$

37. $10xy^2\sqrt{2y}$ [9.7] **38.** 32   [9.5] **39.** $21 - 5\sqrt{2}$ [9.6] **40.** $\{16\}$

CHAPTER 10 QUADRATIC EQUATIONS

SECTION 10.1   (PAGE 580)

EXERCISES   1. C   **3.** A   **5.** B   **7.** true   **9.** true   **11.** According to the square root property, -9 is also a solution, so her answer was not completely correct. The solution set is $\{\pm 9\}$.   **13.** $\{\pm 9\}$ **15.** $\{\pm \sqrt{14}\}$
17. $\{\pm 4\sqrt{3}\}$ **19.** $\emptyset$ **21.** $\{\pm 1.5\}$ **23.** $\{\pm 2\sqrt{6}\}$ **25.** $\{-2, 8\}$ **27.** $\emptyset$ **29.** $\{8 \pm 3\sqrt{3}\}$
31. $\left\{-3, \dfrac{5}{3}\right\}$ **33.** $\left\{0, \dfrac{3}{2}\right\}$ **35.** $\left\{\dfrac{5 \pm \sqrt{30}}{2}\right\}$ **37.** $\left\{\dfrac{-1 \pm 3\sqrt{2}}{3}\right\}$ **39.** $\{-10 \pm 4\sqrt{3}\}$ **41.** $\left\{\dfrac{1 \pm 4\sqrt{3}}{4}\right\}$
43. Johnny's first solution, $\dfrac{5 + \sqrt{30}}{2}$, is equivalent to Linda's second solution, $\dfrac{-5 - \sqrt{30}}{-2}$. This can be verified by multiplying $\dfrac{5 + \sqrt{30}}{2}$ by 1 in the form $\dfrac{-1}{-1}$. Similarly, Johnny's second solution is equivalent to Linda's first one.   **45.** $\{-4.48, .20\}$
47. $\{-3.09, -.15\}$ **49.** $(x + 3)^2 = 100$ **50.** $x + 3 = -10$ or $x + 3 = 10$ **51.** $\{-13\}; \{7\}$ **52.** $\{-13, 7\}$
53. $\{-7, 3\}$ **54.** $\{-3, 6\}$ **55.** .983 foot   **57.** 3.442 feet   **59.** about $\dfrac{1}{2}$ second   **61.** 9 inches   **63.** 5%

SECTION 10.2   (PAGE 587)

CONNECTIONS   Page 586:   1. x^2 **2.** $x^2 + 8x$ **3.** $x^2 + 8x + 16$ **4.** It occurred when we added the 16 squares.

EXERCISES   1. 16   **3.** multiplying $(t + 2)(t - 5)$ to get $t^2 - 3t - 10$ **5.** (d)   **7.** 49   **9.** $\dfrac{25}{4}$ **11.** $\dfrac{1}{16}$

13. $\{1, 3\}$ **15.** $\{-1 \pm \sqrt{6}\}$ **17.** $\{-3\}$ **19.** $\left\{-\dfrac{3}{2}, \dfrac{1}{2}\right\}$ **21.** $\emptyset$ **23.** $\left\{\dfrac{-7 \pm \sqrt{97}}{6}\right\}$ **25.** $\{-4, 2\}$
27. $\{1 \pm \sqrt{6}\}$ **29. (a)** $\left\{\dfrac{3 \pm 2\sqrt{6}}{3}\right\}$ **(b)** $\{-.633, 2.633\}$ **31. (a)** $\{-2 \pm \sqrt{3}\}$ **(b)** $\{-3.732, -.268\}$
35. 75 feet by 100 feet   **37.** 1 second and 5 seconds   **39.** 3 seconds and 5 seconds   **41.** 8 miles

SECTION 10.3   (PAGE 594)

EXERCISES   1. 4; 5; -9 **3.** 2   **5.** $a = 3, b = -4, c = -2$ **7.** $a = 3, b = 7, c = 0$ **9.** $a = 1, b = 1, c = -12$
11. $a = 9, b = 9, c = -26$ **13.** If a were 0, the equation would be linear, not quadratic.   **15.** No, because $2a$ should be the denominator for $-b$ as well. The correct formula is $x = \dfrac{-b \pm \sqrt{b^2 - 4ac}}{2a}$.   **17.** $\{-13, 1\}$ **19.** $\{2\}$
21. $\left\{\dfrac{-6 \pm \sqrt{26}}{2}\right\}$ **23.** $\left\{-1, \dfrac{5}{2}\right\}$ **25.** $\{-1, 0\}$ **27.** $\left\{0, \dfrac{12}{7}\right\}$ **29.** $\{\pm 2\sqrt{6}\}$ **31.** $\left\{\pm \dfrac{2}{5}\right\}$ **33.** $\left\{\dfrac{6 \pm 2\sqrt{6}}{3}\right\}$
35. $\emptyset$ **37.** $\emptyset$ **39.** $\left\{\dfrac{-5 \pm \sqrt{61}}{2}\right\}$ **41. (a)** $\left\{\dfrac{-1 \pm \sqrt{11}}{2}\right\}$ **(b)** $\{-2.158, 1.158\}$ **43. (a)** $\left\{\dfrac{1 \pm \sqrt{5}}{2}\right\}$
(b) $\{-.618, 1.618\}$ **45.** $\left\{-\dfrac{2}{3}, \dfrac{4}{3}\right\}$ **47.** $\left\{\dfrac{-1 \pm \sqrt{73}}{6}\right\}$ **49.** $\{1 \pm \sqrt{2}\}$ **51.** $\emptyset$ **53.** The solution(s) must make sense in the original problem. (For example, a length cannot be negative.)   **55.** $r = \dfrac{-\pi h \pm \sqrt{\pi^2 h^2 + \pi S}}{\pi}$
57. $P = .241$ (about 88 days)   **59.** $P = 1.881$ **61.** $P = 84.130$ **63.** $1^2 = 1^3$, or $1 = 1$. This is a true statement.
65. 4 seconds   **67.** approximately .2 second and 10.9 seconds   **69.** 30 units   **70.** It is the radicand in the quadratic formula.   **71. (a)** 225   **(b)** 169   **(c)** 4   **(d)** 121   **72.** Each is a perfect square.   **73. (a)** $(3x - 2)(6x + 1)$
(b) $(x + 2)(5x - 3)$ **(c)** $(8x + 1)(6x + 1)$ **(d)** $(x + 3)(x - 8)$ **74. (a)** 41   **(b)** -39 **(c)** 12   **(d)** 112   **75.** no
76. If the discriminant is a perfect square the trinomial is factorable.   **(a)** yes   **(b)** yes   **(c)** no

SUMMARY: EXERCISES ON QUADRATIC EQUATIONS (PAGE 598)

1. $\{\pm 6\}$ **3.** $\left\{\pm \dfrac{10}{9}\right\}$ **5.** $\{1, 3\}$ **7.** $\{4, 5\}$ **9.** $\left\{-\dfrac{1}{3}, \dfrac{5}{3}\right\}$ **11.** $\{-17, 5\}$ **13.** $\left\{\dfrac{7 \pm 2\sqrt{6}}{3}\right\}$ **15.** $\emptyset$

17. $\left\{-\dfrac{1}{2}, 2\right\}$ **19.** $\left\{-\dfrac{5}{4}, \dfrac{3}{2}\right\}$ **21.** $\{1 \pm \sqrt{2}\}$ **23.** $\left\{\dfrac{2}{5}, 4\right\}$ **25.** $\left\{\dfrac{-3 \pm \sqrt{41}}{2}\right\}$ **27.** $\left\{\dfrac{1}{4}, 1\right\}$

29. $\left\{\dfrac{-2 \pm \sqrt{11}}{3}\right\}$ **31.** $\left\{\dfrac{-7 \pm \sqrt{5}}{4}\right\}$ **33.** $\left\{\dfrac{8 \pm 8\sqrt{2}}{3}\right\}$ **35.** $\emptyset$ **37.** $\left\{-\dfrac{2}{3}, 2\right\}$ **39.** $\left\{-4, \dfrac{3}{5}\right\}$

41. $\left\{-\dfrac{2}{3}, \dfrac{2}{5}\right\}$ **43.** We need to know other methods because many quadratic equations cannot be solved by elementary factoring methods.

SECTION 10.4 (PAGE 604)

CONNECTIONS **Page 603: 1.** $\langle -1, -4 \rangle, \langle 0, 2 \rangle, \langle -5, 0 \rangle$ **2.** $-1; 2$

EXERCISES **1.** $3i$ **3.** $2i\sqrt{5}$ **5.** $3i\sqrt{2}$ **7.** $5i\sqrt{5}$ **9.** $5 + 3i$ **11.** $6 - 9i$ **13.** $6 - 7i$ **15.** $14 + 5i$
17. $7 - 22i$ **19.** 45 **21.** $18i$ **23.** -4 **25.** $6 + 4i$ **27.** $4 - 3i$ **29.** $3 - i$ **31.** $2 - 6i$
33. $-\dfrac{3}{25} + \dfrac{4}{25}i$ **35.** $-1 + 3i$ **36.** The product of $-1 + 3i$ and $2 + 9i$ is $-29 - 3i$, which *is* the original dividend.
37. $-3 - 4i$ **38.** The product of $-3 - 4i$ and i is $4 - 3i$, which *is* the original dividend. **39.** Because
$(3 - 2i)(4 + i) = 14 - 5i$ is true, the given statement is true. **40.** The product of the quotient and the divisor must equal
the dividend. **41.** $\{-1 \pm 2i\}$ **43.** $\{3 \pm i\sqrt{5}\}$ **45.** $\left\{-\dfrac{2}{3} \pm i\sqrt{2}\right\}$ **47.** $\{1 \pm i\}$ **49.** $\left\{-\dfrac{3}{4} \pm \dfrac{\sqrt{31}}{4}i\right\}$
51. $\left\{\dfrac{3}{2} \pm \dfrac{\sqrt{7}}{2}i\right\}$ **53.** $\left\{\dfrac{1}{5} \pm \dfrac{\sqrt{14}}{5}i\right\}$ **55.** $\left\{-\dfrac{1}{2} \pm \dfrac{\sqrt{13}}{2}i\right\}$ **57.** $\left\{\dfrac{1}{2} \pm \dfrac{\sqrt{11}}{2}i\right\}$ **59.** If the discriminant $b^2 - 4ac$
is negative, the equation will have solutions that are not real. **61.** true **63.** false; For example, $3 + 2i$ is a complex
number but it is not real.

SECTION 10.5 (PAGE 614)

CONNECTIONS **Page 610:** The maximum height is 8 feet at 16 feet from the initial point. The maximum distance is 32 feet.

EXERCISES **1.** vertex **3.** $(2, -5)$ **5.** $(-1, 2)$ **7.** $(-3, 0)$

9. $(3, 4)$ **11.** $(2, 0)$ **13.** one real solution; $\{2\}$ **15.** two real solutions; $\{\pm 2\}$

17. no real solutions; $\emptyset$ **19.** $\{-2, 3\}$ **21.** $\{-1, 1.5\}$ **23.** $\{-4, 2\}$ **25. (a)** the point $(0, -2)$ **(b)** the points
$(-1, 0)$ and $(2, 0)$
In Exercises 27–31, the domain is first and the range is second.
27. $(-\infty, \infty); [0, \infty)$ **29.** $(-\infty, \infty); (-\infty, 4]$ **31.** $(-\infty, \infty); [1, \infty)$ **33.** 3 **35.** 21 **37.** 40 and 40

39. 320 feet by 640 feet; 204,800 square feet **41.** $y = \dfrac{11}{5625}x^2$ **43.** In each case, there is a vertical "stretch" of the

parabola. It becomes narrower as the coefficient gets larger. **45.** The graph of Y_2 is obtained by reflecting the graph of Y_1
across the x-axis. **47.** By adding a positive constant k, the graph is shifted k units upward. By subtracting a positive constant
k, the graph is shifted k units downward.

CHAPTER 10 REVIEW EXERCISES (PAGE 623)

1. $\{\pm 12\}$ **3.** $\{\pm 8\sqrt{2}\}$ **5.** $\{3 \pm \sqrt{10}\}$ **7.** $\emptyset$ **9.** $\{-5, -1\}$ **11.** $\{-1 \pm \sqrt{6}\}$ **13.** $\left\{-\dfrac{2}{5}, 1\right\}$

15. 2.5 seconds   **17.** $\left(\dfrac{k}{2}\right)^2$ or $\dfrac{k^2}{4}$ **19.** $\{1 \pm \sqrt{5}\}$ **21.** $\left\{\dfrac{2 \pm \sqrt{10}}{2}\right\}$ **23.** $\left\{\dfrac{-3 \pm \sqrt{41}}{2}\right\}$ **25.** Multiply both the

numerator and the denominator by the conjugate of the denominator.   **27.** $5 - i$ **29.** 20   **31.** i **33.** $\dfrac{7}{50} + \dfrac{1}{50}i$

35. No, the product $(a + bi)(a - bi) = a^2 + b^2$ will always be the sum of the squares of two real numbers, which is a real

number.   **37.** $\left\{\dfrac{2}{3} \pm \dfrac{2\sqrt{2}}{3}i\right\}$ **39.** $\left\{-\dfrac{3}{2} \pm \dfrac{\sqrt{23}}{2}i\right\}$ **41.** $\left\{-\dfrac{1}{9} \pm \dfrac{2\sqrt{2}}{9}i\right\}$ **43.** vertex: $(1, 4)$

45. two; $\{\pm 2\}$ **47.** two; $\{-1, 4\}$ **49.** none; $\emptyset$ **51.** $\left\{-\dfrac{11}{2}, 5\right\}$ **53.** $\left\{\dfrac{-1 \pm \sqrt{21}}{2}\right\}$ **55.** $\left\{\dfrac{-5 \pm \sqrt{17}}{2}\right\}$

57. $\emptyset$ **59.** $\left\{-\dfrac{5}{3}\right\}$ **61.** $\{-2 \pm \sqrt{5}\}$ **63.** approximately 424   **65.** $\{1 \pm \sqrt{3}\}$ **66.** $(x - (1 + \sqrt{3})) \cdot$

$(x - (1 - \sqrt{3}))$ **67.** $((x - 1) - \sqrt{3})((x - 1) + \sqrt{3})$ **68.** $(x - 1)^2 - \sqrt{3}^2 = x^2 - 2x + 1 - 3 = x^2 - 2x - 2$

69. They are both $x^2 - 2x - 2$.   **70.** $(x - (2 + \sqrt{5}))(x - (2 - \sqrt{5}))$ or $((x - 2) - \sqrt{5})((x - 2) + \sqrt{5})$

CHAPTER 10 TEST (PAGE 626)

[10.1] 1. $\{\pm \sqrt{39}\}$ **2.** $\{-11, 5\}$ **3.** $\left\{\dfrac{-3 \pm 2\sqrt{6}}{4}\right\}$ **[10.2] 4.** $\{2 \pm \sqrt{10}\}$ **5.** $\left\{\dfrac{-6 \pm \sqrt{42}}{2}\right\}$

[10.3] 6. (a) 0   **(b)** one (a double solution)   **7.** $\left\{-3, \dfrac{1}{2}\right\}$ **8.** $\left\{\dfrac{3 \pm \sqrt{3}}{3}\right\}$ **9.** $\left\{-1 \pm \dfrac{\sqrt{7}}{2}i\right\}$ **10.** $\left\{\dfrac{5 \pm \sqrt{13}}{6}\right\}$

[10.1–10.3] 11. $\{1 \pm \sqrt{2}\}$ **12.** $\left\{\dfrac{-1 \pm 3\sqrt{2}}{2}\right\}$ **13.** $\left\{\dfrac{11 \pm \sqrt{89}}{4}\right\}$ **14.** $\{5\}$ **15.** 2 seconds   **16.** 247.3 years

[10.4] 17. $-5 + 5i$ **18.** $-17 - 4i$ **19.** 73   **20.** $2 - i$ **[10.5] 21.** vertex: $(3, 0)$

22. vertex: $(-1, -3)$ **23.** vertex: $(-3, -2)$ **24. (a)** two   **(b)** $\{-3 \pm \sqrt{2}\}$

(c) $-3 - \sqrt{2} \approx -4.414$ and $-3 + \sqrt{2} \approx -1.586$ **25.** 200 and 200

CUMULATIVE REVIEW EXERCISES CHAPTERS 1-10 (PAGE 627)

[1.2] 1. 15   **[1.5] 2.** -2 **[1.6] 3.** 5   **[1.8] 4.** $-r + 7$ **5.** $-2k$ **6.** $19m - 17$ **[2.1] 7.** $\{18\}$

[2.2] 8. $\{5\}$ **9.** $\{2\}$ **[2.4] 10.** $100°, 80°$ **11.** width: 50 feet; length: 94 feet   **12.** $L = \dfrac{P - 2W}{2}$ or $L = \dfrac{P}{2} - W$

[2.7] 13. $(-2, \infty)$ **14.** $(-\infty, 4]$ **[3.2] 15.**   **[3.3] 16.** -1

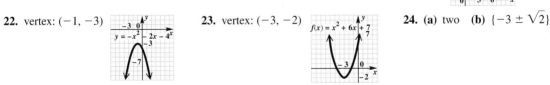

[4.1] **17.** $8x^5 - 17x^4 - x^2$ **18.**

x	y
-2	1
-1	-2
0	-3
1	-2
2	1

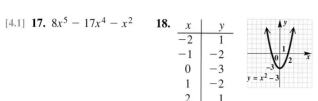

[4.5] **19.** $\dfrac{x^4}{9}$ **20.** $\dfrac{b^{16}}{c^2}$ **21.** $\dfrac{27}{125}$

[4.3] **22.** $2x^2 - 9x - 18$ **23.** $2x^4 + x^3 - 19x^2 + 2x + 20$ [4.6] **24.** $3x^2 - 2x + 1$ [4.7] **25. (a)** 6.35×10^9
(b) $.00023$ [5.1] **26.** $16x^2(x - 3y)$ [5.3] **27.** $(2a + 1)(a - 3)$ [5.4] **28.** $(4x^2 + 1)(2x + 1)(2x - 1)$

29. $(5m - 2)^2$ [5.5] **30.** $\{-9, 6\}$ [5.6] **31.** $2\frac{1}{2}$ seconds   [5.7] **32.** $(-\infty, -1] \cup [6, \infty)$

[6.1] **33.** $\dfrac{x - 3}{x - 2}$; -2 and 2 [6.2] **34.** $\dfrac{4}{5}$ [6.4] **35.** $\dfrac{-k - 1}{k(k - 1)}$ **36.** $\dfrac{5a + 2}{(a - 2)^2(a + 2)}$ [6.5] **37.** $\dfrac{b + a}{b - a}$

[6.6] **38.** $\left\{-\dfrac{15}{7}, 2\right\}$ [6.7] **39.** 3 miles per hour   [7.1, 7.2] **40. (a)** $-\dfrac{1}{3}$ **(b)** $y = -\dfrac{1}{3}x + 6$ **(c)** The line crosses
the y-axis above the origin, indicating that $b > 0$, since $(0, b)$ is the y-intercept.   [7.1] **41.** $2x - y = -3$
[7.2] **42.**

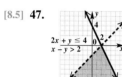

[7.3] **43.** -18 [8.1–8.3] **44.** $\{(-3, 2)\}$ **45.** $\emptyset$ [8.4] **46.** Krystalite: $\$39.99$;

Contempra II: $\$29.99$ [8.5] **47.**   [9.1] **48.** 10 [9.4] **49.** $\dfrac{6\sqrt{30}}{5}$ **50.** $\dfrac{\sqrt[3]{28}}{4}$ [9.3] **51.** $4\sqrt{5}$

52. $-ab\sqrt[3]{2b}$ [9.6] **53.** $\{7\}$ [9.7] **54.** 4 [10.1] **55.** $\left\{\dfrac{-2 \pm 2\sqrt{3}}{3}\right\}$ [10.2] **56.** $\{-1 \pm \sqrt{6}\}$

[10.3] **57.** $\left\{\dfrac{2 \pm \sqrt{10}}{2}\right\}$ [10.2, 10.3] **58.** $\emptyset$ **59.** $3, 4, 5$ [10.4] **60. (a)** $8i$ **(b)** $5 + 2i$

61. $\left\{-\dfrac{1}{2} \pm \dfrac{\sqrt{17}}{2}i\right\}$ [10.5] **62.**   vertex: $(-1, 2)$; domain: $(-\infty, \infty)$; range: $(-\infty, 2]$ **63.** positive

APPENDIX A   (PAGE 633)

1. 117.385   **3.** 13.21   **5.** 150.49   **7.** 96.101   **9.** 4.849   **11.** 166.32   **13.** 164.19   **15.** 1.344
17. 4.14   **19.** 2.23   **21.** 4800   **23.** .53   **25.** 1.29   **27.** .96   **29.** .009   **31.** 80%   **33.** .7%
35. 67%   **37.** 12.5%   **39.** 109.2   **41.** 238.92   **43.** 5%   **45.** 110%   **47.** 25%   **49.** 148.44
51. 7839.26   **53.** 7.39%   **55.** $2760   **57.** 805 miles   **59.** 63 miles   **61.** $2400

APPENDIX B   (PAGE 638)

1. $\{1, 2, 3, 4, 5, 6, 7\}$ **3.** $\{$winter, spring, summer, fall$\}$ **5.** $\emptyset$ **7.** $\{L\}$ **9.** $\{2, 4, 6, 8, 10, \ldots\}$
11. The sets in Exercises 9 and 10 are infinite sets.   **13.** true   **15.** false   **17.** true   **19.** true   **21.** true
23. true   **25.** true   **27.** false   **29.** true   **31.** true   **33.** false   **35.** true   **37.** true   **39.** false
41. true   **43.** false   **45.** $\{g, h\}$ **47.** $\{b, c, d, e, g, h\}$ **49.** $\{a, c, e\} = B$ **51.** $\{d\}$ **53.** $\{a\}$
55. $\{a, c, d, e\}$ **57.** $\{a, c, e, f\}$ **59.** $\emptyset$ **61.** B and D; C and D

Index

Index of Applications